Animal Models of Depression

Animal Models of Depression

Edited by
George F. Koob
Cindy L. Ehlers
David J. Kupfer

With 65 Illustrations

Birkhäuser
Boston · Basel · Berlin

George F. Koob, PhD
Cindy L. Ehlers, PhD
Department of Neuropharmacology
Research Institute of the Scripps Clinic
La Jolla, CA 92037, USA

David J. Kupfer, MD
Department of Research and Psychiatry
University of Pittsburgh Medical School
Pittsburgh, PA 15261, USA

Library of Congress Cataloging-in-Publication Data
Animal models of depression / edited by George F. Koob, Cindy L.
Ehlers, David J. Kupfer.
   p. cm.
   Based on a conference held Sept. 1986 sponsored by the MacArthur
Foundation Research Network on the Psychobiology of Depression.
   Bibliography: p.

   1. Depression, Mental—Animal models—Congresses. I. Koob,
George F. II. Ehlers, Cindy L. III. Kupfer, David J., 1941- .
IV. McArthur Foundation Research Network on the Psychobiology of
Depression.
RC537.A568 1989
616.85'27'00724—dc19   88-7437

   CIP-Titelaufnahme der Deutschen Bibliothek
   Animal models of depression: [based in part on a conference
held in September 1986] / ed. by: George F. Koob—
[Sponsored by the MacArthur Foundation Research Network on
the Psychobiology of Depression]. - Boston ; Basel :
Birkhäuser, 1989

   ISBN-13: 978-1-4684-6764-2   e-ISBN-13: 978-1-4684-6762-8
   DOI: 10.1007/978-1-4684-6762-8

   NE: Koob, George F. [Hrsg.]

Printed on acid-free paper

9 8 7 6 5 4 3 2 1

Contents

III. Pharmacologic Models

Introduction

Animal models represent experimental investigations developed in one species for the purpose of studying phenomena in another species and provide numerous advantages for preclinical research. They allow scientists greater control and isolation of important experimental variables. Animal models are safe, reproducible strategies by which to evaluate and design new pharmacological treatment strategies, while also allowing direct central nervous system intervention to alter the course of the aberrant behavior. Animal models have been developed for a number of mental illnesses; in this particular domain, they hold the promise to shed light on the still obscure etiologies of these illnesses and ultimately to facilitate the development and testing of "cures." Yet, true models of mental illness are difficult to develop, because mental illness may be a uniquely human phenomenon.

It was based on these considerations that the MacArthur Foundation Research Network on the Psychobiology of Depression set out to sponsor a conference to review the status, problems, promises, and relevance of animal models to the clinical conditions of affective disorders. The conference was held in September 1986 and included participants from both within the Network as well as scientists and scholars from various disciplines relevant to the concerns of the conference. After the conference was held, it became clear to the organizers that the material presented could be helpful to a broader field of investigators, since a significant portion of the information has not been presented elsewhere or in the unified context of a monograph. Thus, it was decided to ask the participants to prepare book chapters based on their presentations, and the current volume represents the outcome of these efforts.

The monograph follows the structural outline of the conference: The first section is devoted to the presentation of an overview. Chapter 1 reviews the basis of development of animal models in psychiatry. Dr. William T. McKinney emphasizes that in order to understand and evaluate the rapidly expanding literature on several areas of animal modeling research, a fundamental understanding of the bases and the jurisdiction for using animal models to study affective disorders must first be present. An animal model should simulate and reproduce accurately the human syndrome it is designed to represent, or be considered a framework that dictates the direction that experimentation should pursue.

Chapter 2 focuses on the problems and promises of animal models of mental illness. Dr. Conan Kornetsky addresses several basic questions: Do animal models of mental illness hold the promise to determine the etiology of these complex disorders? If it is possible to clarify the etiology, should it not be relatively easy to also find a cure? Are there true animal models of mental illness or are these conditions unique to humans?

Responding to some of these questions, Chapter 3 discusses the relevance of animal models to the clinical condition. Its author, Dr. Robert Post, highlights their potential utility in paralleling processes in the clinical situation that can be conceptualized in new ways, perhaps leading to a new series of clinical research formulations, testable hypotheses, and even pharmacotherapies.

The second and third sections of the monograph focus on two different approaches to animal models—developmental and pharmacological models. Contributors were asked to emphasize a single focus of study for a particular model, with the idea that each would explore the strength and weaknesses of the model and directions for future research.

Chapters 4 through 9 explore the potential of developmental models. They demonstrate the degree to which development can be altered by early childhood experiences and how it can lead to predictable behavioral and social signs of depressive symptoms. The hypothalamic-pituitary-adrenal system and neuroendocrine models are presented by Dr. Ned Kalin in Chapter 4. He concludes his review with a series of proposed experiments to test these models in primates. In Chapter 5, Dr. Robert Robinson explores the use of animal models to study post-stroke depression and provides insights into the parallels between human and animal studies. Chapter 6 presents a peer separation model of depression by Dr. Cindy Ehlers and co-workers. She proposes a new model of depression involving social zeitgebers and biological rhythms. In Chapter 7, Dr. Jay Weiss and co-workers discuss the electrophysiology of the locus coeruelus and implications for stress-induced depression. Their emphasis is on the applications of such a model for aspects of depression in humans. A proposed approach to categorizing depressive syndromes and their animal models is undertaken in Chapter 8 by Dr. Martin Teicher. He applies the techniques of motor activity to test this nosologic schema. Anhedonia as an animal model of depression is discussed in Chapter 9 by Dr. George Koob. In this chapter, he reviews specific measures in CNS reward, namely intracranial self-stimulation, and applies this model to the neurobiology of reward.

Chapter 10 to 13 focus on pharmacologic models. Dr. James Howard and co-workers review various models of depression used in the pharmacological industry in Chapter 10 and provide an excellent critique of these various models as they are used for new drug development. Dr. Zacharko and Dr. Hymie Anisman present animal models used for the pharmacological, biochemical, and behavioral analyses of depression (with an emphasis on the biogenic amines) in Chapter 11. Dr. R. Frances Schlemmer and co-workers discuss pharmacologic probes in primate sòcial interaction in Chapter 12, focusing on biochemical-behavior relationships. Chapter 13 contains an evaluation of the neuropharmacology of serotonin (5HT) and sleep by Dr. John Fernstrom and Dr. Ross H. Pastel. In this critique, they point out that a simple relationship between 5HT neuronal function and sleep does not exist and that studies must be evaluated as acute or chronic investigations, and that the activity of 5HT neurons changes with the state of vigilance.

What we hope to present with this monograph are examples that demonstrate that animal models have the potential to make a critical difference in the study of affective disorders. Of paramount importance, however, is the fact that, even if the pathogenesis of affective diseases was known, it would still be necessary to utilize animal models to monitor the symptoms or objective measures of depression to predict the efficacy of drugs and other treatments.

We are grateful to the John D. and Catherine T. MacArthur Foundation for supporting a meeting that provided to the participants many insights into the usefulness of animal models for furthering research on affective disorders. We are equally appreciative of their encouragement to share these insights with a broader field of investigators through the publication of this monograph.

George F. Koob
Cindy L. Ehlers
David J. Kupfer

Contributors

Hymie Anisman, Ph.D. Department of Psychology, Carleton University, Ottawa, Ontario, Canada K1S 5B6

Ross J. Baldessarini, M.D. Professor of Psychiatry and Neuroscience, Harvard Medical School, Massachusetts General Hospital and Mailman Research Center, McLean Hospital, Belmont, MA 02178, USA

Natacha I. Barber, B.A. Department of Psychiatry, Harvard Medical School and Mailman Research Center, McLean Hospital, Belmont, MA 02178, USA

Floyd E. Bloom, M.D. Member, Research Institute of Scripps Clinic, 10666 North Torrey Pines Road, BCR1, La Jolla, CA 92037, USA

R. Ian Chaplin, B.S. Research Institute of Scripps Clinic, 10666 North Torrey Pines Road, BCR1, La Jolla, CA 92037, USA

Barrett R. Cooper, Ph.D. Department of Pharmacology, Burroughs Wellcome Co., 3030 Cornwallis Road, Research Triangle Park, NC 27709, USA

John M. Davis, M.D. Research Department, Illinois State Psychiatric Institute, 1601 West Taylor Street, Chicago, IL 60612, USA

Cindy L. Ehlers, Ph.D. Associate Adjunct Professor of Psychiatry, Western Psychiatric Institute Clinic, Pittsburgh, PA 15261, and Assistant Member, Research Institute of Scripps Clinic, 10666 North Torrey Pines Road, BCR1, La Jolla, CA 92037, USA

John D. Fernstrom, Ph.D. Professor of Psychiatry, Department of Psychiatry, University of Pittsburgh, School of Medicine, Pittsburgh, PA 15213, USA

Robert M. Ferris, Ph.D. Department of Pharmacology, Burroughs Wellcome Co., 3030 Cornwallis Road, Research Triangle Park, NC 27709, USA

James L. Howard, Ph.D. Senior Research Scientist, Department of Pharmacology, Burroughs Wellcome Co., 3030 Cornwallis Road, Research Triangle Park, NC 27709, USA

Ned H. Kalin, M.D. Director of Research, Department of Psychiatry, Middleton Memorial Veterans Hospital, University of Wisconsin Medical School, Madison, WI 53705, USA

George F. Koob, Ph.D. Associate Member, Research Institute of Scripps Clinic, 10666 North Torrey Pines Road, BCR1, La Jolla, CA 92037, USA

Conan Kornetsky, Ph.D. Professor, Department of Pharmacology and Psychiatry, Boston University School of Medicine, 80 E. Concord Street, Room L602, Boston, MA 02118, USA

David J. Kupfer, M.D. Professor and Chairman, Department of Psychiatry, University of Pittsburgh Medical School, 3811 O'Hara Street, Pittsburgh, PA 15261, USA

Janet M. Lawrence, M.D. Department of Psychiatry and Neuroscience Program, Harvard Medical School, Massachusetts General Hospital and Mailman Research Center, McLean Hospital, Belmont, MA 02178, USA

William T. McKinney, M.D. Professor of Psychiatry, Clinical Sciences Center, D6/244, University of Wisconsin School of Medicine, 600 Highland Avenue, Madison, WI 53792, USA

Ross H. Pastel, Ph.D. Sleep Research Laboratory, Physiology and Behavior Branch, Department of Medical Neuroscience, Walter Reed Army Institute of Research, Washington, DC 20307, USA

Gerald T. Pollard, Ph.D. Department of Pharmacology, Burroughs Wellcome Co., 3030 Cornwallis Road, Research Triangle Park, NC 27709, USA

Robert M. Post, M.D. Chief, Biological Psychiatry Branch, NIMH, Room 3 N212 Building 10, 9000 Rockville Pike, Bethesda, MD 20892, USA

Robert G. Robinson, M.D. Professor of Psychiatry and Neuroscience, Johns Hopkins University School of Medicine, Traylor Building Room 624, Baltimore, MD 21205, USA

R. Francis Schlemmer, Jr., Ph.D. Department of Pharmacodynamics, College of Pharmacy, University of Illinois at Chicago, 833 South Wood Street, Chicago, IL 60612, USA

Peter E. Simson, Ph.D. Department of Psychiatry, Duke University Medical Center, Box 3829, Durham, NC 27710, USA

Francis E. Soroko, B.S. Department of Pharmacology, Burroughs Wellcome Co., 3030 Cornwallis Road, Research Triangle Park, NC 27709, USA

Martin H. Teicher, M.D., Ph.D. Assistant Professor of Psychiatry, Department of Psychiatry, Harvard Medical School, and Mailman Research Center, McLean Hospital, Belmont, MA 02178, USA

Tamara L. Wall, B.S. Research Institute of Scripps Clinic, 10666 North Torrey Pines Road, BCR1, La Jolla, CA 92037, USA

Ching M. Wang, Ph.D.   Department of Pharmacology, Burroughs Wellcome Co., 3030 Cornwallis Road, Research Triangle Park, NC 27709, USA

Jay M. Weiss, Ph.D.   Professor of Psychology, Department of Psychiatry, Duke University Medical Center, Box 3829, Durham, NC 27710 USA

Susan R. B. Weiss, Ph.D.   Senior Staff Fellow, Biological Psychiatry Branch, NIMH, Building 10 S3S239, 9000 Rockville Pike, Bethesda, MD 20892, USA

Stephen P. Wyss, B.S.   Research Institute of Scripps Clinic, 10666 North Torrey Pines Road, BCR1, La Jolla, CA 92037, USA

Jennifer E. Young, M.A.   Department of Pharmacodynamics, College of Pharmacy, University of Illinois at Chicago, 833 South Wood Street, Chicago, IL 60612, USA

Robert M. Zacharko, Ph.D.   Professor of Psychology, Department of Psychology, Carleton University, Ottawa, Ontario Canada K1S 5B6

Section I   Animal Models

1

Basis of Development of Animal Models in Psychiatry: An Overview

WILLIAM T. MCKINNEY

Introduction

In the last fifteen to twenty years the range of research activities that could be subsumed under the general heading of "animal modeling research" has expanded enormously. There are many approaches to developing animal models of depression in particular, and it will quickly become apparent that there is quite a variety of studies underway. In order to understand and evaluate the rapidly expanding literature in several areas of animal modeling research, there must first be a fundamental understanding of the basis and justification for having animal models in the first place. It is also important that this understanding be combined with a realization of the limitations of animal models in relation to psychiatric syndromes and their role in a comprehensive program of psychiatric research with affective disorders as well as other forms of psychopathology. If the fundamental philosophy, rationale, and advantages of animal models are not understood along with their limitations, there will remain a significant risk of either an uninformed rejection of animal modeling research or, on the other hand, an overacceptance of its direct clinical relevance.

The author's position regarding animal models in psychiatry is one which is equally applicable to discussions of animal models of depression, schizophrenia, anxiety, alcoholism, or any other syndrome. There are a number of approaches to developing animal models of depression, and these are frequently discussed in terms of which is the best and most comprehensive one for meeting some set of validity criteria. Without meaning to disparage such discussions, the author would contend that there is no such thing as a comprehensive animal model for depression, mania, or, for that matter, of any psychiatric syndrome. Furthermore, there never will be. To search in these directions is to ask the wrong questions, and the field of animal modeling research needs to start asking the right questions. Animal models are basically experimental preparations developed in one species for the purpose of studying phenomena occurring in another species. The concern ought to be with developing a variety of experimental paradigms in animals to study selected aspects of human psychopathology rather than attempting to develop a "depressed" animal. Certain paradigms are suitable for studying certain phenomena while others are better for different aspects of depression research. If this observation sounds obvious, it has not been so apparent to workers and others concerned with this area. There have been countless presentations and papers arguing about which model is "best." The answer is that there is no universally "best" model. The model used depends on what questions one wants to investigate. For example, if the primary interest is in developing an animal model of depression which will have high validity in terms of predicting treatment responsiveness across species, one could cite several paradigms with high empirical or predictive validity which are quite useful in this context. Such approaches might be essentially useless as far as mechanistic or developmental studies are concerned. On the other hand, a given animal preparation might be useful for studying the developmental origins of certain sets of behaviors, but have very limited usefulness for preclinical drug screening.

There are at least four different kinds of animal models of depression which will be discussed later. However, as we discuss the further development of this field of animal modeling of depression, it could be an instructive exercise to think through the advantages and limitations of each proposed model in the context of these four purposes. There is as yet no "holy grail" that has been found in this area, and the search for the perfect animal model itself is likely misdirected. This is not to say that there has not been major progress in developing animal models of depression and, to a certain extent, mania; but the work is still in a formative stage and much remains to be done. Having the proper theoretical framework to evaluate animal modeling work is critical.

This chapter will be divided into the following major sections:

Section 1—Selected historical considerations regarding animal modeling in psychiatry with a focus on depression
Section 2—Definitional issues: What are animal models and what is the rationale for their use in depression research?
Section 3—General kinds of animal models
Section 4—Summary

Historical Considerations

Pavlov is often said to be the originator of research relevant to animal modeling of human psychopathology in general (Pavlov, 1941). This chapter will not attempt to review the details of his work or to explain his terminology and how it relates to present day nomenclature. Of central importance, however, is the fact that his work represented one of the first moves away from the correlational method of behavioral analysis to the experimental study of psychopathology. To quote Kimmel (1971):

The significance of this change in direction may best be comprehended in relation to its two most important implications. First, the completely correlational method of behavioral analysis which was the empirical foundation of all earlier systematic efforts to understand psychological abnormality, including everything from Hippocrates' humors and Gall's prominences to the ingenious psychoanalytic theorizing of Freud, could now be supplemented, if not altogether supplanted, by a direct experimental approach which was much less fraught with the dual dangers of loose conjecture and empirical untestability. Second, and historically of possibly greater significance, the continuity of animal morphology, physiology, and behavior, already beginning to assume a position on center stage in man's philosophical thinking, received a new extensive thrust from the early Pavlovian findings since for the first time even such "uniquely human" phenomena as emotional breakdowns were seen to occur in subhuman animals.

Pavlov was followed by a number of other workers (Gantt, 1971; Liddell, 1947; Masserman, 1943; Hebb, 1947), and it is difficult to know what conclusions to draw about this early history of experimental psychopathology research. Some have not seen it as a particularly noteworthy beginning. However, the early pioneers may have been more successful than it appears on first glance in developing certain principles that, at times, are seeming to be rediscovered today. These include, but are certainly not limited to:

1. The demonstration that psychopathology could be studied experimentally in animals in addition to the strictly correlational studies done previously in humans.
2. The demonstration of the importance of both careful behavioral observations and of serendipity. While it is true that most of the early workers did not use the more sophisticated and quantifiable behavioral scoring techniques now available, they were keen observers and literary in their descriptions.
3. The repeated proposal of an interactive model of psychopathology. The role of the temperament of the animals, along with a variety of social and neurobiological variables, was repeatedly stressed in the early literature. The concept of individual variability was part of the early work, and investigation of the sources of such variability continues to be an important area of research today.
4. The development of the principle that there could be a persistent internal response, even after the inducing stimulus is no longer present. This development remains as a major contribution in our understanding of a number of forms of psychopathology, including depression.
5. The recognition of the importance of unpredictability and uncontrollability. Systematic investigations of these phenomena continue today.
6. The experimental paradigms they used led to the development of another basic principle which is still important today, namely, that adaptive behavioral processes provide the foundation on which maladaptive behavior patterns are built in the presence of altered environmental demands. Adaptive mechanisms, of animals and humans, are fragile and share a ten-

uous relationship with the environment. Either internal changes in the organism (e.g., with drugs or other altered neurochemistry) and/or changes in the external environment (e.g., separation, the imposition of uncontrollability, etc.) can lead to serious behavioral changes. These behavioral changes can, in turn, lead to neurobiological changes and the development of a vicious circle. The study of these interactions is becoming a cornerstone of animal modeling research in depression and other forms of psychopathology.

One of the problems with early experimental psychopathology was that clinical terms were applied far too loosely and prematurely to a set of behavioral changes induced by methods that seemed to bear only a faint resemblance to inducing conditions for human syndromes. This problem led clinicians to have a certain amount of skepticism and cynicism about the whole field.

With regard to depression in particular, the history of animal modeling research dates back to at least 1928 with the earliest stage consisting of case reports. For example, Tinkelpaugh (1928) reported the case of a young rhesus monkey who developed marked behavioral changes including self-mutilation, agitation, anorexia, and social withdrawal following separation and subsequent viewing again of a female monkey with whom he had lived monogamously for three years. After reunion with the female, he gradually recovered. Absence did not make the heart grow fonder. Yerkes and Yerkes (1929) described the high death rate of newly captured gorillas and attributed this death rate to the loss of familiar surroundings and the severance of all meaningful bonds as antecedent causes of their deaths. They also reported behavioral changes resembling depression in chimpanzees and suggested separation as a probable cause. There have been a number of other case reports suggesting that a variety of animals might experience something similar to human depression. Such syndromes have been described in natural settings, seminaturalistic settings, and in laboratories.

However, more serious scientific development in the field of animal modeling of depression did not begin until there were controlled experimental ways of inducing such syndromes. There were several lines of research activity in the 1960s relevant to the development of animal models of depression. In a little known study, but one of the very first in the animal modeling of depression area, Senay (1966) worked with three-week-old German shepherd puppies. He formed a relationship with each member of a litter over a nine-month period and observed their behavioral changes when he withdrew from social contact with them. Prior to that time he had been their sole human contact. Separation produced increases in object avoidance and aggressive behavior for animals of avoidance temperaments and increases in object seeking for animals of the approach temperament. In other words, different animals reacted differently to attachment bond disruption depending on their temperament and the preloss levels of gratification.

Several other historical developments in this field can be mentioned. One is the development of separation paradigms in primates and other species (Mineka and Suomi, 1978; McKinny and Bunney, 1969). Several laboratories, more or less independently, began a series of studies about separation of nonhuman primates, mostly mothers and infants. These studies, which began in the 1960s and continue today have provided a considerable amount of data about the development and disruption of attachment bonds and represent one of the approaches to developing animal models of certain aspects of depression.

Another important line of work begun in the 1960s was the "uncontrollability" studies (Seligman and Maier, 1967; Sherman and Petty, 1980; Weiss and Goodman, 1984). This work will be reviewed by others in this book so this chapter will not focus on it in any detail. These studies have obviously spawned an enormous amount of literature and relates closely to both cognitive and neurobiological theories of human depression. This line of work and separation may even be related in that one of the theories about the mechanisms of separation reactions relates to controllability-uncontrollability issues. Additional work is needed to investigate this possibility and to relate this possible mechanism to what is now known about the behavioral, developmental, and neurobiological aspects of separation.

During the time that behavioral models of depression were being pursued the development of pharmacological induced models for depression also continued (Porsolt, 1982). These kinds of approaches, which involve pharmacological induction of altered behavior patterns, have a long history and have been heavily utilized for the evaluation of antidepressant agents. This approach to modeling will be discussed below.

A more recent development in the animal modeling of depression area has involved the relationship between the imposition of multiple stressors, conflict, and depression. Recently, animal paradigms have been developed in which one can study these interrelationships, and these paradigms represent a potentially important development (Katz, 1981). Yet another type of approach involves the manipulation of dominance patterns (Price, 1967). Some have written about the importance of such patterns in relationship to depression and mania, although only limited empirical data are available. Neuroendocrine strategies are also being used in yet another attempt to develop animal models of depression.

It should be apparent from this brief historical overview that the field of animal models of depression has gone from case histories to the use of multiple induction techniques. Each method has advantages and disadvantages but, despite being a relatively new undertaking, the field is in a formative, exciting, and still progressing stage.

What Are Animal Models and What Is the Rationale for Their Use in Depression Research?

The concept of animal models in psychiatry, and in depression research in particular, is now regaining attention after having fallen in popularity for a number of years. Most major conferences on depression will now have a section on animal models, and interest is again picking up with major foundation and governmental conferences scheduled.

Several events have combined to account for this waxing and waning of interest. There was an earlier period of disillusionment in several quarters based on a combination of naive promises by some regarding the perfect fit of their "model" with the human disease as well as unrealistic expectations by others. Workers were presenting one model or another as an animal model of depression in some global sense yet differences from the human syndrome were readily apparent. People were forgetting what animal models are and are not.

As mentioned previously, animal models are basically experimental preparations developed in one species for the purpose of studying specific aspects of the human illness. They are not replicas of human depression in its entirety. There inevitably will be important differences, and the study of both the similarities and differences is important. Naturally, one wishes to produce animal preparations as similar to the human illness as possible, which may include genetic considerations, induction methods, pathophysiology, phenomenology, and treatment responsiveness to mention a few. Some models will be better for one or another of these aspects of the illness and none will be inclusive.

Related to the above is the question: Why have animal models in the first place? Why not do all the studies directly in humans and not bother with animal models and the issues of cross species reasoning? As it was expressed to me recently, "If psychiatrists can't solve their research problems, given the number of sick humans they have to work with, research with animals certainly isn't going to help." It is indeed true that we are not short of human patients on which to do depression research, and this network is obviously majorly involved with this important effort.

There have been several lines of work relevant to affective disorders. The first that should be mentioned is retrospective reasoning and theory development based on work with humans who are manifestly sick with the illness. This stage has been, and continues to be, important in terms of generating hypotheses and providing a framework in which to begin to understand affective disorders. The limitations of this approach have been increasingly recognized in that one bases theory on retrospective data obtained from sick patients and/or family, and the accuracy of recall can be seriously questioned. Etiological phenomena and derivative phenomena have become hopelessly intertwined by that point in time.

Other research methods have been utilized in depression research. One epidemiological method involves various community sampling techniques. This has yielded extremely important data concerning both actuarial data as well as associated events that might serve as stressors or precipitants. In more recent years there has been increased clinical research activity in controlled settings, both inpatient and outpatient, and with a variety of depressed patients. Patients with specified diagnoses are carefully studied with highly developed neurobiological techniques and behaviorally assessed with many rating scales.

The following list of reasons for including animal modeling research as part of a comprehensive research program on affective disorders is illustrative rather than comprehensive.

1. Many of the critical questions about the origins of human psychopathology cannot be studied directly in humans. It is possible, by using animal preparations, to control inducing conditions rather precisely and to study the behavioral and neurobiological effects on both a short-term and long-term basis. For example, in relation to depression, prospective studies examining the effects of developmental events on behavior and on neurobiology can be done much more easily in animals. The timing and exact nature of certain alterations in development can be specified and the short- and long-term consequences studied. This aspect of modeling research is relevant to the question of developmental vulnerability based on early experiences and the mediating mechanisms of this vulnerability.

    A particular line of research where animal models have a special contribution to make is in the conduct of prospective studies examining the effects of developmental events on behavior and on neurochemistry. The interactions between these variables can be studied in a controlled and prospective manner. In the last decade, there have been animal preparations under development which make such investigations feasible and will facilitate the movement beyond correlation and retrospective analysis to cause-and-effect studies (Kraemer et al., 1971).

2. The underlying mechanisms associated with specific behaviors and patterns of behaviors can be studied more directly in certain animal species. Animal models potentially make possible the dissection of underlying mechanisms in a more direct way than is possible in human clinical research and complement ongoing efforts in this regard in human protocols. More direct, and potentially more invasive, studies of neurobiological mechanisms can be done. As mentioned previously, such procedures will need to be suited to both the species and the overall purpose of having the experimental paradigm in the first place. Not all procedures are justified on either ethical or economic grounds in all species. The questions have to be clear and specific, especially in proposing such studies in higher order primates. Nevertheless, the time is ripe, indeed overdue, for a vigorous effort in this area. The effort will need to involve multiple laboratories much like collaborative human studies of depression involve many centers. The area of experimental psychopathology in animals has gotten complex enough that this type of collaborative approach needs to be undertaken. For example, different strategies and approaches need to be applied to several kinds of animal preparations. Specific attention needs to be given to what types of mechanism studies are most appropriate to conduct in a given species. Molecular or submolecular studies may

be indicated in some preparations, but these methods may not be the only reasonable way to approach mechanism studies, for example, in a socially behaving species. This is a very controversial frontier area in that basic neuroscientists sometimes want certain types of mechanistic studies to be done, but yet to do them in the way requested would vitiate the social behavioral studies. There needs to be continued dialogue, hopefully not as acrimonious as it sometimes gets about this topic. The issues are complex but probably solvable with enough discussion and the development of some collaborative protocols across laboratories that take advantage of complementing expertise.

Single variables can be evaluated in terms of their main effects, but, very importantly, in terms of the nature of their interaction with each other. For example, the nature of the interaction among genetic, developmental, social, and biological variables can be studied in various combinations in different species. In human clinical research, multiple variables interact simultaneously, and it typically has been impossible to sort them out in any quantifiable way.

3. The ability to isolate specific behavior patterns in animals and study their origins, pathophysiology, and responsiveness to treatment techniques is important. So often when clinicians work with depressed humans, they are dealing with a broad range of behaviors occurring together; and it is impossible to study any one or two in isolation from the others and to understand them more completely. Many examples could be mentioned. One that comes to mind is anhedonia. Another is social withdrawal. If clinicans can begin to better understand these and other particular aspects of the depressive syndromes, they can perhaps over time expand the understanding of the human depressive syndrome where they typically occur together and are impossible to study in isolation from each other.

4. Animal models have played an important role in the preclinical evaluation of antidepressant drugs. This use relates to the empirical or predictive validity of animal models. It is likely that this role for animal models will continue.

A related aspect of animal models in this context is their use in helping to better understand the mechanism of action of drugs in relationship to alteration of specific behavior patterns. This use goes beyond a mere global prediction of whether drugs work or do not work. It relates to studying the behavioral effects of agents with relatively specific mechanisms of action.

5. Animal models can also be used to help understand the mechanisms of established treatment techniques. Animal models potentially make possible the investigation of the mechanisms, not only in terms of pathogenesis, but also treatment responsiveness. That is: Why do certain drugs work in depression and others do not? What are the mechanisms of action of electroconvulsive therapy in depression? Why do certain behavioral interventions work and others do not?

6. Animal models also permit the understanding of a specific behavior or set of behaviors in terms of their developmental and social context as well as their pathophysiology. Rather than just focusing on global syndromes, clinicians can investigate certain behaviors in terms of their origin, their context, and their responsiveness to certain interventions.
7. Animal modeling research, especially with primates, has led to the development of improved behavioral, ethologically based rating methods which can now be used in clinical research settings to evaluate social interactions (e.g., mother and infant, peers).

The above considerations regarding the rationale for animal models in depression research relate closely to a consideration of the general kinds of animal models that are possible. Naturally, there is some overlap, but it may be useful to review the broad types of approaches to animal modeling.

General Kinds of Animal Models

There are several ways that the general kinds of animal models could be categorized, including the following four categories (McKinney, 1984):

1. Those developed to simulate a specific sign or symptom of the human disorder ("behavioral similarity models")
2. Those developed to evaluate etiological theories of psychopathology ("theory driven models")
3. Those developed with the primary purpose of studying underlying mechanisms ("mechanistic models")
4. Those developed to permit preclinical evaluation of treatment methods ("empirical validity" models)

Obviously, there is overlap among these kinds of models. But it is important to maintain clarity about the primary purpose of developing and studying a given animal model since the evaluation of the usefulness of the model for depression, or any other psychiatric illness, needs to be closely linked to the major purpose of developing an experimental paradigm in the first place.

Traditionally, in the animal modeling of depression area, various proposed models have been presented in a competing context as the "best" model of human depression. Persuasive arguments have been advanced regarding how a given paradigm best meets some set of validity criteria regarding etiology, behavioral similarity, underlying mechanisms, and treatment responsiveness. Unfortunately, one seldom hears another type of question asked, namely: What aspect of human depressions would one like to study in animal preparations and which type of animal and what kind of experimental paradigm is best for conducting the study? Investigators too often have become wedded to one kind of animal, variants of the same experimental paradigm, and defensive presentations regarding the overall validity of their particular model. It is past time to take a hard look at the conceptualizations in this area and to begin to specify human linkages in a more careful and specific way that incorporates fundamental principles of ethology and evolutionary theory.

Behavioral Similarity Models

These are models designed to simulate specific signs or symptoms of human depressions in animals. The primary intent is not to evaluate either a specific etiological theory, to study underlying mechanisms, or to evaluate treatment responsiveness. Obviously, these are important questions even in this kind of modeling but are not the primary focus. The validity of these models is judged by how closely they approximate human depression from a phenomenological standpoint. The idea is to produce a particular symptom or set of symptoms. The methods used to produce the behaviors may or may not be how such behaviors are produced in humans.

Two examples could be cited, one from outside psychiatry and the other from within the field. Atherosclerosis has been studied in monkeys. From one standpoint, the techniques used to produce atherosclerotic blood vessels are secondary. Unless one is doing etiological studies, the final common state, namely atherosclerosis, is the object of study. A drug or diet that is never used in humans may be used in animals to produce the state, and a variety of mechanism and even treatment studies can be done without it being possible to make any etiological statements from such studies.

Perhaps closer to the topic of animal models of depression, a state of behavioral immobility in animals can be produced by several methods (e.g., chronic unpredictable stressors, experiences with uncontrollability, separation from mother or peers, being put in a beaker of water, or with pharmacological agents). From a behavioral similarity standpoint, this state can arguably be said to resemble the psychomotor retardation and/or given up, withdrawn behavior seen in some forms of depression. However, in the general kind of models we are now discussing, it is this behavior per se that is the primary focus of study, and the experimental paradigm is evaluated in this context. Since it is a constellation of behaviors that can be produced by a variety of methods, the linkage with a specific inducing condition becomes somewhat secondary. One focuses the research on the dissection of a certain behavior pattern. These kinds of animal models are important as the field of human depression and animal depression research move to more specificity and understanding of specific behavior patterns rather than global syndromes. Animal modeling research can make seminal contributions from this context in helping to develop some general principles of psychopathology that may even cut across psychiatric syndromes.

Theory-Driven Models

These kinds of models are frequently misunderstood. They are "theory driven" in the sense that a theory drives the development of the experimental paradigm; however, they are not theory driven in the sense that one must assume the validity of the theory to value the research. In these kinds of models, no a priori assumptions are made about the validity of the etiological theory. Rather, the attempt is to operationalize the theory and to develop experimental paradigms to evaluate the effects of such inducing conditions. The misunderstanding takes the form of criticism that the theory driving the development of the model has not been substantiated in humans so why study it in animals. There is considerable irony in this criticism because that is the very reason for developing the paradigm in the first place, that is, to test the theory.

Typically, most etiological theories of psychopathology have been developed from studies of sick humans, and therefore, are retrospective in nature. This is an area where suitable animal preparations can play a particularly useful role. One can evaluate prospectively the effects of paradigms designed to represent certain causative theories. For example, clinical evidence suggests that separation is an important variable accounting for a small, but significant, proportion of the variance in the occurrence of certain kinds of depressions. Recent human studies of chronic depressives would suggest a possible important role for early developmental object losses. The evidence, however, comes mostly from clinical retrospective studies and from population surveys. These are important sources of data for generating hypotheses, but it is also important to evaluate prospectively the role of such separations and to what extent they were primary or derivative occurrences. How do they interact with other sources of vulnerability? In animal preparations, one can study animals which have been subjected to controlled separations and prospectively evaluate the consequences, not just in terms of the immediate responses, but in terms of the long term vulnerability induced by alteration of early rearing conditions. Similar comments could be made about experiences with uncontrollability, being subjected to certain pharmacological manipulations, exposure to chronic stressors, etc. Such studies are worthwhile whether or not the state so induced in animals should properly be labeled "depression" or not. These studies represent the use of animal preparations to test, in a systematic manner, the effects of certain inducing conditions and permit the development of careful descriptions of the behavioral and neurobiological effects. Furthermore, one can also quantify how alterations in one or another parameter influence the response to a particular event.

Models Designed to Study Underlying Mechanisms

There are a number of complex and controversial issues surrounding the use of animal models for mechanism studies. For some, mechanism studies represent the only reason for developing animal models in the first place, and proposed animal models of depression or any other illness are evaluated by how well they lend themselves to mechanistic studies. With the increasing availability of high technology methods for the study of underlying neurobiological mechanisms, some have become preoccupied with the molecular and submolecular basis of the altered behaviors seen in many of the proposed animal models of depression.

One should be careful with the term "mechanisms." To some, the term is synonymous with neurobiological mechanisms, and nowadays this translates molecular and submolecular. Correlative studies of social behavior and neurobiology are no longer acceptable among certain groups. While the application of high technology neuroscience techniques to the study of the mechanisms underlying the different models could hardly be debated as a general goal, certain cautions are important.

First, not all animal models lend themselves easily to the type of high technology, molecular and submolecular, mechanistic studies referred to above, which tend to be the more socially and developmentally oriented models in higher order primates. In this day of high technology neuroscience, it should be remembered that behavior and psychopathology occur in a social and developmental context and therefore the continued study of social behavior is essential. A serious challenge for researchers in this area is the development of noninvasive techniques for mechanism studies in socially behaving animals that will be satisfactory to basic neuroscientists as "mechanism" studies. Unfortunately, CSF studies and pharmacological probes will no longer suffice. There is much to be said for the longitudinal study of social behavior in a developmental context by whatever means are available that will permit the ongoing assessment of social behavior. One might have to compromise on the "directness" of the neurobiological studies and settle for such things as CSF and blood studies, use of selected pharmacological agents as tools, and perhaps limited sacrifice of subgroups. Such studies might help develop more rational justification for direct tissue studies and make possible the formation of some specific hypotheses that could then be tested in more direct studies. In the enchantment with the newer neurobiological investigative tools, there is the risk of neglecting the development basis of social behavior and important correlative neurobiological studies that might rationally point the way to more direct studies.

One cannot necessarily directly transpose techniques of mechanism studies from rodents to monkeys nor, for that matter, from monkeys to humans. The study of mechanisms must be approached from different vantage points in different species and in different kinds of protocols. Each approach has advantages and disadvantages which should be acknowledged. At present, there is a need for research utilizing both approaches, i.e., protocols where the central neurobiological mechanisms can be studied rather precisely and directly (generally in rodents and invertebrates) and others where social behavior can be studied rather precisely and longitudinally; but the neurobiological studies that can be done are not as direct as above. In many ways, these studies resemble human research protocols with the additional feature that it is possible to control development and do specified interventions at certain times.

Empirically Valid Models

The use of empirically valid models represents one of the oldest and best known applications. They mainly involve the use of animal preparations to develop and test clinically effective drugs. In this context, an ideal animal model, is one where there are no false positives and no false negatives. That is, in all instances where a drug is effective in the animal preparation it is also clinically effective. Likewise, whenever a drug is inactive in an animal preparation, it is not active clinically in humans. Thus, there is a 100% correspondence between the effects of a drug in the animal model and in the clinical condition, which is known as empirical validity. Though the concordance between the effects in the animal model and in humans will never be exact, there are a number of models with established high empirical validity for depression, schizophrenia, and anxiety disorders.

If one is primarily interested in developing an animal model system in which drugs can be evaluated, the method of inducing the syndrome and even the behavioral similarity issues become secondary. As mentioned above, a number of models with high empirical validity are available and have been important in the development of clinical and basic psychopharmacology. Some caution is in order in that the empirical validity of a model does not necessarily establish it as valid on other grounds. Drugs can have the same effects in two different species for quite different reasons so one cannot necessarily reason to underlying mechanisms or etiology from such studies. This fact does not make them any less useful but does circumscribe the kinds of conclusions that can be drawn from these approaches.

Summary

The introduction to this chapter defines the concept of animal models and gives an overview of what the general kinds of animal models are. The general historical context for experimental psychopathology research is described next, and the transition into specific attempts to develop animal models of depression is provided.

There are basically four general types of animal models, and the issue of the validity of various models needs to be discussed in the context of the purpose of developing the animal model in the first place.

The rationale for animal models and their relationship to a range of research approaches to understanding depression has been discussed. Animal models of depression have a number of areas in which they can make unique contributions. These areas relate to careful control of inducing conditions and the study of the interactions among etiological variables, the ability to control development, the possibility of doing more direct studies of central neurobiological mechanisms, and the study of the mechanisms by which depression treatments work.

There are a number of challenges that confront this relatively new area of depression research, and there needs to be continuing dialogue between animal modeling researchers and clinical depression researchers. This dialogue is an important interface and one with great potential for helping to clarify many aspects of the depressive syndromes.

Parts of the work described in this chapter, as well as the writing of this chapter itself, were supported by research grants MH21892 and MH40748 from the National Institute of Mental Health and by the Wisconsin Psychiatric Research Institute. In addition, some of the ideas presented in this chapter evolved while the author was a Fellow at the Center for Advanced Study in Behavioral Sciences at Stanford, California. He is grateful for financial support provided during this time by the John D. and Catherine T. MacArthur Foundation.

Further Readings

Gantt W (1971): Experimental Basis for Neurotic Behavior. In: *Experimental Psychopathology: Recent Research and Theory*, Kimmel H, ed. New York: Academic Press 33-47

Hebb D (1947): Spontaneous neurosis in chimpanzees: Theoretical relations with clinical and experimental phenomena. *Psychosom Med* 9:3-16

Katz R (1981): Animal models and human depressive disorders. *Neurosci Biobehav Rev* 5:243-246

Kimmel H, ed. (1971): *Experimental Psychopathology: Recent Research and Theory.* New York: Academic Press

Kraemer G, et al. (1971): Hypersensitivity to d-amphetamine several years after early social deprivation. *Psychopharmacology* 82:266-271

Liddell H (1947): The experimental neurosis. *Annu Rev Psysiol* IX:569-580

Masserman J (1943): *Behavior and Neurosis: An Experimental Psychoanalytic Approach to Psychobiologic Principles.* Chicago: University of Chicago Press

McKinney W (1984): Animal models of depression: an overview. *Psychiatr Dev* 2:77-96

McKinney W, Bunney W (1969): Animal model of depression, review of evidence: implications for research. *Arch Gen Psychiatry* 21:240-248

Mineka S, Suomi S (1978): Social separation in monkeys. *Psychol Bull* 85:1376-1400

Pavlov I (1941): *Lectures on Conditioned Reflexes: Vol. 2 Conditioned Reflexes and Psychiatry.* Gantt W, trans. New York: International Publishers

Porsolt R (1982): Pharmacological models of depression. In: *Proceedings of Dahlem Conference on the Origins of Depression: Current Concepts and Approaches,* Angst J, ed. Berlin: Springer-Verlag

Price J (1967): The dominance hierarchy and the evolution of mental illness. *Lancet* 2:243-246

Seligman M and Maier S (1967): Failure to escape traumatic shock. *J Exp Psychol* 74:1-9

Senay E (1966): Toward an animal model of depression: a study of separation behavior in dogs. *J Psychiatr Res* 4:65-71

Sherman A, Petty F (1980): Neurochemical basis of the action of antidepressants on learned helplessness. *Behav Neural Biol* 35:344-353

Tinkelpaugh O (1928): The self-mutilation of a male macaque rhesus monkey. *J Mammal* 9:293–300

Weiss J, Goodman P (1984): Neurochemical mechanisms underlying stress-induced depression. In: *Stress and Coping.* Vol. 1, McCabe P, Schneiderman N, eds. Hillsdale, New Jersey: Lawrence Erlbaum Associates

Yerkes R, Yerkes A (1929): *The Great Apes.* New Haven: Yale University Press

2

Animal Models: Promises and Problems

CONAN KORNETSKY

The promise that a model of a mental illness holds for us is that once the model has been developed it should be relatively easy to not only determine the etiology of the illness but to effect a cure. The problem is that it is not easy to develop a true model of a mental illness. It could be that animal models are not possible because mental illnesses may be uniquely human disorders.

Models are used when experiments cannot be carried out in humans because of a possible insult to the physical and/or psychological integrity of the subject. Sometimes nature or society provides natural experiments that are so compelling that further modeling may be superfluous or the natural phenomenon clearly points to the animal experiment that must be done. When a child is separated from its parent or parents, an experiment in subhuman species is not needed to confirm that the child is depressed. It is known that the depression can be prevented by not allowing parental separation. It is not known, however, if childhood separation has any relationship to the depression seen in adults.

A good definition of a model was given a number of years ago by Baldessarini and Fischer (1975) at a conference on models in mental illness. They pointed out that a model is an experimental compromise in that a simple experimental system is used to represent a much more complex and less readily accessible system: ". . . the animal to represent the patient, the tissue slice to represent the intact living brain, the isolated nerve ending to represent the intact synapse."

A model may be nothing more than a series of postulates that build one on another to model a particular behavior or disease. Clark Hull (1943) proposed a behavioral theory of learning that was of this type. For Hull the model was a theory of behavior that was so precisely formulated that the models designed to test the theory were clearly mandated. Psychologists, many years before Hull, attempted to develop a science of human behavior that was rooted in the animal laboratory. The narrower the view of behavior presented, the less apparent was the relevance of the model to human behavior. For example, John B. Watson, the conceded father of "Behaviorism" wrote in 1914, "It is possible to write a psychology, to define it . . . as a science of behavior and never go back on the definition; never to use the terms consciousness, mental, state, content, will, imagery, and the like . . . It can be done in terms of stimulus and response . . ." Thus, for Watson, behavior could be explained wholly in terms of stimulus and response. It is unlikely that he would have attempted to model a mental disease unless the abnormal behavior characteristic of the disease could be redefined in his terms of stimulus and response. B.F. Skinner, although a behaviorist much in the tradition of Watson, felt neither constrained in using animals to model the aberrant behavior of humans nor to label the behavior with a uniquely human name; albeit, quotation marks were placed around the name. He labeled as "anxious" the suppression of lever pressing by a rat to an auditory stimulus that had been previously presented just prior to receiving foot-shock (Estes and Skinner 1941). In an often referenced paper, Skinner (1948) described as "superstitious" the increase in normally low frequency behavior in the pigeon resulting from chance reinforcement.

The behaviorists share with Freud a fundamental belief that all behavior is determined. Although Freud's conceptual schema have a systematic logic, the concepts elude the operational definitions necessary for the behaviorists. Thus, although the behaviorists may easily develop animal models and the psychoanalysts can easily develop conceptual models, the former may have little obvious relevance to human behavior, and the latter defy experimental verification.

What then makes a good model for human psychopathology? Many scientists would argue that for a model to be viable it must have heuristic value. Let us briefly examine the concept of heuristic value. The dictionary definition is, in part, "adj: stimulating interest as a means of furthering investigation" (Stein and Urdang, 1967). The author does not believe that there has been an investigator who has submitted a paper for publication who believed that his work did not have heuristic value. The work of Clark Hull and his students (as every graduate student of the period was aware) had a great deal of heuristic value. Each experiment clearly suggested another. Despite this, the system and postulates finally died out. It would be surprising to find a psychology graduate student in the United States who is currently involved in a thesis specifically designed to test a postulate of Clark Hull. Since "heuristic value" is an inflated commodity, it may qualify as a necessary but certainly not a sufficient test of the viability of an animal model. Thus, some of the other essential components of a good animal model must be identified.

The first question is whether a pathological condition that may be uniquely human can be modeled in an animal. If a whole animal model is used, inferences are often made from the antecedent conditions of the ongoing behavior that animals are in pain, hungry, or sexually aroused. These are constructs that cannot always be documented. One can only infer in animals such human states as depression or anxiety. An increase in the number of fecal boli in the bottom of a cage may reflect a state of anxiety but it may also be caused by a parasympathetic discharge due to the direct pharmacological action of a drug. A statement, probably apocryphal, attributed to Freud made the point, "a cigar may only be a cigar." When attempting to model often poorly defined emotional states, there is an increased likelihood that the labeling and the subsequent interpretation may be more in the eye of the beholder than in the brain of the animal. A common error is the assumption that similar overt behavior implies similar mood or experience. Despite these problems, many (if not most) researchers who use subhuman primates will argue that depression is not foreign to these animals and many of the things that make humans depressed also make these primates depressed. As indicated in some of the subsequent chapters, environmental conditions can be created in order to produce behavior that is analogous to some of the behavior seen in depressed human patients.

A model of a disease state need not produce the disease in the animal. It may be sufficient to model only one aspect of the phenomenon which may be a critical component of the disease. Studies of animal behavior designed to elucidate some aspect of normal behavior are of this type. When a behaviorist looks at the acquisition of a behavior by an animal in an operant box, he is not looking at many of the important variables that lead to a child's learning in school. The experimenter is ignoring many organismic and social factors; but by manipulating the environment of the experimental chamber and the schedule of reinforcement, he can learn a great deal about the contingencies that will facilitate or impair learning. He focuses only on those parts of the environment that he can control.

In regard to pathological behavior, some specific aspect of the disease may be the focus. Although not the disease, the pathological behavior focused on could be the critical aspect of the disease process. To build a model of depression of this type would require a specific symptom of depression and then the ability to produce this symptom in the animal. An example might be to focus on sleep disturbance. Animals could be sleep deprived and subsequent performance measured on a variety of procedures. It could then be determined whether the model is predictive in that known antidepressants would reverse the state.

The development of an animal model based on what is believed to be a critical aspect of a disease was carried out by Kornetsky and Eliasson (1969) and Eliasson and Kornetsky (1972) in the study of an animal model of schizophrenia. Based on findings in humans that a significant number of schizophrenic patients had an attentional deficit the investigators built a model on this single major aspect of the disease. Further, since it was also postulated that this deficit was caused by a state of central arousal, this deficit in animals was produced by electrically stimulating the mesencephalic reticular formation of the rat. In these early studies, chlorpromazine, which by itself impaired attentional performance, blocked the impairment caused by electrical stimulation to the mesencephalic reticular formation. The results of one of these studies is shown in Figure 2.1.

A similar attentional deficit as that obtained by electrical stimulation in the rat was achieved by directly applying norepinephrine (NE) to the mesencephalic reticular formation of the rat (Bain, 1980; Kornetsky and Markowitz, 1975). Low concentrations of NE improved performance and as the concentration was increased, impairment in performance was observed. This impairment caused by NE stimulation could be reversed by either chlorpromazine or thioridazine, but not haloperidol. Figure 2.2 shows an example of the results of this experiment. Although, an attentional disorder is still believed to be a major deficit in schizophrenic patients (Nuechterlein and Dawson, 1984), the failure of haloperidol to reverse the attentional deficit not only decreased the utility of the model described above, but also demonstrated the vulnerability of simple one-dimensional models. Thus, although an attentional deficit could be caused, the etiology of the attentional deficit in the rat was not homologous with that seen in the schizophrenic patient. The use of this type of model, in which a single behavioral symptom of the disease is selected, is especially difficult when looking for a single relevant symptom of depression. The deficits of depression are those of the tired, bored, and uninterested.

In classifying types of animal models, the author has defined three types: homologous, isomorphic, and, for a lack of a better name, predictive. The author does not know who first described models in this way; his first use of these terms was in a conference on animal models in 1977 (Kornetsky). An homologous model can be described as something like a description of a duck. If it looks like a duck, and it sounds like a duck, and it smells like a duck, and its mother was a duck, then, if it is not a duck, it is a pretty good homologous model of a duck. The homologous model creates the disease, or

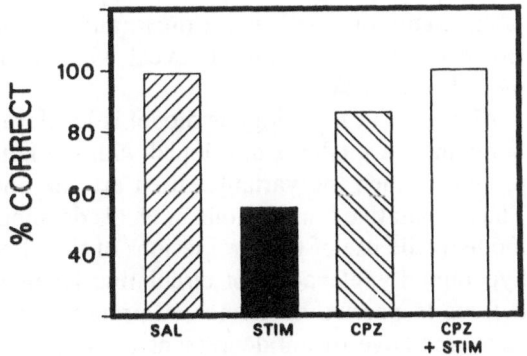

FIGURE 2.1. The mean effects of chlorpromazine (1.0 mg/kg) and intracranial electrical stimulation to the mesencephalic reticular formation, alone and in combination, on correct responses on a simple test of attention in the rat (an animal version of the Continuous Performance Test). Stimulation or chlorpromazine alone significantly differed from saline or chlorpromazine and stimulation. (Abstracted from Kornetsky and Eliasson, 1969.)

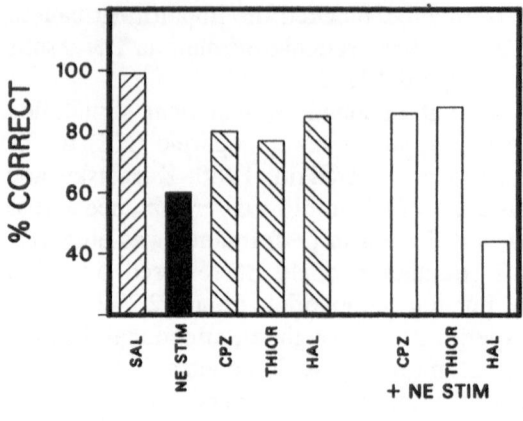

FIGURE 2.2. The mean effects of chlorpromazine (1 to 2 mg/kg), thioridazine (2 to 3 mg/kg), and haloperidol (.05 to 1.0 mg/kg), alone and in combination with intracranial stimulation of the mesencephalic reticular formation by norepinephrine (NE) (2 to 8 μg/ml), on correct responses on a simple test of attention in the rat. NE stimulation was significantly different from saline alone or NE and chlorpromazine or thioridazine. NE stimulation plus haloperidol performance was significantly poorer than stimulation alone. (Abstracted from Bain, 1980.)

at least some aspect of the disease, in the animal; the state in the animal is caused by the same things that cause the state in humans; and therapy that works in humans works in the model. The model of an attentional disorder in the rat, described above, was believed to be an homologous model of one major aspect of the schizophrenia process. The etiology was a central state of arousal that was mimicked by electrical or norepinephrine stimulation of the midbrain.

The isomorphic model is similar to the homologous model except that the etiology of the behavior or state may be quite different than that of the state in humans. It is also predictive of therapeutic outcome. In the model of a duck, it looks, sounds, and smells like a duck; but its mother may have been a goose.

The last category is the predictive model. An example of this type of model in depression might be the blocking of mouse killing behavior by antidepressant drugs (Horowitz et al., 1965). In this model, the duck does not look, sound, or smell like a duck; and nothing is known about its lineage. However, the model might reveal what medication will make a duck act more like a duck.

Although an homologous model is ideal, rarely is such a model available. Sometimes a model is developed with the hope that, if the condition can be produced, then the variables used are the ones responsible for the disease. Thus, a putative homologous model is developed. An example of this type of model is the use of bilateral lesions of the olfactory bulb to cause irritability, hyperactivity, elevation of circulating levels of plasma corticosteroids, and deficits in passive avoidance (Cairncross et al., 1978). Since this model is also predictive of antidepressant activity, the belief is that by determining the neurochemical change as a result of the bulbectomy, the neurochemistry of depression might be understood.

The advantage of a completely homologous model is that it probably would not only contribute to the understanding of the brain biochemistry of depression but would lead to the development of effective drugs for treating the disease. Do such models of depression currently exist? The answer is probably yes but only for some types of human depression. The difficulty in answering the question is that many definitions are probably orthogonal to each other. Depression has been defined behaviorally as motor and cognitive slowing, defined biochemically as serotonergic or noradrenergic, and defined etiologically as endogenous or exogenous. Which definition should be used to build an homologous model? Until the disease can be more accurately described at various levels of behavior and biology, investigators may not know when they actually have an homologous model.

Since depression is a state or disease that manifests itself by altered behavioral states that can be defined as they are in DSM III, there are some models that are certainly isomorphic and may even be homologous with some types of depression. A good example is the separation model (See Chapter 1). The model has face validity and seems to meet the diagnostic criterion of DSM III. It can be said with certainty that a pathological state has been produced in the animal. However, the question is whether it is a model of endogenous depression, depression caused by separation, or, more specifically, anaclitic depression as suggested by Wilner (1984)?

Among the many isomorphic models is learned helplessness (Seligman and Beagley, 1975) (see Chapter 7). This model is predictive in that the effects of the stress can be reversed by tricyclics and MAOI antidepressants. It has a certain amount of face validity in that many depressed patients have many of the symptoms seen in animals exposed to inescapable foot shock, a major method of producing learned helplessness. This type of stressful treatment will also cause a phenomenon called stress-induced analgesia. The author believes that both models have some problems with specificity.

Although these models are modeling some aspect of depression, it is not certain that any of them have revealed very much about depression in humans that was not already known. For instance, it is known that separation from a loved one can cause depression; and that an inescapable, overwhelming, and chronic painful experience will cause depression in many. It is also known that many antidepressants as well as some nonantidepressants will ameliorate these conditions. What else have these models revealed? Has any new theory or treatment been developed that is testable?

If these models are predictive, then they can be used for the discovery of new antidepressant drugs. There may be, however, simpler predictive models than separation and learned helplessness models. These latter models are mostly carried out in infrahuman primates. Some drug models, such as the reversal of the behavioral and physiological effects of reserpine by MAOI and tricyclic antidepressants (Costa et al., 1960), are useful screening procedures that are simple and easy to perform; however, they will produce only drugs that are similar to the ones already available. These drugs probably will not reveal much about depression.

The authors in this volume describe many techniques that have already been described and have pointed out the similarities of these models to depression in people. The authors have each argued for the predictive value of their methods. However, is any more known about depression? Will the authors rush back to their laboratories eager to test new hypotheses?

Despite the author's negativism, some of the models described in this volume have helped in predicting efficacy of new drugs. More important, however, is that although these procedures are models for the investigators, they are not models to the animal. As far as the animals are concerned, the huddling monkeys of Harlow are not modeling anything. The technique allows the understanding of attachment and interaction between animals that has validity in its own right as well as being a model for the sequelae of infant separation in humans. The problem may be that the behavior observed may not be modeling the human condition called endogenous depression, but it is probably a very good homologous model of an exogenous depression. However, to the extent that the same antidepressant drugs are therapeutically useful in the treatment of both exogenous and endogenous depression, the model may be useful for the study of various types of depression.

A literature search conducted in preparation for this chapter surprisingly uncovered no reference to bipolar models of affective disorders. Also the word bipolar does not appear in any of the chapter titles for this volume. However, an induced condition that might resemble the bipolar patient is the drug abuser. George Koob's presentation on anhedonia as a model of depression may make use of this model. Since the research currently being conducted by the author is not directed toward models of depression but models of drug-induced euphoria, the author explored whether this work is a model or even a partial model of human depression.

Among those working in the field of drug abuse, a large number of clinical researchers believe that drug taking is restorative. Thus, the drug user was suffering from some form of pathology prior to use; and the drug is used to restore them to a normal state. Some investigators believe that the pathology is depression; others believe it is uncontrolled feelings of aggression; and still others feel that the pathology might be some form of metabolic disorder.

This point of view was eloquently stated by Beckett (1974): ". . . Heroin addiction in some is symptomatic of an underlying depression in a wounded personality. . . ." Khantzian (1980) argued ". . . that opiates counteracted regressed, disorganized, and dysphoric ego states associated with overwhelming feelings of rage, anger and related depression." Dole and Nyswander (1967), in their introduction of the use of methadone as a treatment modality, argued for the existence of a metabolic disturbance in the narcotic addict and that drug use was restorative.

Whether those who become heroin users are pathological prior to their introduction to drug use will not be argued here. Whether or not there is pathology prior to the person ever using heroin may not be important for purposes of this volume. However, it is clear that there is heroin induced euphoria. Heroin also counteracts the dysphoric feelings that come with chronic use. Thus, some understanding of the heroin action and a model of the affective states produced by these drugs may have some application to the study of depression. Many nonopiates also cause drug induced euphoria, with cocaine use resembling in some respects the high of mania and the crash of depression.

The author's model for the study of drug-induced euphoria is brain-stimulation reward, or as it has often been called, intracranial self-stimulation (ICSS). The author and others have found that opiate drugs as well as cocaine and many other abused substances will facilitate brain-stimulation reward. Some early investigators have suggested that depression may result from some dampening of this brain reward system (e.g., Stein, 1962; Ferster, 1973). If this were so, then antidepressant drugs should make the animal more sensitive to this stimulation, which would be manifested in a lowering of the threshold or an increase in rate of responding. Although there has been some suggestion that there is facilitation in ICSS after a single administration of antidepressant drugs, the most common effect is a depression in response rate (Wauquier, 1976). Figure 2.3 shows the failure to find any increased sensitivity to rewarding brain stimulation caused by an antidepressant drug.

Considering the normal time course of the therapeutic response to the antidepressant drugs, facilitation in ICSS would only be seen after a period of chronic drug administration. Fibiger and Phillips (1981) found such an increase in facilitation of ICSS in rats treated for two weeks with desipramine. The author and colleagues, however, were not successful in effecting such a change after chronic imipramine administration since the animals developed tolerance to the threshold raising effects of imipramine on brain-stimulation reward. This is shown in Figure 2.4. The discrepancy between these findings and those of Fibiger and Phillips could be due to differences in a number of variables, (e.g., dose, frequency of testing); however, a more likely difference probably is related to the methods used. Liebman (1983), in his review of methods for studying the effects of drugs on brain-stimulation reward, indi-

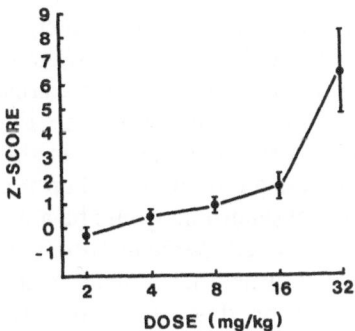

FIGURE 2.3. Mean effects of single doses of imipramine on the reward threshold. Results are expressed as the mean ± standard error of the standard (z) scores for each of the animals.

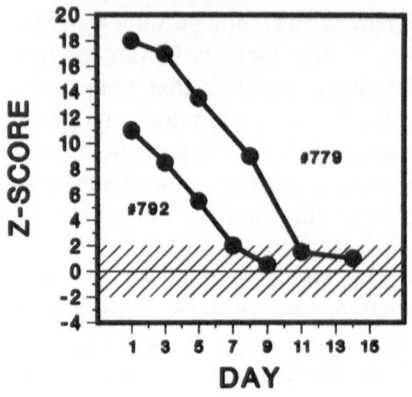

FIGURE 2.4. Effects of daily doses (32 mg/kg) of imipramine on the reward threshold for each of two animals. Results are expressed as standard scores. The shaded area indicates the approximate 95 percent confidence limits.

cated that a major problem is dissociating motor effects from effects on the reinforcing value of the rewarding stimulation. The study on the effects of imipramine reported above, a rate-independent procedure was used that allowed measurement of the threshold of the stimulation independent of motor effects (Esposito and Kornetsky, 1977; Payton and Kornetsky, 1983).

Since a common symptom of depression is an inability to appreciate life's usual rewards (e.g., food, sex, music), it would be expected that antidepressant drugs would enhance these reinforcers as well as rewarding brain stimulation.

In retrospect, it was not surprising that imipramine failed to increase the sensitivity of animals to rewarding brain stimulation. Since imipramine does not elevate mood in normal persons, what was lacking was some manipulation that modeled the depression by raising the reward threshold. Such an experiment was carried out by Leith and Barrett (1976). In this experiment, animals were made putatively depressed by withdrawal from chronic amphetamine administration. This depression, which resulted in a decrease in lever pressing for rewarding brain stimulation, was reversed by both MAO inhibitors as well as tricyclic antidepressants. To test this model further, its specificity must be established. The problem is that non-antidepressants also may reverse the effect. If it is a drug like morphine or a benzodiazepine, it can be argued that morphine is an antidepressant and that benzodiazepines are useful in some types of depression. If a neuroleptic seems to reverse the effect, then it is clear that this is not a particularly good predictive model. If, however, a case can be made that the model is a homologous model, then a great deal can be concluded about the psychology and biology of the disease. Furthermore, if the variables that produced the model result in changes in the brain that are homologous with changes in depression in humans, then it can be argued that any drug that reverses the model syndrome would be a useful therapeutic drug even if it is currently not used as such. More important, if the model is homologous, it will allow for the study of neurochemical events in the brain with some hope of determining the events that are specifically related to depression.

An interesting attempt to understand the role of the brain reward system in depression was reported in 1982 by Cornfeldt et al. and Szewczak et al. They lesioned the internal capsule, at about the level of the telecephalon-diencephalon border. This lesion caused a decrease in lever pressing for rewarding brain stimulation which was reported to be reversed by tricyclics and MAOI as well as atypical antidepressants. Non-antidepressants were not effective. This would seem to be an ideal model. If the effect of the internal capsule lesions could be understood, the phenomenon of depression might also be understood. Although there are no follow-up experiments to these 1982 reports, a recent experiment by Hubner and Koob (1987) offers some explanation as to the role of this area of the brain in affective disorders. They reported that lesions of the ventral pallidum, an area likely to have been affected by lesions of the internal capsule, caused a marked reduction in the self-administration of cocaine as well as heroin, which was interpreted as reduction in the rewarding effects of these drugs. Because drugs of abuse act on the same systems as rewarding brain stimulation, manipulation of this area of the brain may be a useful tool for modeling depression.

Animal model studies that are, at the least, isomorphic for human depression and, in some instances, homologous for some types of exogenous depression as well as relevant for understanding the behavior of the species studied have produced quality insights. The problem is that a good model of exogenous depression may not be helpful in understanding those depressions that are called endogenous and are believed to have their etiology in the neurochemistry of the individual. Furthermore, as a test of their validity many of the models use changes in the biochemistry of the brain that fit current theories, or reversal of the modeled pathology by currently used antidepressants. This type of validation is most likely to result in both the development of more drugs like the ones already available and the reinforcement of current theories concerning the underlying neurochemistry of depression. Current theories are for the most part inferential, and the experiments and the models built to test these theories are for the most part additive. Therefore, the experiments are not designed to disprove current hypotheses, but are either neutral or add more inferential evidence to already entrenched hypotheses. Unfortunately, until more is known about the neuropathology of exogenous and endogenous depression, investigators may be merely involved in a sophisticated game of neurochemical phrenology.

Acknowledgment. Preparation of this manuscript was supported by NIDA grants DA02326 and Research Scientist Award DA00099.

Further Readings

Bain G (1980): The effects of norepinephrine perfusion of the mesencephalic reticular formation on behavior in the rat: interaction with several neuroleptic drugs on a measure of sustained attention. Ph.D. Dissertation, Boston University

Baldessarini RJ and Fischer JE (1975): Biological models in the study of false neurochemical synaptic transmitters. In: *Model Systems in Biological Psychiatry,* Ingle DJ, Shein HM, eds. Cambridge, MA, MIT Press

Beckett HD (1974): Hypotheses concerning the etiology of heroin addiction. In: *Addiction,* Bourne PG, ed. New York, Academic Press

Cairncross KD, (1978): A new model for the detection of anti-depressant drugs: olfactory bulbectomy in the rat compared with existing models. *J Pharmacol Methods,* 1:131–143

Cornfeldt M, Fisher B, Fielding S (1982): Rat internal capsule lesions: a new test for detecting antidepressants. *Fed Proc* 41:1066

Costa E, Garattini S, Valzelli L (1960): Interaction between reserpine, chlorpromazine and imipramine. *Experientia* 16:461–463

Dole VP and Nyswander ME (1967): Addiction—a metabolic disease. *Arch Intern Med* 120:19–24

Eliasson M and Kornetsky C (1972): Interaction effects of chlorpromazine and reticular stimulation on visual attention behavior in rats. *Psychonomic Science* 26:261–262

Esposito R and Kornetsky C (1977): Morphine lowering of self-stimulation thresholds: lack of tolerance with long-term administration. *Science* 195:189–191

Estes WK and Skinner BF (1941): Some quantitative properties of anxiety. *J Exper Psychol* 29:390–400

Ferster CB (1973): A functional analysis of depression. *Am Psychol* 28:857–870

Fibiger HC and Phillips AG (1981): Increased intracranial self-stimulation in rats after long-term administration of desipramine. *Science* 214:683–685

Horowitz ZP, Ragozzino PW, and Leaf RC (1965): Selective block of rat mouse-killing by anti-depressants. *Life Sci* 4:1901–1912

Hubner CB and Koob GF (1987): Ventral pallidal lesions produce decreases in cocaine and heroin self-administration in the rat. *Society for Neuroscience Abstract* 13(3):1717

Hull C (1943): *Principles of Behavior.* New York: Appleton-Century

Khantzian EJ (1980): An ego-self theory of substance dependence. In: *Theories of Addiction,* Lettieri DJ, Sayers M, Wallenstein HW, eds. NIDA Monograph 30, Rockville, Maryland, National Institute on Drug Abuse

Kornetsky C (1977): Animal models: promises and problems. In: *Animal Models in Psychiatry and Neurology,* Hanin I, Usdin E, eds. New York: Pergamon Press

Kornetsky C and Eliasson M (1969): Reticular stimulation and chlorpromazine: an animal model for schizophrenic overarousal. *Science* 165:1273–1274

Kornetsky C and Markowitz R (1975): Animal models and schizophrenia. In: *Model Systems in Biological Psychiatry,* Ingle I, Shein HM, eds. Cambridge, MA: MIT Press

Leith NJ and Barrett RJ (1976): Amphetamine and the reward system: evidence for tolerance and post drug depression. *Psychopharmacology* 46:19–25

Liebman JM (1983): Discrimination between reward and performance: A critical review of intracranial self-stimulation methodology. *Neurosci Biobehav Rev* 7:45–72

Nuechterlein KH and Dawson ME (1984): Information processing and attentional functioning in the developmental course of schizophrenic disorders. *Schizophr Bulletin* 10:160–203

Payton M and Kornetsky C (1983): Brain-stimulation reward: thresholds versus response rates from various brain loci. *Society for Neuroscience Abstract* 9(2):977

Seligman MEP and Beagley G (1975): Learned helplessness in the rat. *J Comp Physiol Psychol* 88:534–541

Skinner BF (1948): "Superstition" in the pigeon. *J Exper Psychol* 83:658–666

Stein J and Urdang L (1967): *The Random House Dictionary of the English Language.* New York: Random House

Stein L (1962): New methods for evaluating stimulants and anti-depressants. In: *Psychosomatic Medicine: The First Hahnemenn Symposium,* Nodine JH, Moyer JH eds. Philadelphia: Lea and Fibiger

Szewczak MR, Fielding S, Cornfeldt M (1982): Rat internal capsule lesion: further characterization of antidepressant screening potential. *Society for Neuroscience Abstracts* 8:465

Watson JB (1914): *Behavior: An Introduction to Comparative Psychology,* Henry Holt, New York. As reviewed in R.S. Woodworth (1948): *Contemporary Schools of Psychology,* New York: Ronald Press

Wauquier A (1976): The influence of psychoactive drugs on brain self-stimulation in rats: a review. In: *Brain-stimulation Reward,* Wauquier A, Rolls ET, eds. Amsterdam: North-Holland Publishing Company

Wilner P (1984): The validity of animal models of depression. *Psychopharmacology* 83:1–16

3

Non-Homologous Animal Models of Affective Disorders: Clinical Relevance of Sensitization and Kindling

ROBERT M. POST and SUSAN R.B. WEISS

Introduction

In many ways this chapter would be better titled "uses and abuses of animal models of affective disorders." The previous two chapters have outlined some of the systematic criteria for the development and use of animal models for various psychiatric disorders, including: similarity of behaviors, inducing agents or principles, biochemical and physiological correlates, and pharmacological responsivity. These criteria have been widely accepted, as have, for example, the criteria for confirming that a given chemical is a neurotransmitter in the central nervous system. Without fulfilling each of several criteria, a putative neurotransmitter candidate cannot be considered "proven." The paradigms discussed in this chapter cannot be considered formal animal models for affective disorder, as they do not meet many of the basic requirements of an animal model. In particular, there is little evidence of (1) parallel inducing procedures, (2) behavioral homology to the disorder studied, or (3) selective response to those pharmacological interventions that are successful in affective illness. Nonetheless, these models are potentially useful for other purposes. Specifically, they help focus on possible parallel processes in the clinical situation that can be conceptualized in new ways, perhaps leading to a new series of clinical-research formulations, testable hypotheses, and even pharmacotherapies.

This chapter focuses on two animal models: (1) behavioral sensitization to the psychomotor stimulant cocaine, and (2) electrophysiological kindling of the amygdala. Although the mechanisms underlying sensitization and kindling appear quite different, both of these models demonstrate a common characteristic: responsivity increases as a function of repeated stimulation. Various aspects of affective disorder also appear to show a pattern of increasing psychopathological responsivity to repeated stimuli over time (Post et al., 1984c, 1986b). The animal models of sensitization and kindling may help focus on analogous processes that may be occurring in the course of affective disorder, even though sensitization processes in affective disorder may be mediated by quite different underlying neurobiological mechanisms from those in behavioral sensitization and kindling. Kindling and sensitization may thus provide useful analogies in considering aspects of the longitudinal course of manic-depressive illness—its ability over time to accelerate, show more rapid onsets of individual episodes, and eventually develop "spontaneous" recurrences.

Throughout the research described here (this use or misuse of animal models) there has been frequent interplay between clinical studies and those in the laboratory. A brief chronological review of these developments will illustrate this interaction and provide a background for consideration of the behavioral sensitization and kindling models for their clinical implications. Drs. W. E. Bunney, Jr. and F. K. Goodwin first suggested in 1970 that one of the authors (RP) explore the clinical utility of cocaine as an antidepressant. Cocaine, as a potent euphoriant and blocker of catecholamine reuptake, appeared to be an excellent test of the catecholamine deficiency hypothesis of depression. If the simple catecholamine hypothesis were valid, cocaine should have proven to be a superb antidepressant. This did not turn out to be the case (Post et al., 1974). Moreover, the literature on the effects of chronic cocaine administration suggested that with chronic use it, like amphetamine, may be dysphorogenic and psychotogenic (Post, 1975). The early literature also indicated that animals showed reverse tolerance or behavioral sensitization to many effects of cocaine, including hypothermia, hyperactivity, stereotypy, cataplexy, and even seizures (see review in Post and Kopanda, 1976; Post and Contel, 1983).

These developments sparked an interest in studying the effects of chronic cocaine in animals in order to explore the phenomena and mechanisms of behavioral sensitization to psychomotor stimulants using relatively low doses of cocaine (with behavioral hyperactivity or stereotypy the endpoint being measured). Attempts to find a conceptual model for the progressive development of seizures in response to repeated high-dose cocaine administration revealed interesting temporal, EEG, and behavioral similarities between cocaine-induced seizures and electrophysiological kindling, in which repeated electrical stimulation of the brain induces progressive changes in seizure thresholds and behavior and ultimately results in the production of seizures to stimulation which had previously been ineffective (Goddard et al., 1969). This concept led to studies of "pharmacological kindling" utilizing high doses of psychomotor stimulants (cocaine) and pure local anesthetics (lidocaine) with seizures as the endpoint to be measured. (Since lidocaine is as potent as cocaine as a local anesthetic but lacks cocaine's psychomotor stimulant properties, it can be used to dissociate cocaine's local anesthetic and stimulant effects.)

Work with seizure models raised the question of whether drugs that might relatively selectively dampen pathological neuronal excitability in the limbic system could be clinically useful. The limbic system has long been implicated in the modulation of affective behavior in animals and humans (Papez, 1937; MacLean, 1954; Gloor et al., 1981 [see review of Post, 1986]). The anticonvulsant carbamazepine was capable of inhibiting limbic system excitability and some types of kindling. These findings as well as the empirical observations of improvement in mood and behavior in epileptic (Dalby, 1971) and affective (Okuma et al., 1973) patients who were administered carbamazepine, indicated that clinical trials of carbamazepine in affective illness might provide an alternative therapy for unresponsive patients (Ballenger and Post, 1978b, 1980). Initial data, as well as a growing literature, now suggest that carbamazepine is clinically effective in the treatment of various phases of manic-depressive illness (Okuma 1984; Post et al., 1984a, 1986d,e).

This evidence of clinical efficacy now led the focus of research back to the laboratory in order to attempt to elucidate possible mechanisms of action of carbamazepine (Post et al., 1984d; Post, 1987). In addition, one can also ask whether some of the initial rationales for the use of carbamazepine are in fact relevant and, in particular, is the ability of carbamazepine to dampen neural activity in the limbic system related to its mechanism of action in manic-depressive illness? The evidence to date is at best equivocal (Post and Uhde, 1985). However, further clinical research with carbamazepine, as well as with a variety of other anticonvulsants that are or are not useful in the treatment of limbic system disorders, should help provide partial answers to this question.

In attempting to examine possible mechanisms of at least the anticonvulsant action of carbamazepine, the authors and colleagues observed that this drug was effective in different seizure models as a function of the stage of development of kindling (Post et al., 1984e; Weiss et al., 1987, 1988; Weiss & Post, 1987). These observations in the animal laboratory again led to the question of whether an analogous phenomenon occurs in the psychopharmacotherapy of affective disorder: i.e., can it also vary as a function of course of affective illness (Post et al., 1987b)? This question is raised not with the idea that the same drugs that are effective in manic-depressive illness will also be useful in kindling or vice versa but with the idea that the phenomenon of differential pharmacology as a function of stage in kindling may be a relevant principle for a parallel phenomenon in manic-depressive illness. The results of several studies that led to this sequence of studies of laboratory animal models and clinical investigations are discussed below.

Behavioral Sensitization

Repeated administration of the same dose of cocaine evokes a progressive increase in the degree of locomotor hyperactivity produced (Figure 3.1). This phenomenon represents a relatively long-lasting change in responsivity (weeks to months) and involves a change in both the rapidity of onset

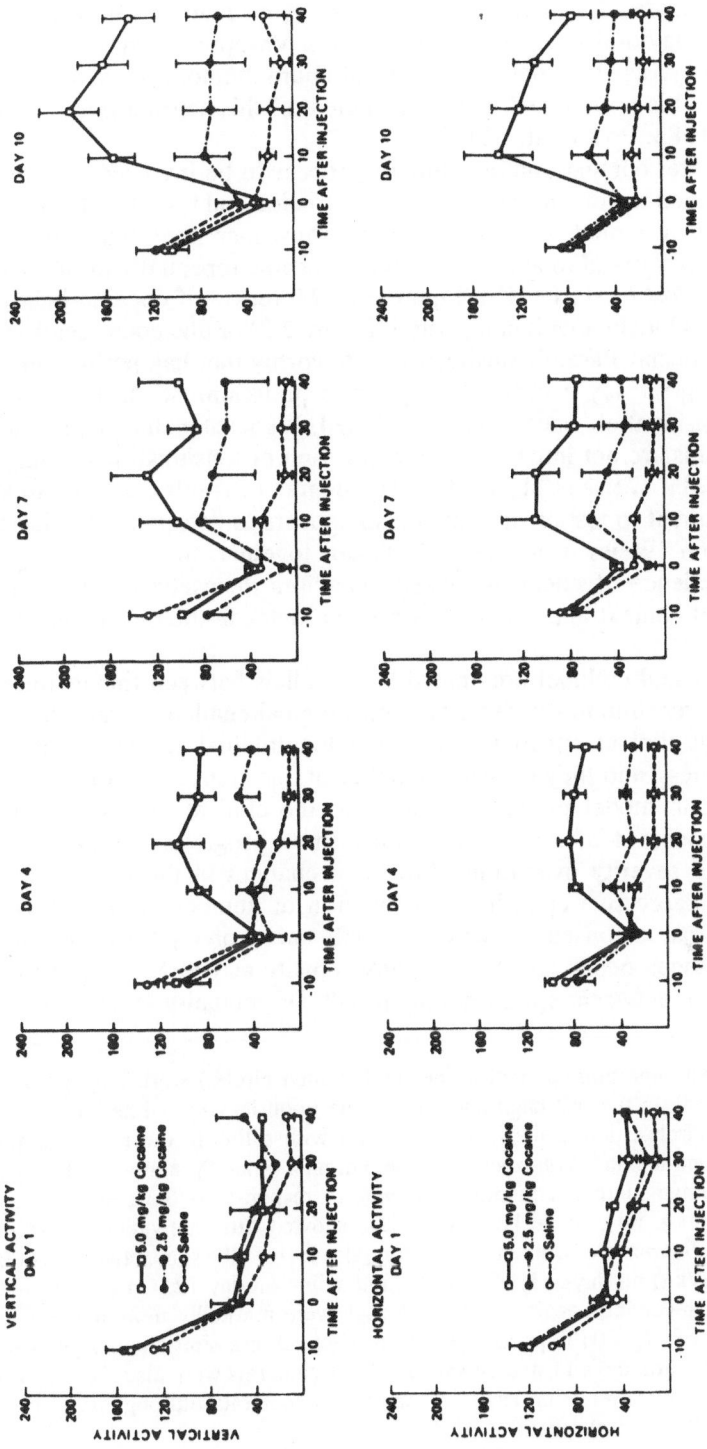

FIGURE 3.1. Female rats are more sensitive to the effects of cocaine than male rats. The progressive increases in motor response to cocaine at 5 mg/kg i.p., in the female approximate that observed in males at 10 mg/kg, i.p. Note increases in both horizontal and vertical activity as a function of day of cocaine administration (i.e., behavioral sensitization). The same once-daily dose of cocaine that had little effect on day 1 produces robust effects by days 7 or 10.

and the magnitude of hyperactivity and stereotypy. Female rats are more sensitive to behavioral sensitization than males. Vasopressin may modulate the responsivity, as it is deficient in Brattleboro homozygotes that lack vasopressin and is normalized with replacement of this neurohormone (Post and Contel, 1983; Post et al., 1982).

Environmental context and conditioning appear to be involved in behavioral sensitization to cocaine, since an animal treated and tested with cocaine in different environments does not generally show increased behavioral responsivity, as compared to animals receiving cocaine repeatedly in the same environment (Post et al., 1981b) (Figure 3.2). Moreover, if cocaine-induced hyperactivity is blocked with haloperidol (Figure 3.3) or diazepam, sensitization does not occur. Parenthetically, it is noteworthy that haloperidol (given prior to testing on day 2 rather than prior to pretreatment on day 1) will not block sensitization once it has developed, suggesting that dopaminergic mechanisms are not integral to the expression of sensitization (Beninger and Hahn, 1983; Weiss et al., 1986a). The degree of similarity of pretreatment environment to test environment also appears to determine the degree of sensitization (Weiss et al., 1986a). Taken together, these data suggest that the experience of cocaine-induced behaviors themselves in the same environmental context appears to be a critical factor in the development of sensitization.

The authors and collaborators noted the parallels between this increased responsivity over time to the same pharmacological challenge and selective aspects of the clinical phenomenology and longitudinal course of manic-depressive illness, and they formed a number of questions accordingly: does the sensitization model provide insights into the clinical observation that repeated episodes of affective illness may show progressive increases in frequency and severity over time? Are there changes in the rate of onset of individual depressive episodes as a function of number of recurrences? That is, do gradual onsets occur early while more precipitous and rapid onsets of episodes occur later, after many repetitions (Post et al., 1981a). Is the interval between episodes important? Do magnitude of response

———————————————————————————————————————→

FIGURE 3.2. Test cage animals (solid lines and shaded circles) were injected once daily with cocaine in the test cage and with saline upon leaving, while home-cage animals (open circles, dotted lines) were injected with saline in the test cage and with cocaine upon leaving. *Top row:* Test-cage animals (male Sprague-Dawley rats) showed progressive increases in motor activity in response to the same dose of cocaine (10 mg/kg, i.p.), while home-cage rats showed little activation following saline injection. *Bottom row:* Both groups of animals received the same challenge drug (cocaine, 10 mg/kg) on days 11, 13, and 14 and saline on day 12. On days 13 and 14, the animals receiving cocaine in the test cage were markedly more active than the home-cage rats ($p < .01$T, $p < .01$, $p < .05$, respectively); a similar nonsignificant trend was observed on days 11 and 14 ($p < .05$). Test-cage rats were also significantly ($p < .001$) more active after saline (day 12) than home-cage rats, although only minor degrees of activity were observed.

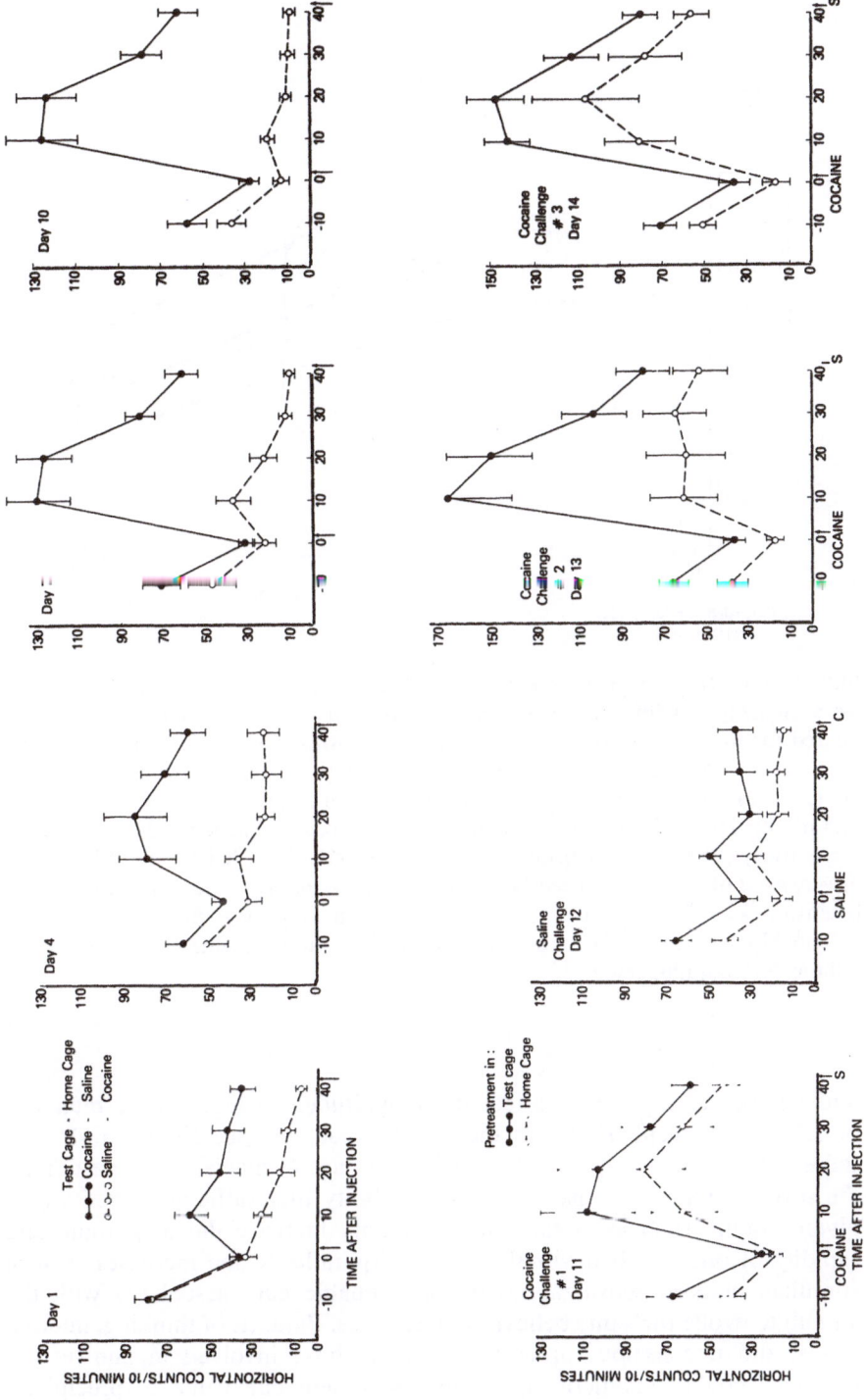

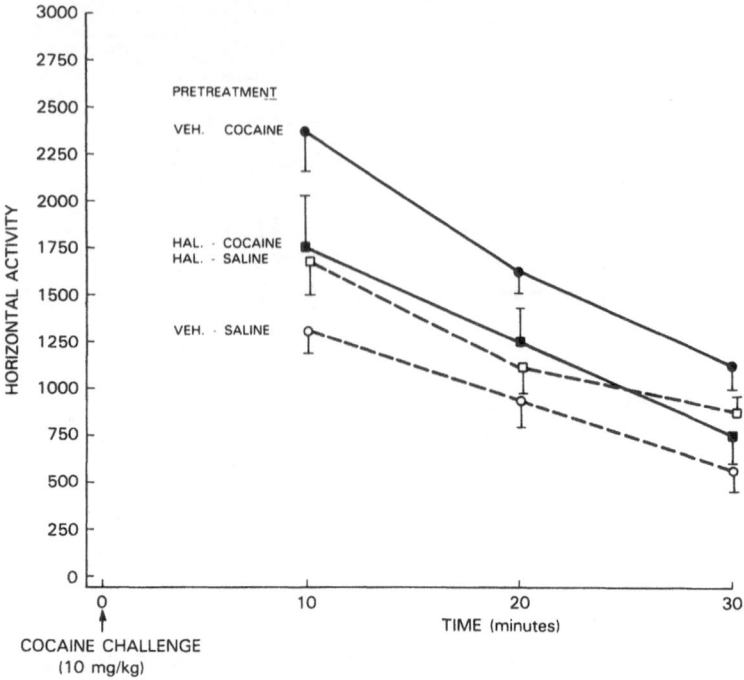

FIGURE 3.3. Day 1 injections are specified in the pretreatment column. All rats were challenged with cocaine (10 mg/kg, i.p.) on day 2 and horizontal activity plotted for the various groups. The behavioral sensitization effect of prior cocaine pretreatment (40 mg/kg, i.p.; shaded circles) on day 1 compared with saline (open circles) is illustrated (VEH = vehicle for haloperidol). In contrast, when haloperidol (0.5 mg/kg, i.p.) is given prior to cocaine or saline on day 1 (shaded and open squares, respectively), the effect of prior cocaine is blocked. Thus, haloperidol blocks the *development* of cocaine-induced behavioral sensitization. However, once the animal is sensitized by day-1 cocaine, haloperidol pretreatment prior to the day-2 challenge will not block the *expression* of cocaine-induced behavioral sensitization (Weiss et al., 1986a; data not illustrated).

and number of repetitions affect the magnitude, quality, and permanence of subsequent responsivity, as they do for cocaine (see Figure 3.4)? Are these changes over time related to conditioning mechanisms? In particular, the investigators have discussed the possibility that sufficient repetition of given stimuli, such as stress, could not only increase the magnitude and rapidity of onset of behavioral responses (parallel to the increases seen in stimulant-induced sensitization) but also enable cues associated with the stimuli to evoke the same behavioral responses. Patterns of thinking, internal states, and unconscious processes such as those involved in anniversary reactions could hypothetically acquire sufficient cue value to precipitate

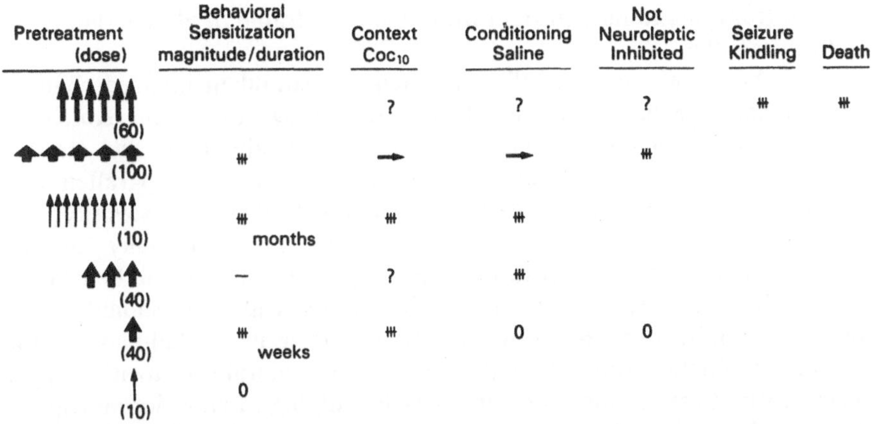

Pretreatment (dose)	Behavioral Sensitization magnitude/duration	Context Coc_{10}	Conditioning Saline	Not Neuroleptic Inhibited	Seizure Kindling	Death
↑↑↑↑↑↑ (60)		?	?	?	#	#
◆◆◆◆◆ (100)	#	→	→	#		
↑↑↑↑↑↑↑↑↑ (10)	# months	#	#			
↑↑↑ (40)	—	?	#			
↑ (40)	# weeks	#	0	0		
↑ (10)	0					

FIGURE 3.4. A dose/frequency matrix for the effects of cocaine is illustrated. Effects become more robust, more long-lasting, and less conditioned as dose and number of repetitions are increased. For example, a single injection of cocaine (10 mg/kg; small arrow) is not sufficient to produce increased behavioral activity to a rechallenge with the same dose the next day, while repeated treatment with this dose (multiple small arrows) induces a robust sensitization effect (see Figure 3.2) that lasts for several months. A single dose of cocaine (40 mg/kg; large arrow) will produce sensitization that persists for only a short period of time and in which there is no conditioned effect upon rechallenge with saline. The effects of this single dose of cocaine are blocked by neuroleptics; this is indicated by the zero in the "not neuroleptic inhibited" column. In contrast, repeated subcutaneous administration of large doses of cocaine (short arrows, row 2) produce behavioral sensitization that is not dependent on conditioning and is not blocked by neuroleptics (Gale, 1984). Finally, repeated high doses of cocaine (60 mg/kg, i.p.) result in kindling and a high probability of lethality associated with the seizures.

some of the behavioral and/or biochemical changes associated with the initial stimuli (for details, see Post et al., 1984c, 1986b).

This analysis obviously uses paralogical reasoning on one level in order to use some of the principles derived from the animal model on another level. For example, it disregards the fact that cocaine-induced behaviors in the rat are excessive and, like those of amphetamine, are perhaps a better model for mania than depression (Jacobs and Silverstone, 1986). The conditioning literature in animals does provide a model for considering how opposite physiological and behavioral responses might be triggered following the same stimulus or its withdrawal (Siegel, 1979). Such a mechanism may be relevant to some of the phenomena in manic-depressive illness, in which largely opposite behaviors (depression and mania) may occur following similar precipitants (Post et al., 1984c, 1986b). This analysis also makes the leap from cocaine-induced processes to stress-induced processes, not to mention from rats to humans and from one time-frame of effects (hours) to another (months).

While the author (Post, 1975) suggested that stimulant-induced changes may, in some instances, model those induced by stress, and Antelman has extensively documented cross-sensitization of stimulants and stresses (Antelman et al., 1980; Antelman and Chiodo, 1983), the parallels and dissimilarities are in need of further clarification. The type, severity, and timing of the stressor, as well as context and conditioning, may each be important variables in determining the degree of subsequent behavioral responsivity. Thus, simulant-induced behavioral sensitization should not be considered a mirror of stress-sensitization. However, it may help answer the question of whether some of the principles of behavioral sensitization apply to stress-sensitization and whether either is edifying in considering aspects of manic-depressive illness which appear to show sensitization in humans. The potential linkage between cocaine-induced behavioral sensitization and stress has been strengthened by the recent findings that cocaine not only potentiates effects of catecholamines and indoleamines involved in the stress response but also releases corticotropin-releasing factor (CRF) (Rivier and Vale, 1987; Calogero et al, 1988), the paradigmatic neuropeptide mediator of the stress response.

Kindled Seizures

The major characteristics of amygdala kindling are summarized in Table 3.1. A point of great clinical interest is that, after many repetitions of previously ineffective stimuli, increasing behavioral and electrophysiological effects occur which may finally emerge in a full-blown or explosive fashion. The intermittency of stimulation is critical for amygdala kindling; continuous stimulation does not evoke amygdala-kindled seizures. Once an animal has demonstrated amygdala-kindled seizures, this measure of amygdala excitability and responsivity appears to be permanently changed (Goddard et al., 1969; Racine, 1978). Moreover, if one "overkindles" an animal—i.e., produces hundreds of amygdala-kindled seizures—a phase of "spontaneity" will develop in which exogenous electrophysiological stimulation is no longer required in order to induce seizures and the animal demonstrates spontaneous epilepsy (Pinel and Rovner, 1978a, 1978b).

Given these characteristics of amygdala kindling as they relate to its specified endpoint of seizures, it is obvious that this model is inadequate and not behaviorally homologous to any aspects of the clinical phenomenology of manic-depressive illness which, by definition, does not involve a seizure disorder. Even though in kindling one can demonstrate changes in neural excitability and afterdischarge thresholds short of producing full-blown seizures, any direct extrapolation of a model of amygdala kindling to affective illness should be undertaken with caution. Nonetheless, various aspects

TABLE 3.1. Electrical kindling: major characteristics*

1. Repeated stimulations
2. Local afterdischarges progressively develop
   a. Increases in amplitude, frequency
   b. Increase in duration
   c. Increase in complexity of wave form
   d. Increase in anatomical spread
3. Replicable sequence of seizure stages
   (I) Behavioral arrest; (II) blinking, masticatory movements, head nodding; (III) contralateral, then (IV) bilateral forelimb, clonus with; (V) rearing and falling
4. Limbic system kindles more readily than cortex
5. History of kindled convulsion development is recapitulated as seizure builds
6. Transfer effects to secondary sites; kindling facilitated in other sites even after primary site destroyed
7. Interference: A secondary kindled site interferes with primary site rekindling
8. No toxic or neuropathological changes evident: kindling is a trans-synaptic process
9. Permanent change in neural excitability
10. Seizure develops spontaneously in chronically kindled animals.
11. Stage of kindling (development, completed, spontaneous) is differentially pharmacologically responsive

*See Goddard et al. (1969); Wada and Sato (1974); Wada et al. (1974); Racine (1978); Pinel and Rovner (1978a, 1978b); Pinel (1980); Post et al. (1984c).

of the kindling model may be relevant in the re-examination of longitudinally developing clinical phenomena in manic-depressive illness.

For example, the authors and colleagues are using the kindling model as a conceptual bridge in approaching the problem of how an organism may manifest a sudden eruption of behavioral and/or neurophysiological dysfunction to a stimulation which had previously been without effect. Moreover, once this threshold has been exceeded, the dysfunction can recur in a highly reproducible fashion upon each subsequent stimulation, and with sufficient repetition "spontaneity" can occur. The model may thus help focus on: (1) the processes preceding, and mechanisms for, the emergence of the first major affective episode, (2) the increasing vulnerability to relapse in manic-depressive illness, (3) the relative similarity of successive recurrences, and finally, (4) the development of spontaneous episodes.

The concept of spontaneity as it is most clearly delineated in the amygdala-kindling model may have its analogues in various neuropsychiatric syndromes including manic-depressive illness. Here the suggestion is that if there are sufficient recurrences of episodes that are either psychosocially or pharmacologically precipitated (such as with a tricyclic antidepressant or monoamine oxidase inhibitor [see discussion of Pickar et al. (1982)], episodes may begin to occur spontaneously without these apparent precipitants (see Figure 3.5 and Table 3.2). Clearly, many patients with manic-depressive illness have such rapidity of cycling that it is difficult to conceptualize how definite psycho-social precipitants could any longer play a role in the etiology of individual episodes, even though a wealth of data (although still controversial) suggests that psychosocial precipitants and

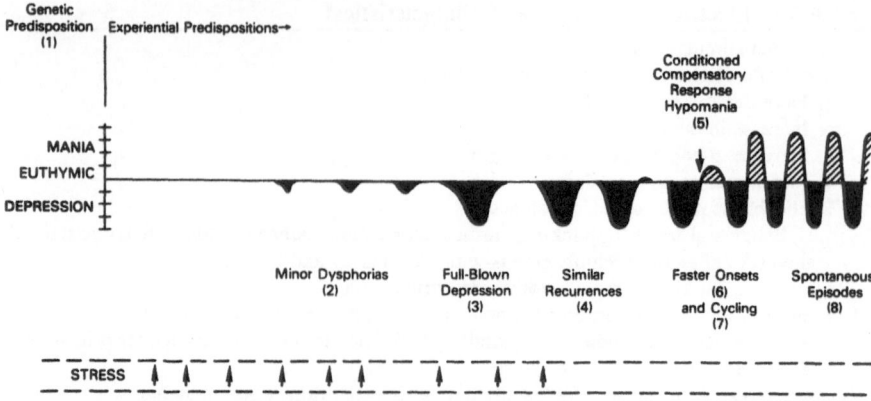

FIGURE 3.5. This figure illustrates one pattern in the evolution of recurrent affective illness in which, over time, episodes become more severe, rapid in onset, and faster in cycling (with shorter well interval). Linkage to environmental events (deprivations, stresses, losses) also evolves from episodes that are triggered (arrows) to those that are "spontaneous." See text for discussion.

stresses may be important in some, perhaps earlier, stages of affective illness (Paykel, 1979; Dunner and Hall, 1980; Brown et al., 1975; Amelas, 1979; Lloyd, 1980).

Thus, manic-depressive illness is phenomenologically very distant from anything resembling amygdala-kindled seizures. Yet, there are some interesting links between the seizures of temporal lobe epilepsy and affective disorders (see discussion of Post et al., 1986c) as well as electroconvulsive

TABLE 3.2. Parallels in the phenomenology of affective illness to kindling and sensitization

Characteristic of Affective Illness	Characteristic of Kindling	Sensitization
1. Genetic component	+	+
2. Early experience predisposes	+	+
3. Mild alterations emerge as full-blown episodes	+	+
4. Episodes reach plateau and are similar in content and behavior over repeated recurrences	+	+
5. Repeated episodes may lead to emergence of opposite phase (conditioned compensatory reactions)	+	+
6. Onset of maximum disturbance occurs earlier in episode	+	+
7. Vulnerable to recurrences and cycles may accelerate	+	?
8. Early episodes may be precipitated, later appear spontaneously	+	?
9. Lithium carbonate effective	−	(+)
10. Carbamazepine effective	(+)	−

1–8: See corresponding numbers in Figure 3.5.
+ Evidence positive.
− Evidence negative.
(+) Evidence inconsistent.

seizures and antidepressant effects (Post et al., 1984b; 1986a). The relationships appear to be complex, and any direct equation between seizures and affective illness remains full of pitfalls. It is perhaps easier to conceptualize kindling in a literal fashion as a mechanism in alcohol withdrawal syndromes (Ballenger and Post, 1978a) or with chronic cocaine intoxication (Post et al., 1987b), in which actual seizures can be observed as an endpoint. That is, repeated episodes of alcohol withdrawal or repeated administration of cocaine may eventually result in the production of full-blown seizures following bouts of alcohol withdrawal or cocaine intoxication that had previously been well tolerated. It is also of interest that in both of these models, a range of behavioral disorders may show a sensitization or kindling-like time-course. For example, one can observe delirium tremens in the absence of seizures and/or one can observe the late occurrence of cocaine-induced dysphoria, psychosis, and panic attacks, also without a seizure endpoint. It is of particular interest that cocaine-induced panic attacks can eventually evolve into a spontaneous phase; that is, following sufficient repetitions of cocaine-related panic attacks, the panic attacks may begin to occur in the absence of cocaine administration (Uhde et al., unpublished data; Post et al., 1987b).

Thus, although cocaine can "kindle" to a seizure endpoint, behavioral changes (such as panic attacks) follow a kindling-like time-course and can also be observed short of seizures. One of the advantages of using seizures as an endpoint in the kindling model is their ease of production, their reproducibility, quantifiability, and amenability to psychopharmacological dissection. Thus, one can ask a series of questions about the ability of pharmacological agents to interfere with limbic seizure thresholds, seizure and after discharge duration, or seizure stage. Further, one can question whether relative efficacy of anticonvulsants on a measure such as amygdala-kindled afterdischarges is related to degree of clinical efficacy in manic-depressive illness (Table 3.3).

At the moment, the data are quite sparse as to whether the ability to act as an anticonvulsant at limbic system sites is in fact associated with efficacy in manic-depressive illness. While carbamazepine, valproic acid, clonazepam, and phenytoin each have been reported to be effective in some aspects of manic-depressive illness (Post and Uhde, 1986), clonazepam and valproic acid are not generally considered first-line treatments for psychomotor epilepsy. Moreover, although phenytoin and carbamazepine appear to show approximately equal efficacy in the treatment of psychomotor epilepsy, carbamazepine is more potent than phenytoin in inhibiting amygdala-kindled versus cortical-kindled seizures in animals (Albright and Burnham, 1980). The assessment of the relative efficacy of these two anticonvulsants may be important in elucidating their possible mechanisms of action on a biochemical level. For example, carbamazepine and phenytoin show equal ability to stabilize sodium channels in physiological and biochemical laboratory preparations (MacDonald et al., 1985; Willow et al., 1985). Thus, if both drugs show equal efficacy in manic-depressive illness, one might consider sodium

TABLE 3.3. Anticonvulsants in manic-depressive illness

Inhibition of Amygdala-Kindled After-Discharge		Mania	Depression	Prophylaxis
++	Electroconvulsant therapy	++	++	±
++	Carbamazepine	++	+	++
++	Valproic acid (+ lithium)	++	±	++
	Clonazepam	++		±
	Acetazolamide	+		+
	Alprazolam	——	+	——
+	Diazepam (i.v. only)	±		
	Progabide	±	+	
±	Phenytoin	±	±	

++ = Robust effects, well documented.
+ = Positive effects, less well documented.
± = Equivocal effect, requires further study.
—— = Negative effect or exacerbation.
Blank = Data not available.

channel stabilization as a candidate for this effect. If, however, as initial data suggest, carbamazepine emerges as a more effective antimanic agent than phenytoin, one would have to consider other potential mechanisms of carbamazepine as critical to this differential response (i.e., mechanisms that are not shared by phenytoin). Included among these are the abilities: to enhance firing of the locus coeruleus, to increase plasma tryptophan, to bind to adenosine receptors, to act as a vasopressin agonist, to bind and physiologically interact with "peripheral-type" benzodiazepine receptors, and to act preferentially at limbic- versus cortical-kindled loci (Post et al., 1984d; Post, 1987).

In contrast to the antimanic and antidepressant effects of carbamazepine, in which the critical biochemical substrates remain elusive, current evidence is quite strong that noradrenergic (alpha$_2$), "peripheral-type" benzodiazepine receptor and sodium channel stabilization mechanisms may be importantly involved in the anticonvulsant effects of carbamazepine. Consequently, kindled seizures may serve as a first-line screening device for examining possible mechanisms of action of this potent psychotropic drug. Once probable anticonvulsant mechanisms of carbamazepine are elucidated, one can ask whether these mechanisms are also involved in its psychotropic effects. A second generation of animal models, as well as clinical studies, may then be conducted in order to further assess these possibilities, once the field of interest has been focused with this particular use of distant animal modeling (i.e., seizures). Since the anticonvulsant and antinociceptive effects of carbamazepine are apparent almost immediately, while the antimanic and antidepressant effects may be delayed for several weeks, different mechanisms may underlie the efficacy of carbamazepine in these different syndromes (Post, 1987, 1988). Thus, development of an animal seizure model that requires chronic carbamazepine administration to demonstrate anticonvulsant effects might be one step closer to identifying the mechanisms underlying the psychotropic effects of carbamazepine, which also require chronic administration.

Recent data from the authors' laboratory (Weiss et al., 1987, 1988) suggest that chronic carbamazepine is required in order to inhibit the development of lidocaine- and cocaine-kindled seizures and their associated lethality. In these instances, chronic oral carbamazepine is effective, but repeated, intermittent (i.p.) administration is ineffective. Elucidating biochemical effects of chronic versus repeated-intermittent carbamazepine may thus provide new insights into the mechanisms related to the psychotropic effects of carbamazepine.

Pharmacotherapy of Kindling Varies as a Function of Stage of Development

In attempting to examine the range of effects of carbamazepine on amygdala-kindled seizures, the author and collaborators discovered that carbamazepine, as well as other anticonvulsant agents, was effective in some stages of amygdala kindling but not others. For example, carbamazepine is unable to block the early developmental phases of amygdala kindling in the rat (although it is effective in the cat and monkey in these early phases), while it is highly effective (if not the most effective) for completed amygdala-kindled seizures in the rat (Weiss and Post, 1987) (Figure 3.6). In contrast pharmacologically kindled seizures, induced by the local anesthetic lidocaine, were differentially responsive to carbamazepine as a function of kindled seizure stage, but in the opposite direction (Weiss et al., 1986b). That is, carbamazepine was potent in inhibiting the development of lidocaine- and cocaine-kindled seizures but was without effect on these seizures once they were fully developed.

As summarized in Figure 3.7, differential responsivity as a function of stage of evolution appears to be a general principle among a variety of anticonvulsants. Pinel (1983) reported that diazepam was highly effective on completed kindled seizures but ineffective in the treatment of spontaneous seizures. In contrast, phenytoin, which is relatively ineffective in the developmental and completed phases of kindling, was quite potent in inhibiting spontaneous seizures. These observations of a differential pharmacotherapy of amygdala kindling as a function of kindling stage are particularly interesting from the perspective that kindling, on the surface, appears to represent a continuous and homogeneous process induced by the same one-second stimulation of the amygdala per day, yet displays very different pharmacological responsivity over time. By inference, differential biochemical mechanisms must underlie these different stages of amygdala kindling development, evolution, and eventual spontaneity.

Amygdala kindling has been considered by Goddard and Douglas (1975)

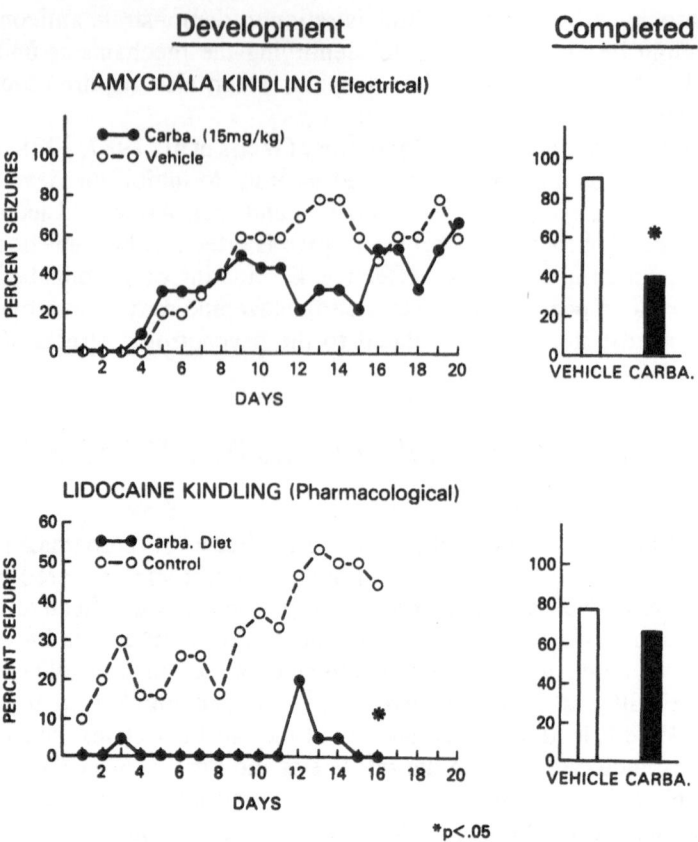

FIGURE 3.6. The early developmental phase of amygdala kindling in the rat is not responsive to carbamazepine treatment (top left), while the mid phase, wherein seizures are regularly triggered by amygdala stimulation, is very responsive to carbamazepine (also 15 mg/kg, i.p. top right, * = p<.01). Conversely, the early developmental phase of pharmacological kindling with lidocaine (bottom; * = p<.01) or cocaine (not illustrated) is very responsive to carbamazepine chronically administered in the diet, but no dose of carbamazepine is able to reverse completed lidocaine-kindled seizures (bottom right) or high-dose acute cocaine seizures (not illustrated). Thus, carbamazepine is effective in some stages of the kindling process but not others, and this varies according to different types of kindling.

as a model of neural learning and memory. Perhaps in an analogous way, it is not surprising that different phases of short- and long-term memory consolidation have been described which appear to have very different time constants, permanence of memory traces, and responsiveness to psychopharmacological intervention. Other medical syndromes also appear to show differential pharmacotherapy as a function of stage of evolution of the illness. For example, in parkinsonism, early stages of the illness may be exquisitely

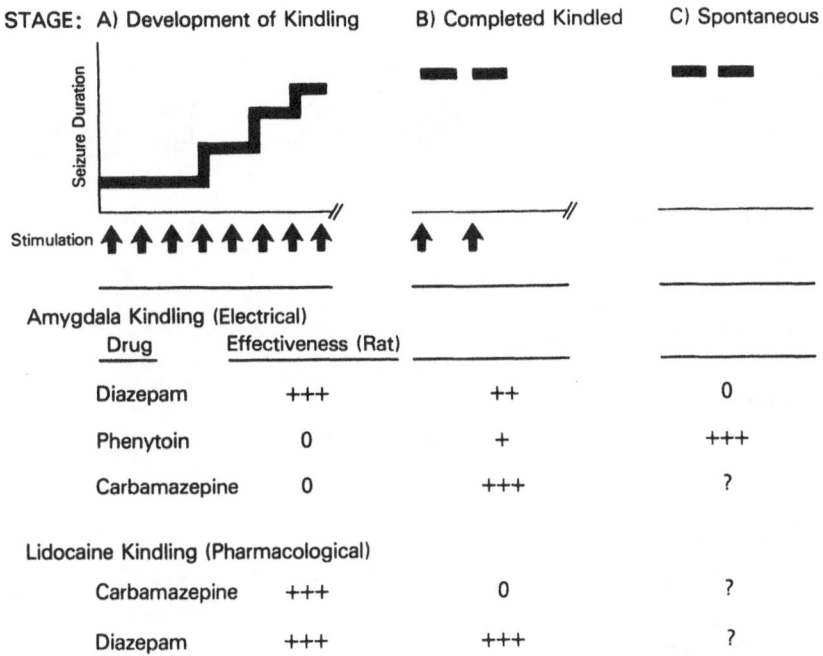

STAGE: A) Development of Kindling	B) Completed Kindled	C) Spontaneous

Amygdala Kindling (Electrical)

Drug	Effectiveness (Rat)		
Diazepam	+++	++	0
Phenytoin	0	+	+++
Carbamazepine	0	+++	?

Lidocaine Kindling (Pharmacological)

Carbamazepine	+++	0	?
Diazepam	+++	+++	?

FIGURE 3.7. This figure summarizes the data in Figure 3.6 and adds data of Pinel (1983) regarding "C," the spontaneous phase of kindling during which seizures occur without exogenous stimulation. Diazepam is very effective in inhibiting seizures (+++, ++) during (A) the development and (B) completed phase of kindling, but is ineffective in (C) the spontaneous phase. Conversely, phenytoin is ineffective (D) or weakly effective (+) in the early phases, but highly effective in the late spontaneous phase.

responsive to either DOPA or dopamine agonist treatment, but later stages may require both drugs in combination and, eventually, neither may be effective.

In a similar fashion, the differential effects of carbamazepine on amygdala kindling have recrystallized the question of whether the pharmacoresponsivity of manic-depressive illness may also vary as a function of stage in the course of its evolution. As illustrated in Figure 3.8, one might ask whether psychotherapy and benzodiazepines are more effective in the treatment of early and mild dysphoric reactions while later, more severe depressive episodes are often optimally treated with tricyclics and monoamine oxidase inhibitor antidepressants. Lithium carbonate, likewise, appears to be effective in the treatment of early and mid phases of recurrent unipolar and bipolar illness. However, a robust literature suggests that lithium is less effective in patients with rapid-cycling manic-depressive illness, particularly in those patients with more than four episodes per year (Dunner and Fieve, 1974; Kukopulos et al., 1980; Himmelhoch, 1984; Hanus and Zapletalek,

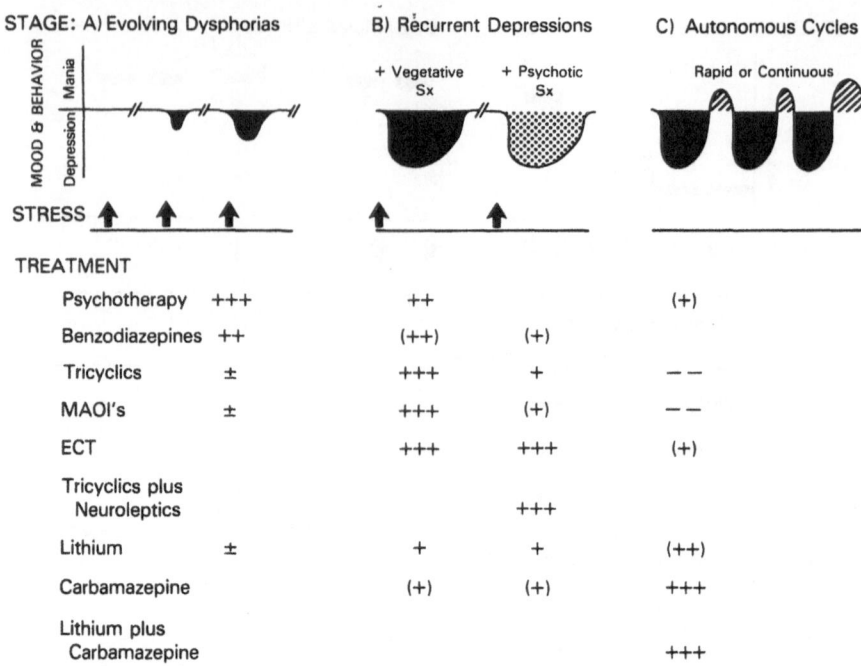

FIGURE 3.8. A provisional schema based on existing data is organized to illustrate the possible relationship of treatment efficacy to longitudinal course of illness organized roughly into stages A, B, C in parallel with stages of kindling in Figure 3.7—i.e., (A) early development, (B) completed or full-blown, and (C) spontaneous or autonomous. Efficacy ranges from +++ (strong or well-documented) to ± (equivocal) to— (inefficacy or relative contraindication), where tricyclics and MAOI have the liability of inducing rather than treating rapid-cycling bipolar illness. Parentheses represent areas where data are weak or findings are inconsistent and in need of further study. Blanks represent the relative absence of data. This schema is presented as provisional and for its heuristic value, not with the perspective that all would agree with the rough classifications of efficacy which may change considerably with further study. Nonetheless, across many rows (different types of treatments) the efficacy does appear to change as a function of course-of-illness.

1984; Prien, 1985). As this rapid-cycling phase of the illness often appears as a late manifestation of affective illness, systematic studies directly testing the hypothesis that lithium is more effective in early and mid phases of the illness but less so in the later phases should provide insight into this question.

Preliminary data from three laboratories (Post et al., 1987a; Kishimoto and Okuma, 1985; and Joffe, personal communication, September 1987) suggest that rapid- or continuous-cycling manic-depressive illness is a positive predictor of response to carbamazepine. Thus, carbamazepine and related anticonvulsants such as valproic acid might be particularly effective in the late stages of manic-depressive illness, which is often characterized by more rapid cycling. The schema in Figure 8 is presented for its heuristic value and in order to generate specific tests of this hypothesis, not with the assumption that the initial data unequivocally support this conceptualization. In fact, there appears to be a range of possibilities that might help explain the preliminary evidence of differential predictors of response to lithium versus carbamazepine (Post et al., 1987a). These might include: (1) initial clinical severity of mania, (2) type of mania (dysphoric), (3) pattern of illness (rapid or continuous cycling with sharp onsets and offsets), (4) genetic subtype (family history negative); as well as (5) the latter suggestion that response to carbamazepine may vary as a function of course-of-illness and be more effective in later stages. Direct tests of carbamazepine's efficacy in early phases of illness with and without rapid cycling should help clarify and unravel several of the possibilities mentioned.

Clearly, much clinical work remains to be performed in order to choose among these various alternatives. However, the clarity of the phenomenon of differential pharmaco-responsivity as a function of stage-of-development of kindling helps to raise and focus on the question of course-of-illness as a possible factor in pharmaco-responsivity in manic-depressive illness with a new perspective and cogency. Again, these observations do not support a literal extrapolation from the drugs that work in one phase or another of amygdala kindling to a parallel phenomenon in manic-depressive illness, particularly in light of the differential effects of carbamazepine in different stages of kindling, depending on whether one is considering electrical kindling of the amygdala or pharmacological kindling with lidocaine. However, they do suggest that the principle derived (pharmaco-responsivity as a function of stage-of-illness) is worthy of further clinical investigation.

Similarly, the sensitization and kindling models offer possible direct tests of this perspective on the progressive course of manic-depressive illness. One would postulate that, if these conceptual models were valid, adequate pharmaco-prophylaxis of recurrent affective episodes with either lithium or carbamazepine should result in a less malignant course-of-illness than that seen in patients whose recurrent episodes are only acutely treated and where no prophylaxis is employed. That is, the prevention of affective episodes themselves should prevent the development of some aspects of behavioral sensitization if such sensitization occurs in manic-depressive illness. The differential prediction in this model is illustrated in Figure 3.9.

Conclusions

This chapter has been an attempt to illustrate, with a variety of examples drawn primarily from experience, the utility of crossing studies and conceptual models back and forth between the clinic and the laboratory. Clinical observations often raise questions that can be adequately answered only in

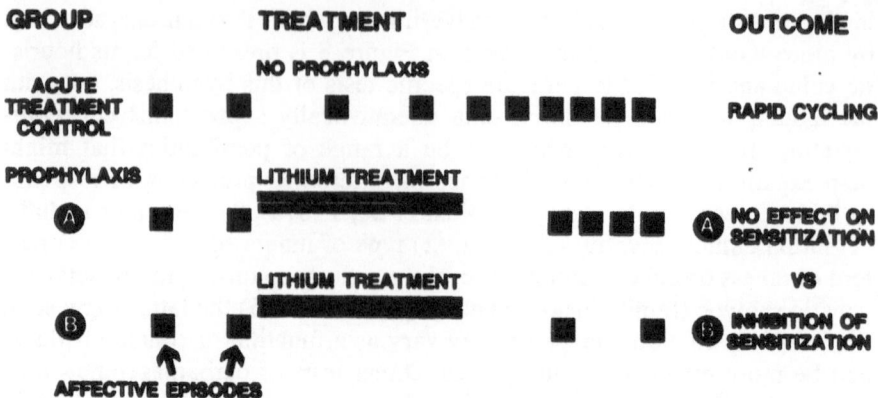

FIGURE 3.9. Differential outcome of successful lithium prophylaxis is illustrated in A and B in comparison with natural course of illness, with its propensity for showing shorter well intervals as a function of repeated episodes (top line). If lithium did not block the sensitization process, episodes at A would reappear at the expected baseline frequency. If lithium blocked the development of sensitization (as in the last line, B), episodes would reemerge at a slower frequency and longer interval (i.e., similar to the interval between the second and third and the third and fourth episodes in the no-prophylaxis group).

the controlled laboratory setting. Reciprocally, use of laboratory models may propel a new series of clinical questions. The work of this research team with behavioral sensitization and electrophysiological and pharmacological kindling has, if nothing else, sensitized its members to a variety of clinical issues relevant to the longitudinal course of manic-depressive illness. The longitudinal perspective has been emphasized by some of the earliest workers in the field, including Kraepelin (1921), but has often been overlooked in current psychological and biochemical theories of affective illness, which have tended to focus on abnormalities in acute episodes and their acute treatment.

Perhaps the time is also ripe for a conceptual approach to the longitudinal course of affective illness because of the rather new development of a range of treatment options in addition to lithium carbonate for bipolar manic-depressive patients (Post and Uhde, 1986). It is thus of increasing importance to ask whether the course of illness and other variables are critical determinants of pharmacological responsivity. If pharmacological efficacy does change as a function of course-of-illness, different underlying psychological and biological mechanisms are implied. It also remains very much an open question as to whether some of the initial theoretical rationales that sparked this interest in the amygdala-kindling concept prove to be correct in relation to carbamazepine's mechanism of action in affective illness (Post and Uhde, 1985). Carbamazepine has been proven effective in a variety of neuropsychiatric syndromes, including trigeminal neuralgia and related paroxysmal pain syndromes, which do not necessarily represent either a convulsive disorder or a disorder of limbic system function. Therefore, the emerging positive empirical observations of carbamazepine's clinical efficacy in manic-depressive illness do not alone support the argument that kindling is occurring, nor do they firmly uphold some of the investigators' initial theoretical rationales regarding limbic mechanisms in affective illness. In fact, the evidence that directly relates carbamazepine's efficacy in manic-depressive illness to its ability to stabilize limbic system excitability is quite sparse. This hypothesis, nonetheless, remains an area of active investigation and one that is likely to be rewarding, whether or not the answer is positive.

In this chapter, the authors have attempted to provide a number of caveats on the use and interpretation of animal models in relation to clinical phenomena in affective illness while emphasizing the utility of considering animal models that are less than specific or homologous for the illness being considered.

Conceptualizations of the theoretical underpinnings of manic-depressive illness have often been polarized between those emphasizing psychological and psychodynamic principles and those touting biological substrates and causes. Consideration of the two animal models discussed here suggests the possibility that neither approach is mutually exclusive, that the two may be interactive, and that a more prominent role of one or the other may even occur as a function of stage in the course of evolution of the illness. In fact, these concepts lead to the testable postulate that psychosocial precipitants may be more clear-cut early rather than later in the course of manic-depressive illness, although the role of psychological variables, anniversary reactions, and conditioning may nonetheless be prominent factors even late in the course of the illness. If behavioral response to the psychomotor stimulant cocaine can vary as a function of prior exposure, experience, and environmental context in the lowly rat, it perhaps becomes less difficult to argue that, in humans, a series of psychobiological variables might also be important determinants of behavioral responsivity to stresses and endogenous perturbations that are pertinent to manic-depressive illness.

Altered secretion of a stress-related peptide, corticotropin-releasing hormone (CRH), has been postulated in affective illness (Gold et al., 1984; 1986). Excessive secretion of CRH could account for some of the mood and vegetative symptoms of depression and its endocrine concomitants, such as cortisol hypersecretion. CRH administered intracerebroventricularly in rats produces dose-related effects on motor activity, inhibition of feeding and sexual behavior, "anxiety-like" behaviors, and, in high doses, the late onset of aggression and limbic seizures (Ehlers et al., 1983; Weiss et al., 1986b; Koob and Bloom, 1985; Koob, 1985). While Ehlers has postulated that CRH "kindles" limbic seizures, Weiss et al. (1986b) found tolerance to the effects of repeated CRH. Nonetheless, it is of interest that a putative stress-related hormone is capable of inducing both limbic seizures and a variety of behaviors that are altered in affective illness. Cain and Corcoran (1984) reported that several endogenous opiate peptides related to enkephalin could also produce limbic seizures when injected into discrete areas of the amygdala. This paradigm more closely represents a typical kindled phenomenon than that observed after repeated CRF—i.e., increasing response and eventual convulsions to a previously subconvulsant dose.

Thus, two peptides that can be released during stress, CRH into CSF (Koob and Bloom, 1985; Koob, 1985) and endogenous opiates into the amygdala (Seeger et al., 1984), may be capable of eliciting a variety of phenomena altered in affective illness and inducing limbic seizures and, perhaps, kindling. Some of the "rules" that apply to the development and maintenance of behavioral sensitization to cocaine and various types of kindling may thus be relevant to the effects of endogenous peptides, especially since cocaine has recently been shown to release CRF (Rivier and Vale, 1987; Calogero et al., 1988). Clearly, many studies can be performed which would more directly test this proposition and the utility of these paradigms as indirect models of affective disorders.

The kindling and sensitization models raise a series of new questions and new perspectives in examining the longitudinal course of manic-depressive illness. While limited as models of affective illness, they have already been productive in generating new approaches and therapies. The development of better animal models (several of which are presented in following chapters) will enable investigators to dissect the critical behavioral principles involved and the underlying neural substrates of the affective disorders.

References

Albright P, Burnham W (1980): Development of a new pharmacological seizure model: effects of anticonvulsants on cortical- and amygdala-kindled seizures in the rat. *Epilepsia* 21:681-689

Amelas A (1979): Psychologically stressful events in the precipitation of manic episodes. *Br J Psychiatry* 135:15-21

Antelman SM, Chiodo LA (1983): Amphetamine as a stressor. In: *Stimulants: Neurochemical, Behavioral and Clinical Perspectives,* Creese I, ed. New York: Raven Press

Antelman SM, Eichler AJ, Black CA, Kocan D (1980): Interchangeability of stress and amphetamine in sensitization. *Science* 207:329-331

Ballenger JC, Post RM (1980): Carbamazepine (Tegretol) in manic-depressive illness: a new treatment. *Am J Psychiatry* 137:782-790

Ballenger JC, Post RM (1978a): Kindling as a model for the alcohol withdrawal syndromes. *Br J Psychiatry* 133:1-14

Ballenger JC, Post RM (1978b): Therapeutic effects of carbamazepine in affective illness: a preliminary report. *Commun Psychopharmacol* 2:159-178

Beninger RJ, Hahn BL (1983): Pimozide blocks establishment but not expression of amphetamine-produced environment-specific conditioning. *Science* 220:1304-1306

Brown GW, Bhrolchain MN, Harris T (1975): Social class and psychiatric disturbance among women in an urban population. *Sociology* 9:225-254

Cain DP, Corcoran ME (1984): Intracerebral beta-endorphin, met-enkephalin and morphine: kindling of seizures and handling-induced potentiation of epileptiform effects. *Life Sci* 34:2535–2543

Calogero AE, Kling MA, Bernardini R, Gallucci WT, Post RM, Chrousos GP, Gold PW (1988): Cocaine stimulates rat hypothalamic corticotropin releasing hormone secretion *in vitro. Clin Res Abstracts,* April 1988

Dalby MA (1971): Antiepileptic and psychotropic effects of carbamazepine (Tegretol) in the treatment of psychomotor epilepsy. *Epilepsia* 12:325–334

Dunner DL, Fieve RR (1974): Clinical factors in lithium prophylactic failure. *Arch Gen Psychiatry* 30:229–233

Dunner DL, Hall KS (1980): Social adjustment and psychological precipitants in mania. In: *Mania: An Evolving Concept,* Belmaker RH, van Praag HM, eds. New York: Spectrum Publications

Ehlers CL, Henriksen SJ, Wang M, Rivier J, Vale W, Bloom FE (1983): Corticotropin releasing factor produces increases in brain excitability and convulsive seizures in rats. *Brain Res* 278:332–336

Gale K (1984): Catecholamine-independent behavioral and neurochemical effects of cocaine in rats. In: *Mechanisms of Tolerance and Dependence, NIDA Research Monograph Series 54,* Sharp CW, ed. Washington DC: US Government Printing Office

Gloor P, Olivier A, Quesney LF (1981): The role of the amygdala in the expression of psychic phenomena in temporal lobe seizures. In: *The Amygdaloid Complex, INSERM Symposium #20,* Ben-Ari Y, ed. Amsterdam: Elsevier/North Holland Biomedical Press

Goddard GV, Douglas RM (1975): Does the engram of kindling model the engram of normal long term memory? *Can J Neurol Sci* 2:385–394

Goddard GV, McIntyre DC, Leech CK (1969): A permanent change in brain function resulting from daily electrical stimulation. *Exp Neurol* 25:295–330

Gold PW, Chrousos G, Kellner C, Post RM, Roy A, Avgerinos P, Schulte H, Oldfield E, Loriaux DL (1984): Psychiatric implications of basic and clinical studies with corticotropin-releasing factor. *Am J Psychiatry* 141:619–627

Gold PW, Loriaux DL, Roy A, Kling MA, Calabrese JR, Kellner CH, Nieman LK, Post RM, Pickar D, Gallucci W (1986): Responses to corticotropin-releasing hormone in the hypercortisolism of depression and Cushing's disease: Pathophysiologic and diagnostic implications. *N Engl J Med* 314:1329–1335

Hanus H, Zapletalek M (1984): The prophylactic lithium treatment in affective disorders and the possibilities of the outcome prediction. *Sb Ved Pr Lek Fak Univ Karlovy* 27:5–75

Himmelhoch J (1984): Carbamazepine: an alternative to lithium? New Clinical Drug Evaluation Unit Annual Meeting, Key Biscayne, Florida, May 20–23, 1984

Jacobs D, Silverstone T (1986): Dextroamphetamine-induced arousal in human subjects as a model for mania. *Psychol Med* 16:323–329

Kishimoto A, Okuma T (1985): Antimanic and prophylactic effects of carbamazepine in affective disorders. Abstr, IVth World Congress of Biol Psychiatry, Philadelphia, pg 363, #506.4

Koob GF (1985): Stress, corticotropin releasing factor and behavior. In: *Perspectives in Behavioral Medicine, Neuroendocrine Control and Behavior,* Williams RB, ed. New York: Academic Press

Koob GF, Bloom FE (1985): Corticotropin-releasing factor and behavior. *Fed Proc* 44:259–263

Kraepelin E (1921): *Manic-Depressive Insanity and Paranoia.* Translated by Barclay RM, Robertson GM, ed. Edinburgh: E & S Livingstone

Kukopulos A, Reginaldi D, Laddomada P, Floris G, Serra G, Tondo L (1980): Course of the manic-depressive cycle and changes caused by treatments. *Pharmacopsychiatria* 13:156-167

Lloyd C (1980): Life events and depressive disorder reviewed. II. Events as precipitating factors. *Arch Gen Psychiatry* 37:541-548

MacDonald RL, McLean MJ, Skerritt JH (1985): Anticonvulsant drug mechanisms of action. *Fed Proc* 44:2634-2639

MacLean PD (1954): The limbic system and its hippocampal formation: studies in animals and their possible application to man. *J Neurosurg* 11:29-44

Okuma T (1984): Therapeutic and prophylactic efficacy of carbamazepine in manic-depressive psychosis. In: *Anticonvulsants in Affective Disorders,* Emrich HM, Okuma T, Muller AA, eds. Amsterdam: Excerpta Medica

Okuma T, Kishimoto A, Inoue K, Matsumoto H, Ogura A, Matsushita T, Naklao T, Ogura C (1973): Anti-manic and prophylactic effects of carbamazepine on manic-depressive psychosis. *Folia Psychiatr Neurol Jpn* 27:283-297

Papez JW (1937): A proposed mechanism of emotion. *Arch Neurol Psychiatry* 38:725-743

Paykel ES (1979): Causal relationship between clinical depression and life events. In: *Stress and Mental Disorder,* Barrett JE, Rose RM, Klerman GL, eds. New York: Raven Press

Pickar D, Murphy DL, Cohen RM, Campbell IC, Lipper SC (1982): Selective and non-selective MAO inhibitors: behavioral disturbances in depressed patients. *Arch Gen Psychiatry* 39:535-540

Pinel JP (1983): Effects of diazepam and diphenylhydantoin on elicited and spontaneous seizures in kindled rats: a double dissociation. *Pharmacol Biochem Behav* 18:61-63

Pinel JPJ, Rovner LI (1978a): Electrode placement and kindling-induced experimental epilepsy. *Exp Neurol* 58:335-346

Pinel JPJ, Rovner LI (1978b): Experimental epileptogenesis: kindling-induced epilepsy in rats. *Exp Neurol* 58:190-202

Post RM (1975): Cocaine psychoses: a continuum model. *Am J Psychiatry* 132:225-231

Post RM (1986): Does limbic system dysfunction play a role in affective illness? In: *The Limbic System: Functional Organization and Clinical Disorders,* Doane BK, Livingston KE, eds. New York: Raven Press

Post RM (1987): Mechanisms of action of carbamazepine and related anticonvulsants in affective illness. In: *Psychopharmacology: A Third Generation of Progress,* Meltzer H, Bunney WE Jr, eds. New York: Raven Press

Post RM (1988): Time course of clinical effects of carbamazepine. Implications for mechanisms of action. *J Clin Psychiatry* (Suppl), 49:35-46

Post RM, Ballenger JC, Rey AC, Bunney WE Jr (1981a): Slow and rapid onset of manic episodes: implications for underlying biology. *Psychiatry Res* 4:229-237

Post RM, Ballenger JC, Uhde TW, Bunney WE Jr (1984a): Efficacy of carbamazepine in manic-depressive illness: implications for underlying mechanisms. In: *Neurobiology of Mood Disorders,* Post RM, Ballenger JC, eds. Baltimore: Williams & Wilkins

Post RM, Contel NR (1983): Human and animal studies of cocaine: implications for development of behavioral pathology. In: *Stimulants: Neurochemical, Behavioral, and Clinical Perspectives,* Creese I, ed. New York: Raven Press

Post RM, Contel NR, Gold PW (1982): Impaired behavioral sensitization to cocaine in vasopressin deficient rats. *Life Sci* 31:2745-2950

Post RM, Kopanda RT (1976): Cocaine, kindling, and psychosis. *Am J Psychiatry* 133:627-634

Post RM, Kotin J, Goodwin FK (1974): Effects of cocaine in depressed patients. *Am J Psychiatry* 131:511-517

Post RM, Lockfeld A, Squillace KM, Contel NR (1981b): Drug-environment interaction: context dependency of cocaine-induced behavioral sensitization. *Life Sci* 28:755-760

Post RM, Putnam FW, Contel NR, Goldman B (1984b): Electroconvulsive seizures inhibit amygdala kindling: implications for mechanisms of action in affective illness. *Epilepsia* 25:234-239

Post RM, Putnam F, Uhde TW, Weiss SRB (1986a): ECT as an anticonvulsant: implications for its mechanism of action in affective illness. In: *Electroconvulsive Therapy: Clinical and Basic Research Issues. Annals of the New York Academy of Sciences, Vol 462*, Malitz S, Sackeim HA, eds. New York: New York Academy of Sciences

Post RM, Rubinow DR, Ballenger JC (1984c): Conditioning, sensitization, and kindling: implications for the course of affective illness. In: *Neurobiology of Mood Disorders*, Post RM, Ballenger JC, eds. Baltimore: Williams & Wilkins

Post RM, Rubinow DR, Ballenger JC (1986b): Conditioning and sensitization in the longitudinal course of affective illness. *Br J Psychiatry* 149: 191-201

Post RM, Uhde TW (1985): Are the psychotropic effects of carbamazepine in manic-depressive illness mediated through the limbic system? *Psychiatr J Univ Ottawa* 10:205-219

Post RM, Uhde TW (1986): Anticonvulsants in non-epileptic psychosis. In: *Aspects of Epilepsy and Psychiatry*, Trimble MR, Bolwig TG, eds. New York: John Wiley & Sons Ltd.

Post RM, Uhde TW, Ballenger JC (1984d): Efficacy of carbamazepine in affective disorders: implications for underlying physiological and biochemical substrates. In: *Anticonvulsants in Affective Disorders*, Emrich HM, Okuma T, Muller AA, eds. Amsterdam: Excerpta Medica

Post RM, Uhde TW, Joffe RT, Bierer L (1986c): Psychiatric manifestations and implications of seizure disorders. In: *Medical Mimics of Psychiatric Disorders*, Extein I, Gold M, eds. Washington DC: APA Press

Post RM, Uhde TW, Roy-Byrne PP, Joffe RT (1986d): Antidepressant effects of carbamazepine. *Am J Psychiatry* 143:29-34

Post RM, Uhde TW, Roy-Byrne PP, Joffe RT (1987a): Correlates of antimanic response to carbamazepine. *Psychiatry Res* 21: 71-83

Post RM, Uhde TW, Rubinow DR, Weiss SRB (1986e): Antimanic effects of carbamazepine: mechanisms of action and implications for the biochemistry of manic-depressive illness. In: *Mania: New Research and Treatment*, Swann A, ed. Washington DC: APA Press

Post RM, Weiss SRB, Pert A (1984e): Differential effects of carbamazepine and lithium on sensitization and kindling. *Prog Neuropsychopharmacol Biol Psychiatry* 8:425-434

Post RM, Weiss SRB, Pert A, Uhde TW (1987b): Chronic cocaine administration: sensitization and kindling effects. In: *Cocaine: Clinical and Biobehavioral Aspects*, Fisher S, Raskin A, Uhlenhuth EH, eds. New York: Oxford University Press

Prien RF (1985): Research overview of drug therapy in recurrent and chronic depression. In: *Chronic Treatments in Neuropsychiatry,* Kemali D, Racagni G, eds. New York: Raven Press

Racine R (1978): Kindling: the first decade. *Neurosurgery* 3:234–252

Rivier C, Vale W (1987): Cocaine stimulates adrenocorticotropin (ACTH) secretion through a corticotropin-releasing factor (CRF)-mediated mechanism. *Brain Res* 422:403–406

Seeger TF, Sforzo GA, Pert CB, Pert A (1984): *In vivo* autoradiography: visualization of stress-induced changes in opiate receptor occupancy in the rat brain. *Brain Res* 305:303–311

Siegel S (1979): The role of conditioning in drug tolerance and addiction. In: *Psychopathology in Animals: Research and Clinical Applications,* Keehn JD, ed. New York: Academic Press

Wada JA, Sato M (1974): Generalized convulsive seizures induced by daily electrical stimulation of the amygdala in cats. *Neurology* 24:565–574

Wada JA, Sato M, Corcoran ME (1974): Persistent seizure susceptibility and recurrent spontaneous seizures in kindled cats. *Epilepsia* 15:465–478

Weiss SRB, Costello M, Woodward R, Nutt DJ, Post RM (1987): Chronic carbamazepine inhibits the development of cocaine-kindled seizures. *Abstracts, Society for Neuroscience 17th Annual Meeting, New Orleans, Nov 16–21, 1987,* Abstr #262.20, p 950

Weiss SRB, Post RM (1987): Carbamazepine and carbamazepine-10, 11-epoxide inhibit amygdala-kindled seizures in the rat but do not block their development. *Clin Neuropharm* 10:272–279

Weiss SRB, Murman D, Post RM, Pert A (1986a): Conditioning in cocaine-induced behavioral sensitization. *Abstracts, Society for Neuroscience 16th Annual Meeting, Washington DC, Nov 9–14, 1986,* Abstr #249.13, p 914

Weiss SR, Post RM, Gold PW, Chrousos G, Sullivan TL, Walker D, Pert A (1986b): CRF-induced seizures and behavior: interaction with amygdala kindling. *Brain Res* 372:345–351

Weiss SRB, Post RM, Seele F, Woodward R, Nierenberg J (1988): Chronic carbamazepine inhibits the development of local anesthetic seizures kindled by cocaine and lidocaine. *Brain Res,* in press

Willow M, Gonoi T, Catterall WA (1985): Voltage clamp analysis of the inhibitory actions of diphenylhydantoin and carbamazepine on voltage-sensitive sodium channels in neuroblastoma cells. *Mol Pharmacol* 27:549–558

Section II Developmental Models

4

The HPA System and Neuroendocrine Models of Depression

NED H. KALIN

For years, clinicians have known that depression is associated with specific endocrine abnormalities. Perhaps the most frequent examples are found in patients who suffer from primary disorders of the pituitary-adrenal system. In 1913, Harvey Cushing published a landmark paper in the *American Journal of Insanity* titled "Psychiatric Disturbances Associated with Disorders of the Ductless Glands" (Cushing, 1913). In addition to reporting case histories where cognitive, affective, and vegetative symptoms occurred in individuals with hypercortisolism (Cushing's syndrome), he astutely commented on the complex relationship between the central nervous system and altered endocrine function:

Psychic conditions profoundly influence the discharges from the glands of internal secretion, but we are on a much less secure footing when we come to the reverse, namely the effect on psyche and nervous system of chronic states of glandular overactivity or underactivity. However, . . . it is fair to assume that each of the resultant clinical types will exhibit more or less characteristic mental deviations; for the influence of the somatic condition on the mind is certainly as great as that of mind on body. (Cushing, 1913)

Research over the last decade has supported Cushing's early observations and notions. Numerous reports confirm a high incidence of psychiatric symptoms in patients with Cushing's syndrome and in patients receiving high doses of glucocorticoids (Cohen, 1980; Jeffcoate et al., 1979; Starkman et al., 1981; Hall et al., 1979). Almost always, successful treatment of the primary endocrine abnormality is accompanied by remission of the psychiatric symptoms. Conversely, a high percentage of patients with primary depression develop hypercortisolism and other hypothalamic-pituitary-adrenal (HPA) system abnormalities, and successful antidepressant treatment brings about normalization of these HPA abnormalities (Kalin and Dawson, 1986).

Evidence from inheritance, pharmacological, neurochemical, electrophysiological, and neuroendocrine studies strongly suggests that the depressive syndrome is in part biologically based. While the etiology of depression remains to be established, a number of studies have demonstrated altered function of brain norepinephrine, serotonin, and acetylcholine systems in depressed patients (Kalin and Dawson, 1986). Traditionally, interest in neu-roendocrine alterations associated with depression has been based on the fact that CNS neurotransmitter systems regulate endocrine function. Therefore endocrine alterations associated with depression are thought of as markers of changes in specific CNS neurotransmitter systems. However, in understand-ing the pathophysiology of depression, it is of greater interest that hormonal systems directly modulate CNS neurotransmitter function and regulate mo-tivational and behavioral systems. This chapter will present a model for the involvement of endocrine systems in the pathophysiology of depression and, based on data from animal studies, will speculate on the role of the HPA system in mediating certain aspects of the depressive syndrome. The HPA system has been selected as a focus because altered function of this sys-tem frequently results in depressive symptoms and occurs secondary to the depressive syndrome.

CNS Regulation of the Pituitary-Adrenal System

Selye (1946) provided early evidence of the role of the adrenal cortex in the stress response. It was then shown that ACTH secreted from the ante-rior pituitary was necessary to stimulate the release of glucocorticoids from the adrenal gland. From the work of Harris (1948), Saffran and Schally (1955), and Guillemin and Rosenberg (1955), the existence of a substance in the hypothalamus was hypothesized—a "corticotropin releasing factor" that would stimulate the anterior pituitary's release of ACTH. Other work established that ACTH is co-released with β-endorphin (βE) from the an-terior pituitary gland as a product of the precursor proopiomelanocortin (POMC) (Eipper and Mains, 1978; Guillemin et al., 1977). In 1981, Vale et al. identified the corticotropin-releasing hormone (CRH). Glucocorticoids modulate the release of ACTH and CRH by inhibiting effects at the levels of the pituitary, hypothalamus, and hippocampus (Krieger, 1979; Sapolsky et al., 1984). Secretion of hypothalamic CRH appears to be under the in-hibitory control of brain noradrenergic systems, while central cholinergic and serotonergic systems promote CRH release. At the level of the pituitary, outside the blood-brain barrier, regulation of ACTH secretion is also com-plex, with numerous substances besides CRH (norepinephrine, epinephrine, vasopressin, oxytocin, and somatostatin) affecting the corticotrope's release of ACTH (Axelrod and Reisine, 1984).

Extrahypothalamic Brain CRH, ACTH, and Glucocorticoid Systems

A recent major advance was the discovery that peptide hormones thought to originate only in the pituitary (e.g., ACTH and βE) occur in brain systems independent of the pituitary system. In addition, releasing hormones such as CRH, thought to exist only in the hypothalamus, have been found in extrahypothalamic brain regions. Receptor studies have demonstrated specific binding for CRH throughout the brain. Glucocorticoid receptors also occur in specific brain regions, where they mediate many of the effects of glucocorticoids on brain function. McEwen (1982) has demonstrated specific corticosterone binding sites on hippocampal, septal, and amygdaloid neurons of the rat. Within the hippocampus, high densities occur in the pyramidal neurons of CA1 and CA2 of Ammon's horn and in the granule neurons of the dentate gyrus (McEwen, 1982).

In the brain, POMC-containing cell bodies are found predominantly in the arcuate region of the hypothalamus. Axonal fibers originating from these cell bodies contain ACTH and project to other areas of the hypothalamus as well as to the amygdala, periaqueductal gray, and reticular formation (Watson et al., 1978). A considerably smaller group of ACTH-containing cell bodies exists in the nucleus tractus solitarius (Romagnano and Joseph, 1983). Brain ACTH systems are regulated differently than those in the anterior pituitary. To date, no specific receptors binding ACTH have been found in the brain.

CRH has an extensive extrahypothalamic distribution. CRH-containing neurons occur in the central nucleus, stria terminalis, central gray, lateral dorsal tegmental nucleus, locus coeruleus, parabrachial nucleus, dorsal vagal complex, regions containing the A1-A5 cell groups, dorsal horn of the spinal cord, and cerebral cortex (Swanson et al., 1983). High-affinity CRH receptors occur in laminae I and IV of the cortex, the median eminence, lateral nucleus of the amygdala, and the striatum (DeSouza et al., 1984; Wynn et al., 1984).

A General Model for the Neuroendocrine Mediation of Depression

At this point it is useful to speculate on how neuroendocrine systems might be involved in the pathophysiology of depression. As Figure 4.1 illustrates, I have conceptualized the depressive syndrome as one consequence of an individual's maladaptive response to external stressors or changes in internal physiological systems. The general model proposed here is based on concepts of behavioral endocrinology established by the early work of Beach (1975), Goy et al. (1964), and Leshner (1978). The model suggests that neuroendocrine alterations may be involved in two different ways. They may (1) affect the development of neural systems, resulting in a neurobiological vulnerability to becoming depressed, and (2) acutely affect the function of specific neuronal systems involved in the mediation of the depressive syndrome.

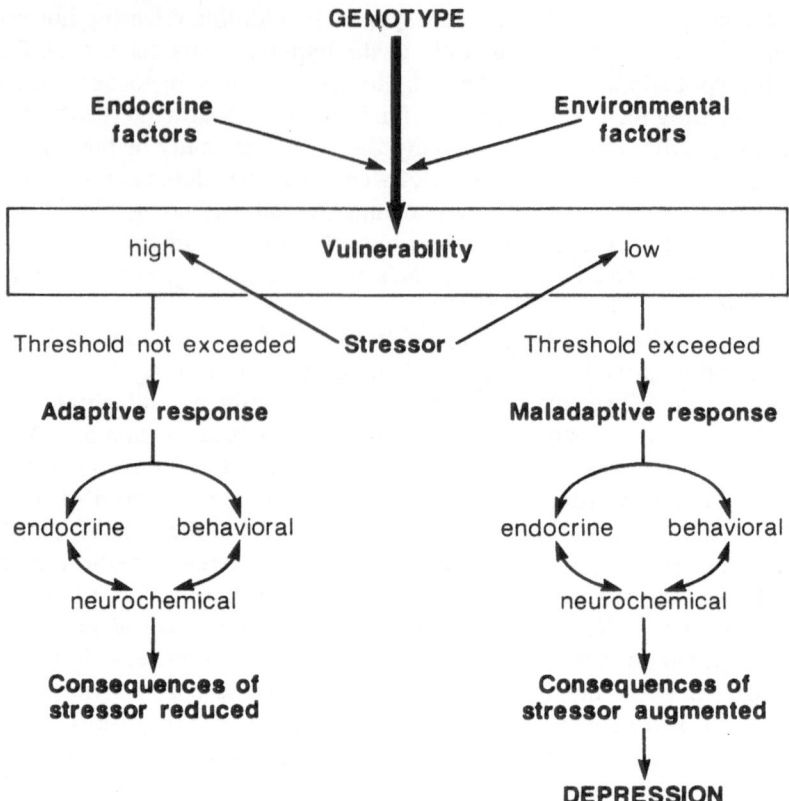

FIGURE 4.1. A model proposing how neuroendocrine systems might be involved in mediating depression. The model suggests that neuroendocrine activation early in development affects maturation of other neural systems involved in the stress response and the vulnerability to depression. When the threshold of vulnerability to depression is high, most exposures to stress will not exceed it, and an adaptive response will be evoked in which endocrine, behavioral, and/or neurochemical mechanisms reduce the consequences of the stressor. When the threshold for vulnerability to depression is low, however, it is readily exceeded. The result is a maladaptive response which augments the consequences of the stressor and may lead to depression. The model also proposes that activation of neuroendocrine hormones in the adult could acutely influence neuronal systems that mediate the depressive syndrome.

Hypothetical Role of the HPA System in Developing a Neurochemical Vulnerability

Numerous factors probably play a role in the development of a neurobiological vulnerability to depression (Whybrow et al., 1984). Genetic influences are important, as demonstrated by family and twin studies of depressed patients. It is also likely that early behavioral and physiological events are involved, as much work has established that events occurring during various critical developmental phases influence the phenotypic expression of the genotype. My hypothesis is that changes in the normal function of the HPA system early in life can influence the development of certain neuronal pathways, rendering an individual more vulnerable to the maladaptive behavioral, endocrine, and neurochemical consequences of stressors. One possible outcome of such a maladaptive response may be depression. To support this hypothesis it is necessary to demonstrate that stress-related HPA alterations occurring during development can cause anatomical and/or functional alterations of CNS systems necessary for the successful adaptation to stressors.

Glucocorticoids and Neuronal Development

Evidence suggests that glucocorticoids play a role in the postnatal development of neurons. Administration of glucocorticoids to infant rats causes an acute reduction of the normal increase in cell number in the cerebrum and cerebellum. When these animals mature the decrease in cell number persists—20% fewer cells in the cerebrum and 30% fewer in the cerebellum (Howard and Granoff, 1968; Balazs and Cotterrell, 1972). Glucocorticoid treatment also alters development of hippocampal neurons. However, these changes are reversed as the animal matures (Bohn, 1980). Of possible relevance to the finding that glucocorticoids affect neuronal development is the suggestion that they modulate the developmental effects of nerve growth factor (Otten and Thoenen, 1977).

It is also likely that glucocorticoids play a role in the development of catecholamine systems. Rats treated neonatally with glucocorticoids develop increased phenylethanolamine N-methyltransferase (PNMT) activity in the brain stem, hypothalamus, and adrenal medulla (Doupe and Patterson, 1982). This enzyme is responsible for catalyzing the methylation of norepinephrine to epinephrine. Evidence suggests that glucocorticoids also influence the development of other catecholamine-synthesizing enzymes, such as tyrosine hydroxylase and dopamine β-hydroxylase (Doupe and Patterson, 1982). Bohn (1983) has shown that rats treated with glucocorticoids at birth have, as adults, steroid-responsive PNMT-containing cells in their superior cervical ganglia. This finding is of interest because these cells are present only in adult rats that were exposed to glucocorticoids as infants, not in control animals.

Glucocorticoids and the Development of the
Pituitary-Adrenocortical System

In the rat, regulation of the HPA system is not completely developed until several weeks after birth (Sapolsky and Meaney, 1986; Walker et al., 1986). This period of development suggests that environmental and hormonal events early in life may influence the development of this stress-responsive system with consequences for the adult animal. Mature rats exhibit decreased basal concentrations of corticosterone and a blunted circadian variation of corticosterone after neonatal administration of glucocorticoids (Krieger, 1972, 1974; Turner and Taylor, 1976). Adult rats that were regularly handled as neonates have a more adaptive adrenocortical response to stressors than do unhandled rats (Levine, 1962; Levine et al., 1967; Haltmeyer et al., 1967). Similarly, animals reared with their mothers have more adaptive endocrine stress responses than animals reared in incubators (Thoman et al., 1968). A recent finding suggests a mechanism by which environmental effects on the adaptive response to stress may be mediated (Meaney et al., 1985). Daily handling of rat pups during their first 22 days of life leads to increased concentrations of glucocorticoid receptors in the hippocampus and frontal cortex of adult rats. This phenomenon is important because hippocampal glucocorticoid receptors play a role in feedback regulation of the HPA system and may modulate behavior.

Glucocorticoids and the Development of Organized Behaviors

Because of the marked effects of glucocorticoids on development of neuronal systems, it is not surprising that behavioral systems are also affected. Studies in mice show that corticosterone administration during infancy produces adults with impaired fine-adjustment mechanisms of motor control (Howard and Granoff, 1968). This deficit is compatible with the effects of corticosterone on development of cerebellar neurons. Other work reveals that exogenous corticosterone retards the normal maturation of swimming behavior in rats (Horne, 1970). It is likely that this effect is mediated by the decrease in neuronal development of the cerebellum and cerebral cortex.

Early treatment with corticosterone appears to heighten emotional reactivity in mature animals. For example, adult mice treated as infants with corticosterone show less exploratory behavior and defecate more when exposed to a novel open field (Howard and Granoff, 1968). Using operant paradigms, it has been shown that rats treated neonatally with corticosterone are hyperresponsive when working for food and are less able to adapt to schedule changes (Howard, 1973). In addition, these rats demonstrate increased water intake—thought to be a sign of anxiety—when working for food in a variable-interval reinforcement schedule (Howard et al., 1974). Adult rats that received corticosterone as neonates have an impaired ability to acquire two-way active avoidance but not one-way active avoidance (Zadina et al., 1985). This has been interpreted as indicative of increased emotionality.

Developmental Effects of ACTH and CRH

Much less work has been done on the effects of ACTH and CRH on CNS development. Both ACTH (Van der Helm-Hylkema and De Wied, 1976) and CRH (Zadina et al., 1985) accelerate eye opening when administered to infant rats, suggesting that these peptides promote earlier brain maturation. This effect of ACTH is independent of its effects on corticosterone secretion. However, since CRH induces ACTH release, it is possible that the effects of

CRH are mediated by increased release of ACTH. ACTH (independent of its effects on corticosterone) has been shown to accelerate the maturation of motor units in the rat (Saint-Come et al., 1983).

In some studies (Nyakas et al., 1981; Champney et al., 1976), neonatal treatment with ACTH has enhanced passive avoidance and reversal learning in adults. ACTH treatment of duckling embryos enhances the behavioral imprinting that occurs in newborn ducks (Martin, 1978); corticosterone appears to have an opposite effect. ACTH given to infant rats results in their increased activity as adults in a novel environment (Frankova and Jakoubek, 1974). Similar effects are seen with CRH administration (Zadina et al., 1985).

Hypothetical Role of the HPA System in the Acute Mediation of Depression

Glucocorticoids, ACTH, and CRH have been shown to have numerous acute effects on brain function. That primary HPA abnormalities frequently result in depressive symptoms and primary depression is commonly associated with altered HPA regulation suggests that these hormones are involved in the acute mediation of the depressive syndrome. This chapter will not attempt to review all of the effects of these hormones on CNS function, but rather will discuss those findings of potential relevance to depression (see Table 4.1).

Glucocorticoids

Glucocorticoids modulate the function of numerous neurotransmitter systems. Hypothalamic norepinephrine, epinephrine, and dopamine systems are likely influenced by glucocorticoids (de Kloet and Veldhuis, 1985). In addition, glucocorticoids alter the number and function of β-adrenergic receptors in various tissues (Stiles et al., 1984) including hippocampus (Roberts and Bloom, 1981). Glucocorticoids affect serotonin metabolism, as demonstrated by their requirement for the stress-induced increase in tryptophan hydroxylase activity (Azmitia and McEwen, 1974). Evidence suggests that

TABLE 4.1. Effects of acute administration of HPA hormones on behaviors potentially relevant to depression

CRH	ACTH	Cortisol
Alters locomotion	Affects learning and memory; increases grooming (agitation)	Decreases ability to selectively attend (difficulty concentrating; altered cognition)
Decreases sleep	Decreases aggression	
Decreases eating	Decreases sociability	
Decreases sex		
Increases huddling		

glucocorticoids are also involved in the regulation of hippocampal GABA systems (Miller et al., 1978).

Since the limbic system is thought to mediate emotional and adaptive behaviors and the highest density of glucocorticoid receptors is in the hippocampus, it is reasonable to hypothesize that the effects of glucocorticoids relevant to depression are mediated through hippocampal pathways. Lesion studies have suggested that the hippocampus is involved in behavioral inhibition and selective attention. For example, one of the effects of hippocampal damage is to retard extinction of an appetitively motivated task (McEwen, 1982). That is, the hippocampus is needed to behaviorally inhibit the response. Micco et al. (1979) showed that this behavior, thought to be mediated by the hippocampus was influenced by glucocorticoids. They found that adrenalectomy facilitates the extinction of an appetitively motivated task. This result is opposite to the effects of hippocampal damage. Corticosterone replacement, and not dexamethasone, reversed these effects, suggesting that the effects are mediated through glucocorticoid receptors (Micco and McEwen, 1980). It thus appears that corticosterone affects behavioral inhibition in the same way that hippocampal ablation does. Therefore, excess corticosterone may release certain behaviors that are normally inhibited by the hippocampus. Data from Pfaff et al. (1971), demonstrating that corticosterone inhibits hippocampal electrical activity, are consistent with this. These findings led to the speculation that corticosterone acting through the hippocampus promotes behavioral persistence in the face of frustrative nonreward. It is of interest that removal of locus coeruleus input to the hippocampus by dorsal bundle lesioning causes increased distractibility. Corticosterone, then, may oppose the effects of locus coeruleus input on selective attention (McEwen, 1982).

Recent work by Sapolsky (Sapolsky, 1985; Sapolsky et al., 1985) has shown that glucocorticoids not only may alter the function of hippocampal cells but actually can promote the death of hippocampal neurons. He hypothesized that basal levels of glucocorticoids play a role in the natural loss of hippocampal neurons that occurs with aging and suggested that chronic stress with resultant increases in glucocorticoids could accelerate this process.

These basic data raise the question: To what degree could glucocorticoids play a role in depression? Animal data suggest that glucocorticoids diminish the ability of organisms to attend selectively to stimuli. Certainly a characteristic of depressed patients is their altered cognition, exemplified by difficulty in concentrating and by dwelling on negative events. In a recent study, Rubinow et al. (1984) provided evidence in depressed humans for a direct relationship between an individual's degree of HPA overactivity and the severity of cognitive dysfunction. Reus et al. (1985) showed that HPA overactivity in depressed patients was associated with impairment of response specificity to a novel stimulus, suggestive of a role of the HPA in selective attention processes.

ACTH

Although the data are less consistent, there is a consensus that exogenous ACTH modifies the function of catecholamine, serotonin, GABA, and acetylcholine systems (Gispen and Zwiers, 1985). ACTH-containing fibers occur in the locus coeruleus and ACTH excites noradrenergic neurons in the locus coeruleus (Olpe and Jones, 1982). It is of interest that after chronic treatment with ACTH[4-9], increased glucose utilization is seen only in selected areas—hippocampus, anterior nucleus of the thalamus, and the cingulate cortex (McCulloch et al., 1982). These findings suggest that the behavioral effects of such treatments may be mediated by these brain regions.

Behavioral studies involving ACTH have demonstrated numerous effects on behavior, learning, and memory. Conditioned avoidance paradigms have been used to assess the role of ACTH in modulation of adaptive behavior. ACTH delays extinction in an active avoidance paradigm, and this effect is unrelated to effects on corticosterone secretion (Gispen and Zwiers, 1985; de Wied, 1966). Thus rats treated with either peripheral or central ACTH continue to make the avoidance response to the conditioned stimulus long after untreated controls have extinguished their response. Subsequent structure-activity studies demonstrated that the necessary sequence for behavioral activity is ACTH[4-7]. It is important to recognize that this sequence has no steroidogenic effects. ACTH also appears to be involved in passive avoidance behavior, although here the results are less consistent (Gispen and Zwiers, 1985).

Another result of ACTH administration is stereotypic behavioral patterns. Administered centrally, ACTH elicits the stretch-yawning syndrome, which is accompanied by EEG and behavioral arousal (Gispen and Zwiers, 1985). However, its significance is not known. Exogenous ACTH also causes excessive grooming (Gispen and Isaacson, 1981). This has been an area of intense investigation, because grooming behavior occurs naturally when animals are exposed to novelty and in other stressful situations. It is thought that under conditions of potential stress, grooming may be a displacement behavior. Dunn et al. (1979) have shown that brain ACTH systems are essential for activation of stress-induced grooming. Studies suggest that the grooming effects of ACTH may be mediated by dopamine systems (Guild and Dunn, 1982).

Only a few studies have assessed the role of ACTH systems in social behavior. Long-term treatment of mice leads to reduced aggressiveness (Leshner et al., 1973; Brain and Poole, 1974), and this effect is independent of ACTH's effect on corticosterone secretion (Gispen and Isaacson, 1981). If a defeated mouse is treated with ACTH immediately after an aggressive attack, it will be more submissive in subsequent encounters up to 48 hours after treatment (Roche and Leshner, 1979). File and Clarke (1980) found that ACTH administration reduces levels of social interaction in rats and this effect appears to be centrally mediated. Other work suggests that ACTH has anxiogenic effects, some of which antagonize the effects of benzodiazepines (Vellucci and Webster, 1982). The author and his colleagues studied the neuroendocrine changes associated with the response of peer-housed rhesus monkeys to separation. They found that, after four days of separation, significant increases in CSF levels of ACTH occur, while peripheral ACTH remains at preseparation levels (Kalin, 1986); and they demonstrated that the ACTH measured in CSF originates from brain ACTH systems. These findings are of interest because social separation in rhesus monkeys has been suggested as an animal model of depression.

CRH

Evidence suggests that extrahypothalamic CRH functions as a neurotransmitter or neuromodulator (Smith et al., 1986). Neurophysiological studies reveal that CRH increases the frequency of spontaneous discharge and decreases the hyperpolarizations after burst discharges in CA1 and CA3 pyramidal neurons of rat hippocampal slices (Aldenhoff et al., 1983). Valentino et al. (1983) reported that ICV-CRH increased the discharge rate of locus coeruleus neurons in anesthetized rats. This is of interest because locus coeruleus neurons become more active during behavioral states that involve heightened vigilance, and it has been argued that these neurons play a major role in anxiety. Consistent with these findings, Ehlers et al. (1983) reported that low doses of ICV-CRH in awake rats produced electroencephalographic changes suggestive of increased arousal. Higher doses of CRH induced seizure activity similar to that which occurs after "kindling" of the amygdala.

ICV-CRH also affects the sympathetic nervous system, causing significant increases in mean arterial pressure and heart rate (Brown et al., 1982; Fisher and Brown, 1984). These changes were associated with prolonged elevation of plasma epinephrine, norepinephrine, and glucose. Increased oxygen consumption occurred without a significant change in body temperature (Brown et al., 1982). Increases in mean arterial pressure and heart rate appear to be secondary to sympathetic nervous system activity, as they were blocked by peripheral ganglionic blockade with chlorisondamine (Fisher and Brown, 1984). Kalin et al. (1983a) found that in monkeys, ICV-CRH causes significant increases in plasma ACTH and cortisol concentrations. This finding has been confirmed in sheep (Donald et al., 1983).

ICV-CRH produces behavioral changes that may be characterized as stress-related. It inhibits sexually receptivity in the female rat (Sirinathsinghji et al., 1983). Also in rats, it elicits behavioral changes similar to those observed in the novel open-field test—namely, increased grooming and decreased ingestive behavior (Britton et al., 1982; Morley and Levine, 1982)—and ICV-CRH has anxiety-like effects in a conflict model of anxiety (Britton et al., 1986). ICV-CRH also potentiates the acoustic startle response, and this effect is blocked by benzodiazepines (Swerdlow et al., 1986). The acoustic startle is a reflexive response sensitive to stress or fear.

Work with rats in the author's laboratory has shown that while ICV-CRH has potent activating effects, it does not selectively enhance the effects of exposure to novelty (Sherman and Kalin, 1987). However, the investigator's more recent studies demonstrate that ICV-CRH potentiates freezing behavior induced by foot-shock and that administration of a CRH antagonist attenuates shock-induced freezing (Sherman and Kalin, 1988: Kalin et al., 1988). That CRH potentiates fear-related freezing but not the effects of novelty suggests that endogenous CRH systems are involved in the activation of only certain stress-related behaviors. Sherman and Kalin (1986 and 1987) also found that ICV-CRH does not alter antinociceptive responding. This finding is of interest because the amount of freezing induced is directly related to pain sensitivity, and many stressors cause enhanced analgesia.

In chair-restrained monkeys, ICV-CRH evokes behaviors including vocalization, head-shaking, and struggling (Kalin et al., 1983a). However, when ICV-CRH is administered to unrestrained monkeys, they exhibit huddling and lying-down behavior (Kalin, 1985). These behaviors are seen in rhesus monkeys undergoing the despair response after attachment bond disruption. It is important to note that the similarities in behavioral posture do not necessarily imply that central CRH mediates huddling in separated monkeys. It is interesting that these CRH-induced behaviors are also associated with pituitary-adrenal activation, because some rhesus monkeys undergoing separation (Kalin et al., 1983b) and 50% of depressed humans have HPA dysregulation. These findings led us to question whether altered central CRH in humans might mediate some of the behavioral and neuroendocrine aspects of depression (Kalin, 1985). Consistent with this, Nemeroff has reported increased levels of CRH in the CSF of depressed patients (Nemeroff et al., 1984).

Experiments Proposed to Test the Neuroendocrine Model of Depression

The model proposed by the author predicts that specific HPA alterations early in life would make animals more likely to respond maladaptively to stressors later on. Therefore it would be expected that in adults the intensity of the stressor required for induction of maladaptive behaviors would be less. For example, brief separations usually produce behavioral agitation and it is not until the length of the separation has progressed that monkeys begin huddling and exhibiting behavioral despair. Demonstrating that adult monkeys treated with HPA hormone as neonates are more likely to respond to brief separations with an intense despair response would validate this model.

Similar results would be predicted for rats treated with HPA hormones as neonates and then exposed as adults to uncontrollable foot-shock. It would be expected that the threshold for the behavioral outcome of such a treatment (learned helplessness) would be lower in prenatally treated rats.

To test the role of HPA hormones in the acute expression of depressive behavior, experiments could be designed to alter HPA hormones before and during environmental manipulations known to result in behavioral outcomes associated with depression. Again using the monkey separation paradigm, it would be expected that a specific HPA alteration would intensify the despair response. Once it has been established that specific HPA manipulations either developmentally or acutely predispose an animal to behaviors associated with depression, then it is important to determine whether these effects are specific for stressors that induce depressive behaviors. Will animals with HPA alterations respond maladaptively only to manipulations that produce depressive behaviors, or will they have more intense responses to all types of environmental stressors? Establishing which of the HPA manipulations are specific for animal models of depression would lend insight into the systems specifically linked to the pathophysiology of depression. Therefore it would be expected that antidepressants would attenuate the hormonally induced intensification of depressive behaviors.

Acknowledgment. The author would like to thank Carol Steinhart for her editorial assistance.

References

Aldenhoff J, Gruol D, Rivier J, Vale W, Siggins G (1983): Corticotropin releasing factor decreases postburst hyperpolarizations and excites hippocampal neurons. *Science* 221:875–877

Axelrod J, Reisine T (1984): Stress hormones: their interaction and regulation. *Science* 224:452–459

Azmitia E Jr, McEwen B (1974): Adrenalcortical influence on rat brain tryptophan hydroxylase activity. *Brain Res* 78:291–302

Balazs R, Cotterrell M (1972): Effect of hormonal state on cell number and functional maturation of the brain. *Nature* 236:348–350

Beach F (1975): Behavioral endocrinology: an emerging discipline. *Am Scientist* 63:178–187

Bohn M (1980): Granule cell genesis in the hippocampus of rats treated neonatally with hydrocortisone. *Neuroscience* 5:2003–2012

Bohn M (1983): Role of glucocorticoids in expression and development of phenylethanolamine N-methyltransferase (PNMT) in cells derived from the neural crest: a review. *Psychoneuroendocrinology* 8:381–390

Brain P, Poole A (1974): The role of endocrines in isolation-induced intermale fighting in albino laboratory mice. *Aggressive Behav* 1:39–69

Britton K, Koob G, Rivier J, Vale W (1982): Intraventricular corticotropin-releasing factor enhances behavioral effects of novelty. *Life Sci* 31:363–367

Britton K, Lee G, Vale W, Rivier J, Koob G (1986): Corticotropin releasing factor (CRF) receptor antagonist blocks activating and 'anxiogenic' actions of CRF in the rat. *Brain Res* 360:303–306

Brown M, Fisher L, Rivier J, Spiess J, Rivier C, Vale W (1982): Corticotropin-releasing factor: effects on the sympathetic nervous system and oxygen consumption. *Life Sci* 30:207-210

Champney T, Sahley T, Sandman C (1976): Effects of neonatal cerebral ventricular injection of ACTH 4-9 and subsequent adult injections on learning in male and female albino rats. *Pharmacol Biochem Behav* 5(Suppl 1):3-9

Cohen S (1980): Cushing's syndrome: a psychiatric study of 29 patients. *Br J Psychiatry* 136:120-124

Cushing H (1913): Psychiatric disturbances associated with disorders of the ductless glands. *Am J Insanity* 69:965-990

de Kloet E, Veldhuis H (1985): Adrenocortical hormone action. In: *Handbook of Neurochemistry*, Vol 8, Lajtha A, ed. New York: Plenum Press, pp 47-82

DeSouza E, Perrin M, Insel T, Rivier J, Vale W, Kuhar M (1984): Corticotropin-releasing factor receptors in rat forebrain: autoradiographic identification. *Science* 224:1449-1451

de Wied D (1966): Inhibitory effect of ACTH and related peptides on extinction of conditioned avoidance behavior in rats. *Proc Soc Exp Biol Med* 122:28-32

Donald R, Redekopp C, Cameron V, Nicholls M, Bolton J, Livesey J, Espiner E, Rivier J, Vale W (1983): The hormonal actions of corticotropin-releasing factor in sheep: effect of intravenous and intracerebroventricular injection. *Endocrinology* 113:866-870

Doupe A, Patterson P (1982): Glucocorticoids and the developing nervous system. In: *Adrenal Actions in Brain*, Ganten D and Pfaff D, eds. Berlin: Springer-Verlag, pp 23-27

Dunn A, Green E, Isaacson R (1979): Intracerebral adrenocorticotropic hormone mediates novelty-induced grooming in the rat. *Science* 203:281-283

Ehlers C, Henriksen S, Wang M, Rivier J, Vale W, Bloom F (1983): Corticotropin-releasing factor produces increases in brain excitability and convulsive seizures in rats. *Brain Res* 278:332-336

Eipper B, Mains R (1978): Analysis of the common precursor to corticotropin and endorphin. *J Biol Chem* 253:5732-5744

File S, Clarke A (1980): Intraventricular ACTH reduces social interaction in male rats. *Pharmacol Biochem Behav* 12:711-715

Fisher L, Brown M (1984): Corticotropin-releasing factor and angiotensin II: comparison of CNS actions to influence neuroendocrine and cardiovascular function. *Brain Res* 296:41-47

Frankova S, Jakoubek (1974): Long term behavioural effects of diazepam and ACTH, administered early in life. *Act Nerv Super (Praha)* 16:247-249

Gispen W, Isaacson R (1981): ACTH-induced excessive grooming in the rat. *Pharmacol Ther* 12:209-246

Gispen W, Zwiers H (1985): Behavioral and neurochemical effects of ACTH. In: *Handbook of Neurochemistry*, 2nd Ed, Vol 8, Lajtha A, ed. New York: Plenum Press, pp 375-411

Goy R, Bridson W, Young W (1964): Period of maximal susceptibility of the prenatal female guinea pig to masculinizing actions of testosterone propionate. *J Comp Physiol Psychol* 57:166-174

Guild A, Dunn A (1982): Dopamine involvement in ACTH-induced grooming behavior. *Pharmacol Biochem Behav* 17:31-36

Guillemin R, Rosenberg B (1955): Humoral hypothalamic control of anterior pituitary: a study with combined tissue cultures. *Endocrinology* 57:599-607

Guillemin R, Vargo T, Rossier J, Minick S, Ling N, Rivier C, Vale W, Bloom F (1977): Beta-endorphin and adrenocorticotropin are secreted concomitantly by the pituitary. *Science* 213:1367–1369

Hall R, Popkin M, Stickney S (1979): Presentation of the steroid psychoses. *J Nerv Ment Dis* 167:229–236

Haltmeyer G, Denenberg V, Zarrow M (1967): Modification of the plasma corticosterone response as a function of infantile stimulation and electric shock parameters. *Physiol Behav* 2:61–64

Harris G (1948): Neural control of the pituitary gland. *Physiol Rev* 28:139–179

Horne R (1970): Hormonal effects on ontogeny of swimming ability in the rat: assessment of central nervous system development. *Science* 168:147–151

Howard E (1973): Increased reactivity and impaired adaptability in operant behavior of adult mice given corticosterone in infancy. *J Comp Physiol Chem* 85:211–220

Howard E, Granoff D (1968): Increased voluntary running and decreased motor coordination in mice after neonatal corticosterone implantation. *Exp Neurol* 22:661–673

Howard E, Olton D, Taylor M (1974): Polydipsia in adult mice and rats given corticosterone in infancy. *J Comp Physiol Psychol* 120:120–125

Jeffcoate WJ, Silverstone JT, Edwards CR, Besser GM (1979): Psychiatric manifestations of Cushing's syndrome: response to lowering of plasma cortisol. *J Med* 48:465–472

Kalin N (1985): Behavioral effects of ovine corticotropin-releasing factor administered to rhesus monkeys. *Fed Proc* 44:433–441

Kalin N (1986): ACTH in plasma and CSF in the rhesus monkey. *Biol Psychiatry* 21:124–140

Kalin N, Dawson G (1986): Neuroendocrine dysfunction in depression: hypothalamic-anterior pituitary systems. *Trends Neurosci* 9:261–266

Kalin N, Shelton S, Kraemer G, McKinney W (1983a): Corticotropin-releasing factor administered intraventricularly to rhesus monkeys. *Peptides* 4:217–220

Kalin N, Shelton S, McKinney W, Kraemer G, Scanlon J, Suomi S (1983b): Stress alters the dexamethasone suppression test in rhesus monkeys. *Psychopharmacol Bull* 19:542–544

Kalin N, Sherman J, Takahashi L (1988): Antagonism of endogenous CRH systems attenuates stress-induced freezing behavior in rats. *Brain Res* 457:130–135.

Krieger D (1972): Circadian corticosteroid periodicity: critical period for abolition by neonatal injection of corticosteroid. *Science* 178:1205–1207

Krieger D (1974): Effect of neonatal hydrocortisone on corticosteroid circadian periodicity, responsiveness to ACTH and stress in prepuberal and adult rats. *Neuroendocrinology* 16:355–363

Krieger D (1979): Plasma ACTH and corticosteroids. In: *Endocrinology*, Vol 2, DeGroot L et al., eds. New York: Grune and Stratton, pp 1139–1156

Leshner A (1978): *An Introduction to Behavioral Endocrinology*. New York: Oxford University Press, pp 1–34

Leshner A, Walker W, Johnson A, Kelling J, Kreisler S, Svare B (1973): Pituitary adrenocortical activity and intermale aggressiveness in isolated mice. *Physiol Behav* 11:705–711

Levine S (1962): Plasma-free corticosteroid response to electric shock in rats stimulated in infancy. *Science* 135:795–796

Levine S, Haltmeyer G, Karas G, Denenberg V (1967): Physiological and behavioral effects of infantile stimulation. *Physiol Behav* 2:55–59

Martin J (1978): Embryonic pituitary adrenal axis, behavior development and domestication in birds. *Am Zool* 18:489–499

McCulloch J, Kelly P, Van Delft A (1982): Alterations in local cerebral glucose utilization during chronic treatment with an $ACTH_{4-9}$ analog. *Eur J Pharmacol* 78:151–158

McEwen B (1982): Glucocorticoids and hippocampus: receptors in search of a function. In: *Current Topics in Neuroendocrinology,* Ganten D and Pfaff D, eds. Berlin: Springer-Verlag, pp 1–22

Meaney M, Aitken D, Bodnoff S, Iny L, Tatarewica J (1985): Early postnatal handling alters glucocorticoid receptor concentrations in selected brain regions. *Behav Neurosci* 99:765–770

Micco D Jr, McEwen B (1980): Glucocorticoids, the hippocampus, and behavior: interactive relation between task activation and steroid hormone binding specificity. *J Comp Physiol Psychol* 94:624–633

Micco D Jr, McEwen B, Shein W (1979): Modulation of behavioral inhibition in appetitive extinction following manipulation of adrenal steroids in rats: implications for involvement of the hippocampus. *J Comp Physiol Psychol* 83:323–329

Miller A, Chaptal C, McEwen B, Peck E Jr (1978): Modulation of high affinity GABA uptake into hippocampal synaptosomes by glucocorticoids. *Psychoneuroendocrinology* 3:155–164

Morley J, Levine A (1982): Corticotropin-releasing factor, grooming and ingestive behavior. *Life Sci* 31:1459–1464

Nemeroff C, Widerlöv E, Bissette G, Walleus H, Karlsson I, Eklund K, Kilts C, Loosen P, Vale W (1984): Elevated concentrations of CSF corticotropin-releasing factor-like immunoreactivity in depressed patients. *Science* 226:1342–1344

Nyakas C, Levay G, Viltsek J, Endröczi E (1981): Effects of neonatal $ACTH_{4-10}$ administration on adult adaptive behavior and brain tyrosine hydroxylase activity. *Dev Neurosci* 4:225–232

Olpe H, Jones R (1982): Excitatory effects of ACTH on noradrenergic neurons of the locus coeruleus in the rat. *Brain Res* 251:177–179

Otten U, Thoenen H (1977): Effect of glucocorticoids on nerve growth factor-mediated enzyme induction in organ cultures of rat sympathetic ganglia: enhanced response and reduced time requirement to initiate enzyme induction. *J Neurochem* 29:69–75

Pfaff D, Silva M, Weiss J (1971): Telemetered recording of hormone effects in hippocampal neurons. *Science* 172:394–395

Reus V, Peeke H, Miner C (1985): Habituation and cortisol dysregulation in depression. *Biol Psychiatry* 20:980–989

Roberts D, Bloom F (1981): Adrenal steroid-induced changes in β-adrenergic receptor binding in rat hippocampus. *Eur J Pharmacol* 74:37–41

Roche K, Leshner A (1979): ACTH and vasopressin treatments immediately after a defeat increase future submissiveness in male mice. *Science* 204:1343–1344

Romagnano M, Joseph S (1983): Immunocytochemical localization of ACTH in the brainstem of the rat. *Brain Res* 276:1–16

Rubinow D, Post R, Savard R, Gold P (1984): Cortisol hypersecretion and cognitive impairment in depression. *Arch Gen Psychiatry* 41:279–283

Saffran M, Schally A (1955): The release of corticotropin by anterior pituitary tissue *in vitro. Can J Biochem Physiol* 33:408–415

Saint-Come C, Acker G, Strand F (1983): Development and regeneration of motor systems under the influence of ACTH peptides. *Psychoneuroendocrinology* 10:445–459

Sapolsky R (1985): Glucocorticoid toxicity in the hippocampus: temporal aspects of neuronal vulnerability. *Brain Res* 359:300–305

Sapolsky R, Meaney M (1986): Maturation of the adrenocortical stress response: neuroendocrine control mechanisms and the stress hyporesponsive period. *Brain Res Rev* 11:65–76

Sapolsky R, Krey L, McEwen B (1984): Glucocorticoid-sensitive hippocampal neurons are involved in terminating the adrenocortical stress response. *Proc Natl Acad Sci USA* 81:6174–6177

Sapolsky R, Krey L, McEwen B (1985): Prolonged glucocorticoid exposure reduces hippocampal neuron number: implications for aging. *J Neurosci* 5:1222–1227

Selye H (1946): The general adaptation syndrome and the diseases of adaptation. *J Clin Endocrinol* 6:117–230

Sherman J, Kalin N (1986): ICV-CRH potently affects behavior without altering antinociceptive responding. *Life Sci* 39:433–441

Sherman J, Kalin N (1987): The effects of ICV-CRH on novelty-induced behavior. *Pharmacol Biochem Behav* 26:699–703

Sherman J, Kalin N (1988): ICV-CRH alters stress-induced freezing behavior without affecting pain sensitivity. *Pharmacol Biochem Behav* (in press)

Sirinathsinghji D, Rees L, Rivier J, Vale W (1983): Corticotropin-releasing factor is a potent inhibitor of sexual receptivity in the female rat. *Nature* 305:232–235

Smith M, Bissette G, Slatkin T, Knight D, Nemeroff C (1986): Release of corticotropin-releasing factor from rat brain regions *in vitro*. *Endocrinology* 118:1997–2001

Starkman M, Schteingart D, Schork A (1981): Depressed mood and other psychiatric manifestations of Cushing's syndrome: relationship to hormone levels. *Psychosom Med* 43:3–17

Stiles G, Caron M, Lefkowitz R (1984): β-Adrenergic receptors: biochemical mechanisms of physiological regulation. *Physiol Rev* 64:661–743

Swanson L, Sawchenko P, Rivier J, Vale W (1983): Organization of ovine corticotropin-releasing factor immunoreactive cells and fibers in the rat brain: an immunohistochemical study. *Neuroendocrinology* 36:165–186

Swerdlow N, Geyer M, Vale W, Koob G (1986): Corticotropin-releasing factor potentiates acoustic startle in rats: blockade by chlordiazepoxide. *Psychopharmacology* 88:147–152

Thoman E, Levine S, Arnold W (1968): Effects of maternal deprivation and incubator rearing on adrenocortical activity in the adult rat. *Dev Psychobiol* 1:21–23

Turner B, Taylor A (1976): Persistent alteration of pituitary-adrenal function in the rat by prepuberal corticosterone treatment. *Endocrinology* 98:1–9

Vale W, Spiess J, Rivier C, Rivier J (1981): Characterization of a 41-residue ovine hypothalamic peptide that stimulates secretion of corticotropin and beta-endorphin. *Science* 213:1394–1397

Valentino R, Foote S, Aston-Jones G (1983): Corticotropin-releasing factor activates noradrenergic neurons of the locus coeruleus. *Brain Res* 270:363–367

Van der Helm-Hylkema H, de Wied D (1976): Effect of neonatally injected ACTH and ACTH analogues on eye-opening of the rat. *Life Sci* 18:1099–1104

Vellucci S, Webster R (1982): Antagonism of the anticonflict effects of chlordiazepoxide by β-carboline carboxylic acid ethyl ester, Ro 15-1788 and ACTH$_{(4-10)}$. *Psychopharmacology* 78:256–260

Walker C-D, Perrin M, Vale W, Rivier C (1986): Ontogeny of the stress response in the rat: role of the pituitary and the hypothalamus. *Endocrinology* 118:1445-1451

Watson S, Richard C III, Barchas J (1978): Adrenocorticotropin in the rat brain: immunocytochemical localization in cells and axons. *Science* 200:1180-1192

Whybrow P, Akiskal H, McKinney W (1984): Toward a psychobiological integration: affective illness as a final common path to adaptive failure. In: *Mood Disorders— Toward a New Psychobiology.* New York: Plenum Press, pp 173-203

Wynn P, Hauger R, Holmes M, Milan M, Catt K, Aguilera G (1984): Brain and pituitary receptors. *Peptides* 5:1077-1084

Zadina J, Kastin A, Coy D, Adinoff B (1985): Developmental, behavioral, and opiate receptor changes after prenatal or postnatal β-endorphin, CRF, or Tyr-MIF-1. *Psychoneuroendocrinology* 10:367-383

5

The Use of an Animal Model to Study Post-Stroke Depression

ROBERT G. ROBINSON

Introduction

The development of an animal model of a medical disorder found in humans provides many research advantages over a purely clinical investigation. The specificity of anatomical, biochemical, or physiological changes can be assessed while controlling for numerous other variables which may or may not be associated with the medical condition. There are numerous examples in literature of the value of investigating animal models. Thus it is not surprising that clinically oriented investigators are always in search of animal models of important human disorders.

When the search for animal models turns to psychiatric conditions such as depression, the problem is more difficult because the etiology of most depressive disorders is not known. Thus, a human condition diagnosed by clinical interview cannot be directly validated in the animal by production of the underlying etiology or by clinical examination. Most animal models are validated only by behavioral observation or response to treatment (the specificity of treatment for the etiology of depression is largely uncertain). These validation techniques leave both clinician and researcher dissatisfied with the relevance of the findings from the animal model to the human condition.

Stroke is a focal neurological deficit occurring over seconds to hours produced by the diminution of blood circulation to a specific area of brain. Stroke can produce any of numerous motor, sensory or behavioral deficits, depending on the particular area of brain involved. For many years, clinicians have recognized that depressive disorders frequently accompany stroke (Bleuler, 1951; Kraeplin, 1921). Since stroke can be experimentally produced in animals, it seems that this would be a possible avenue for studying depressive disorders in animals.

The problem, however, has been the uncertainty of the nature of the relationship between stroke and depression. Many clinicians have assumed that the depressive disorder represents an understandable psychological response to the physical or cognitive impairment. Fisher (1961) stated that the brain was "the most cherished organ in the body" and that injury to this organ would understandably lead to depression. Systematic studies, however, have tended to refute the assumption that depression could be entirely attributed to the physical or cognitive impairment. Folstein et al. (1977) found that 25 stroke patients were significantly more depressed than 15 orthopedic patients even though the two groups had equal degrees of physical impairment. Epidemiological studies by Kay (1962) found a frequent association between cerebrovascular disease and first episodes of severe endogenous depression in the elderly. Post (1962) found cerebral ischemic lesions in a high proportion of geriatric patients who had been admitted to hospital for treatment of endogenous depression. Recently, Finkelstein et al. (1982) found that depression and failure to suppress serum cortisol following dexamethasone administration was significantly more common in stroke patients when compared with a group of control patients with other chronic medical illnesses; and Sinyor et al. (1986) reported that the severity of depression was not significantly correlated with the severity of physical or cognitive impairment.

Thus, systematic studies have suggested that depression may be a specific consequence of focal cerebral damage just as hemiparesis or visual field defect are the specific consequence of injury to localized brain regions. Because stroke can be experimentally produced in animals, this disorder might lend itself to the production of an animal model of a depressive disorder. Before the effectiveness or usefulness of an animal model can be evaluated, however, it is important to detail the course, symptoms, response to treatment, and associated variables in the clinical condition of post-stroke depression.

Clinical Syndromes

In systematic studies of stroke patients conducted by the author and colleagues, three types of mood disorders were recognized. The first mood disorder is a severe depression that meets the Diagnostic and Statistical Manual III (DSM-III) (American Psychiatric Association, 1980) symptom criteria for major depression. The percentage of patients showing various symptoms of major depression from a consecutive series of 41 patients meeting the diagnostic criteria are shown in Table 1. These disorders are strongly associated with left frontal brain injury. The second mood disorder is a depressive syndrome which meets the DSM-III symptom criteria for dysthymic disorder (termed here minor depression). The percent of patients with minor depression showing various depressive symptoms found in a consecutive series of 41 patients is shown in Table 1. Minor depressions are frequently associated with posterior hemisphere (parietal-occipital lobe) lesions of either hemisphere. The third mood disorder associated with stroke is an indifferent, apathetic mental state associated with inappropriate cheerfulness. This

TABLE 5.1. Frequency of symptoms in patients with post-stroke depressions

Clinical Symptoms	Major Depression (% with symptom)	Dysthymic Depression (% with symptom)
	(N = 41)	(N = 41)
Depressed mood	100	100
Diurnal mood variation	76	46
Loss of energy	73	43
Anxiety, restless, worried	73	67
Weight loss, decreased appetite	63	15
Early morning awakening	56	26
Delayed sleep onset	56	33
Social withdrawal	54	23
Irritability	54	—
Hopelessness	—	46
Associated lesion location	Left frontal lobe	Right or left
	Left basal ganglia	Posterior parietal and occipital regions

Reprinted with permission from *Psychosomatics,* 26:771 (1985) ©1985 Academy of Psychosomatic Medicine.

cheerfulness, although not euphoria, is greater than would be expected from an acutely ill patient. This syndrome which has been called the "indifference reaction" by Hecaen (1962) and Denny-Brown (1957), is characterized in our population by anxiety, inappropriate cheerfulness, slowness, agitation, loss of interest, and worrying. The syndrome is found almost exclusively with right hemisphere injury and usually right frontal brain injury.

Prevalence and Duration of Depressive Disorders

In a prospective study of 103 stroke patients, almost half of the patients had a clinically significant mood disorder during the acute post-stroke period which met the diagnostic criteria for either major depression (27%) or minor depression (20%) (Robinson et al., 1983). In addition, 9 patients (9% of the total) had the symptom of inappropriate cheerfulness (indifference reaction) without meeting the diagnostic criteria for a psychiatric diagnosis (Robinson et al., 1983). The percent of patients having major depression at 6 months follow-up was 34%; at one year follow-up, 14%; and at two year follow-up, 21%; while the percentage of patients with minor depression at 6 months, 12 months, and 24 months follow-up was 26%, 19% and 21% respectively. Although this prevalence of major and minor depression remained fairly constant throughout the two year follow-up, this constancy in prevalence was the result of many individual patients getting better during the follow-up period, while others developed late onset depressions.

Figure 5.1 shows the outcome at 1 and 2 years follow-up for each in-hospital diagnostic category. Although not all patients in each diagnostic category were followed up at both 1 and 2 years, none of the patients

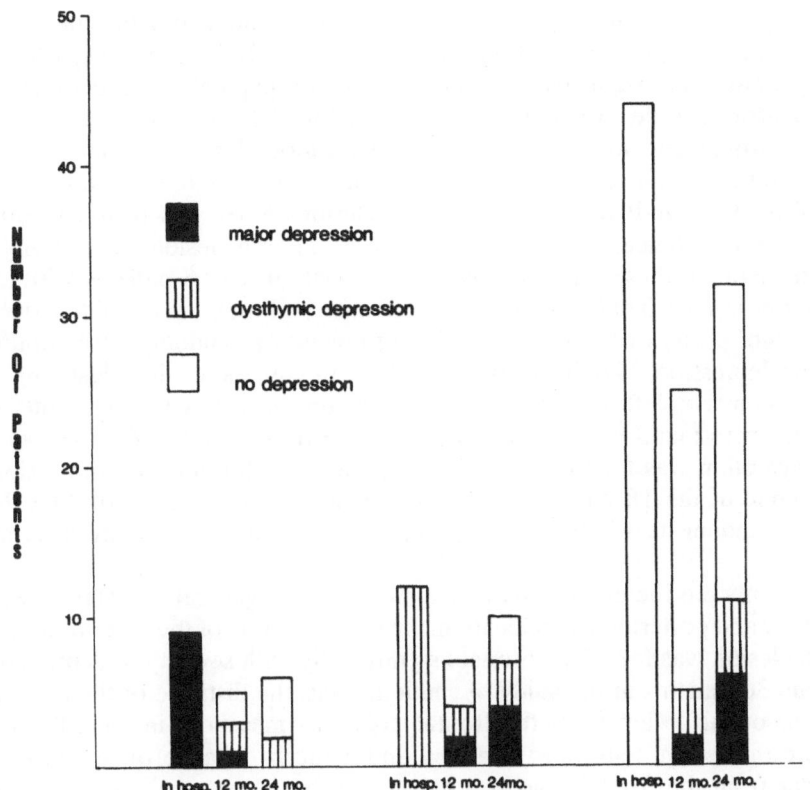

FIGURE 5.1. Diagnostic status at 12- and 24-month time points for the patients grouped according to the in-hospital diagnoses. From Parikh R and Robinson RG, Mood and cognitive disorders following stroke. *Animal Models of Dementia, Neurology and Neurobiology*, Vol. 33, 1987. Reprinted by permission of Alan R. Liss, Inc.

with in-hospital major depression who were reevaluated at 2 years follow-up continued to meet the diagnostic criteria for major depression, while only 30% of the patients with minor depression in-hospital were no longer depressed by 2 years follow-up (i.e., 70% of the patients seen at 2 years had diagnoses of major or minor depression) (Robinson et al., 1987). These findings suggest that major depression has a natural course of about one year duration while the prognosis of minor depression is less certain, and many of these depressions continue for more than two years.

Relationship of Depression to Lesion Location and Size

In an effort to determine whether injury to specific brain regions would be associated with different types of mood disorders, a group of 36 acute stroke patients who had CT verified single infarcts and no previous history of neurological or psychiatric disorder were selected (Robinson et al., 1984a). All patients were right-handed and constituted a population at low risk for psychiatric disorder. Patients were divided into left and right hemisphere lesion groups and also into anterior and posterior lesion location groups. An anterior lesion was defined as one whose anterior border was rostral to 40% of the A-P distance, and the posterior border was rostral to 60% of the A-P distance. Of 10 patients with left anterior lesions, 6 had major depression; but these symptoms were found only in 1 of 8 patients with left posterior lesions, 0 of 6 patients with right anterior lesions, and 1 of 6 patients with right posterior lesions ($p < .05$ that this was a random distribution of major depression) (Robinson et al., 1984a). In contrast to this clustering of depression with left hemisphere lesion location, patients with right anterior lesions were found to be inappropriately cheerful in 5 out of 6 cases, while this symptom was found in only 1 of 6 patients with right posterior lesions and none of the 18 patients with left hemisphere injury ($p < .05$ that this was a random distribution of inappropriate cheerfulness) (Robinson et al., 1984a).

Not only did the author and colleagues find this general association with anterior and posterior lesion location, the exact position of the anterior border of the lesion was found to correlate significantly with severity of depression. Of the 36 patients in the study of single lesions, the distance of the anterior border of the lesion from the frontal pole was measured in all CT slices where the lesion could be visualized and a mean distance of the anterior border from the frontal pole was calculated. In the group of patients with left anterior strokes, depression scores were highly significantly correlated with the proximity of the lesion to the frontal pole (Figure 5.2) (Robinson et al., 1984a). In the right hemisphere lesion group, there was a similar correlation, but in the opposite direction (i.e., the further the lesion was from the frontal pole, the more severe the depression) (Figure 2).

In contrast to this strong correlation with anterior-posterior lesion location, lesion size has not been as consistently correlated with severity of mood disorders in stroke patients (Robinson and Szetela). In the previously described study of 36 patients with single stroke lesions, lesion volume in the left anterior hemisphere group correlated significantly with severity of depression ($r = 0.72$). In the left posterior or right anterior and posterior lesion groups, however, lesion volume did not correlate significantly with severity of depression (Robinson et al., 1984a).

In summary, left frontal lesions are strongly associated with severe depression while right frontal lesions are associated with undue cheerfulness and apathy. Lesion location appears to be more important than lesion size or number of strokes in determining severity of depressive disorder.

The Effect of Bilateral Brain Injury on Mood

This study was undertaken to determine whether undue cheerfulness associated with right hemisphere injury or depression associated with left hemi-

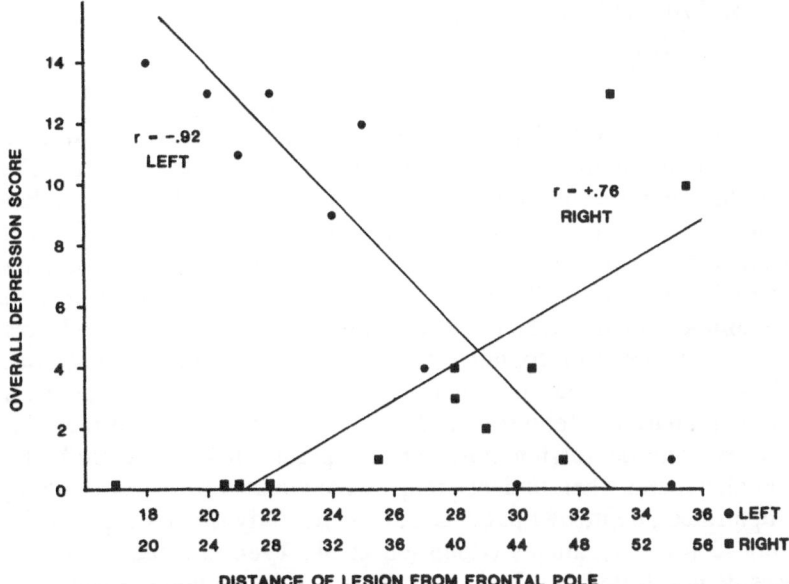

FIGURE 5.2. Relationship between overall depression score and distance of the anterior border of the lesion from the frontal pole for patients with either left anterior hemisphere infarcts or right hemisphere infarcts. The distance from the frontal pole is expressed as a percentage of the total anteroposterior distance. The correlation coefficients are indicated: r left, p < .001; r right, p < .01. From Robinson RG, et al., Mood disorders in stroke patients, *Brain* 107:81–93, 1984. Reprinted by permission of *Brain*.

sphere injury would dominate the clinical picture in patients with bilateral brain injury. The study was also undertaken to determine whether the expression of depression required an intact right hemisphere. Fifteen patients with bilateral hemisphere brain injury secondary to thromboembolic stroke or trauma were examined for mood disorders (Lipsey et al., 1983). Patients with left anterior brain injury were found to be significantly more depressed than patients without such injury. Depression severity also correlated significantly with proximity of the left hemisphere lesion to the frontal pole. Time since stroke, temporal sequence of lesions, right hemisphere lesion location, degree of cognitive impairment, or degree of functional physical impairment did not significantly correlate with depression.

Results of this study led to the conclusion that the depressive symptomatology associated with single left frontal lobe infarction does not depend upon the release of the right hemisphere, but seems to be generated within the left hemisphere itself.

Relationship of Depression to Intellectual Impairment

In the prospective study of 103 patients previously described, the correlation between severity of depression and degree of cognitive impairment as measured by the mini-mental state examination (Folstein et al., 1975) was found to be relatively low ($r = -.29$) (Robinson et al., 1983a). The relationship, however, was statistically significant. Then, in order to assess whether intellectual impairment led to depression or whether depression influenced intellectual impairment, 30 patients with left cerebral hemisphere strokes were examined for depression and cognitive impairment. In non-depressed patients, the severity of cognitive impairment was significantly correlated with both lesion volume and anterior-posterior lesion location as assessed by CT scan analysis (Robinson et al., 1986b). Cognitive impairment in patients with major depression, however, was greater than predicted by lesion volume alone; and when patient groups were equated for severity of cognitive impairment, depressed patients had significantly smaller lesion volumes than non-depressed patients (Robinson et al., 1984b). Multiple regression analysis demonstrated that both depression and lesion volume were significantly and independently related to severity of cognitive impairment.

These findings suggest that in patients who are not depressed the degree of cognitive impairment after stroke is dependent upon both where the brain is injured and how much brain tissue has been injured, while in depressed patients, intellectual impairment is influenced not only by the lesion but also by the depression. Post-stroke depressions may thus produce a "pseudo-dementia" or dementia of depression. These findings suggest that in some patients the treatment of post-stroke depression might improve cognitive functioning following stroke.

Relationship of Depression to Neurological Findings

The neurological examination of stroke patients was done blind to the psychopathological findings and was a standardized examination (Kunitz et al., 1984). Patients were divided into those with mild-to-moderate weakness (i.e., the patient was able to move the limb against gravitational resistance or better) and severe weakness (i.e., the patient was unable to move against gravitational resistance). Touch pain deficits were similarly divided into those who had some touch pain sensation and those who had none. There was no significant difference between the depression scores of patients with mild or moderate motor or sensory deficits of any extremity as compared with those having severe motor or sensory deficits (Robinson et al., 1983). In addition, activities of daily living as measured by the Johns Hopkins Functioning Inventory correlated with severity of depression with a correlation coefficient of .36 (Robinson et al., 1983). This magnitude of correlation coefficient, although significant, could explain only about 10%

of the variance in depression score as compared with lesion location which explained 50% or more of the variance in depression.

In summary, the severity of depression does not appear to be significantly related to the severity of motor or sensory deficit or functional physical impairment. Although there is some association between severity of functional physical impairment and severity of depression, only a relatively small portion of depression could be explained by this variable; and it certainly does not support the conclusion that post-stroke depressions are simply a psychological response to severity of impairment.

Relationship of Depression to Time since Stroke

In an out-patient follow-up study of 103 patients attending a stroke clinic, the prevalence and severity of depression was related to the time that had elapsed since their stroke (Robinson and Price, 1982). This study is different from the prospective study of 103 patients cited earlier and included patients at various time intervals since stroke. Both the severity and prevalence of depression were significantly greater for those patients who were between 6 months and 2 years post-stroke as compared with patients who were between 3 and 9 years post-stroke. Although it did not reach the level of significance ($p < .1$), the data also suggested that there may be an increase in the prevalence of depression in those patients who were more than 10 years post-stroke.

We concluded from this study that the risk period for developing depressive disorders lasts for about 2 years post-stroke. This study also emphasized that time elapsed since stroke is an important variable influencing the prevalence and severity of depression and needs to be considered in the evaluation of mood disorders following stroke.

Treatment of Post-Stroke Depression

As demonstrated in both of the longitudinal studies conducted on post-stroke mood disorders, post-stroke depressive disorders can be longlasting without treatment. Most patients, however, do not receive treatment (Feibel and Springer, 1982). A few anecdotal reports on the benefits of antidepressant medications have been published in the literature (Ross and Rush, 1981), but until 1984 no controlled treatment trials had been done.

We conducted a randomized double-blind treatment trial of post-stroke depression using the tricyclic antidepressant nortriptyline in a group of 34 stroke patients (Lipsey et al., 1984). The drug schedule was 20 mg for one week, 50 mg for two weeks, 70 mg for one week, and 100 mg for two weeks. There was a significantly greater improvement in depression scores in patients treated with nortriptyline ($n = 14$) than in the placebo treated group ($n = 20$) (Figure 5.3). The required blood levels for optimal improvement were in the same range as those needed for patients with functional (i.e.,

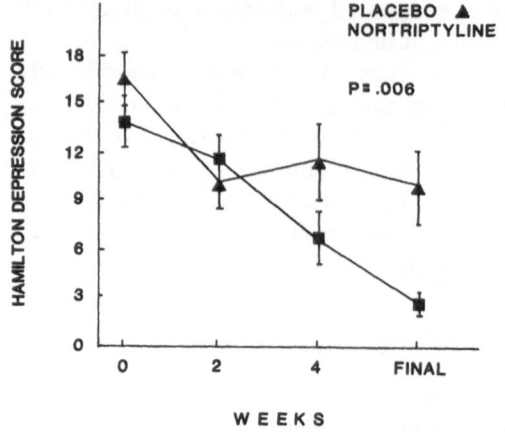

FIGURE 5.3. Hamilton depression scores for nortriptyline and placebo groups over time. Higher scores indicate more severe depression. Error bars represent ± SEM Probability (p) values shown are derived from repeated measures analysis of variance of treatment and time interaction. Reprinted by permission from Robinson RG, et al., *The Lancet* i:1984:297–300.

non-brain injured) depression, and the response to antidepressants required more than two weeks (Figure 5.3). The only potentially serious side effect that occurred more frequently with nortriptyline as compared to placebo was a reversible delirium which occurred in 3 patients.

Although we did not divide the patients into major and minor depression in this study, both groups appeared to respond to antidepressant treatment.

Reding et al. (1986) has also recently reported, based on a double blind treatment trial, that ludiomil was effective in the treatment of post-stroke depression in patients who failed to suppress serum cortisol after dexamethasone administration. Thus, in the majority of cases, post-stroke depressions which last more than one year can be successfully treated with antidepressant medications.

Animal Model of Post-Stroke Depression

Utility of Animal Models to Study Human Disorders

Primitive organisms with simple nervous system organizations can be used to study very basic neural mechanisms of behavior. To study the human brain and its control of complicated behaviors and emotions, however, requires a more highly developed nervous system. The rat is generally the species of choice because of its neuroanatomical similarity to human and its low cost. Virtually all of the anatomical and chemical systems that exist in the human brain can be found in some form in the rat.

Many variables that confound the systematic study and analysis of human data such as age, genetic loading, and environmental effects can be carefully controlled in animal models. Katz (1981) cited some of the advantages of using animal models to study human psychiatric disorders. These advantages include (1) the ability to select dependent variables and then control for multiple independent variables; (2) relatively large numbers of animals can be studied; (3) variables such as receptor binding or biochemical content can be examined, which would not be possible in humans; (4) the short life span allows longitudinal and intergenerational studies; (5) animal models are cost- and time-effective.

The major disadvantage of using animal models, particularly for the study of psychiatric disorders, is the uncertainty that the animal model accurately corresponds to the human condition in its etiology, pathopsychological mechanisms, or behavioral manifestations. Depression has been modelled using a number of different species including monkeys (McKinney, 1974), dogs (Seligman and Maier, 1967), and rats (Porsolt et al., 1978). The methods used for induction of depression have included pharmacological or environmental manipulation, and the models have been validated either by observation of the animals' behavior or response to antidepressant treatment. The difficulty in validating these models has been the uncertainty of correspondence between animals' behavior and depressive symptoms in humans and an uncertainty that antidepressant medications are specifically addressing the etiological abnormality which leads to depressive disorder. One of the advantages using an animal model to investigate post-stroke depressive disorders, however, is that the neural pathology can be used as one means of validating the model; and this validation does not depend so heavily on the animals' behavior looking identical to human depression.

Types of Stroke Models

Peterson and Evans (1937) were the first investigators to demonstrate that cerebral infarction could be produced in animals by occlusion of the middle cerebral artery. Levine and Payan (1966) reported that carotid artery ligation in Mongolian gerbils frequently leads to unilateral cerebral infarction because of an incomplete circle of Willis due to insignificant posterior communicating artery (approximately 40% of Mongolian gerbils have an incomplete circle of Willis). Although carotid artery ligation is technically easier to perform than middle cerebral artery ligation, carotid ligation leads to marked variability in both the size of the cerebral ischemia and the survival rate of the animal because of variation in arterial patency. Stroke models have also been developed in cats (O'Brien and Waltz, 1973), primates (Symon, 1975) and rats.

There are at least four commonly used rat models of stroke. These models include stroke-prone spontaneously hypertensive rats (Yamori and Okamoto, 1974), microembolization of the middle cerebral artery with silicone cylinders or microspheres (Kogrue et al., 1975), carotid vessel of occlusion associated with lower oxygen intake (Levy et al., 1975), and middle cerebral artery ligation (Robinson et al., 1975). Although other lesion techniques were utilized, such as suction lesions (Pearlson and Robinson, 1981), cortical undercutting (Kubos and Robinson, 1984), and intracortical injections of neurotoxins (Kubos et al., 1984), to study specific aspects of the neural mechanisms involved in the behavioral response to stroke, this chapter will deal primarily with the middle cerebral artery ligation model and its relationship to post-stroke depression.

Middle Cerebral Artery Ligation Model

In this study male Sprague Dawley rats approximately 12 weeks of age and weighing 300 grams were utilized in the stroke model. Under chloral hydrate anesthesia (350 mg/kg), rats were placed in a stereotaxic apparatus and a craniotomy was made in the dorsal lateral skull. The borders for the craniotomy extend from the coronal suture posteriorly to the preorbital area anteriorly and from the zygomatic arch inferiorly to the ridge separating the dorsal and lateral aspects of the skull superiorly. Using a dissecting microscope, a semicircular ophthalmic needle with 6-0 suture is passed through the dura behind the middle cerebral artery out through the dura and the artery and overlying dura are ligated. To ensure the occlusion of the artery, a small incision is made in the dura distal to the ligature; and the exposed artery is severed with dural scissors.

Middle cerebral artery ischemia leads to loss of cortical tissue in the lateral aspect of the frontal parietal cortex dorsal to the rhinal fissure and lateral to the cingulate cortex without any postoperative mortality. The lesion extends through a variable extent of the cortex, but never involves subcortical tissue (Figure 5.4). The lesion in either hemisphere has a similar appearance. It is generally circular when viewed from above and varies in diameter from 1 to 5 mm with very little histological evidence of inflammatory reaction.

Effects of Experimental Stroke on Behavior

Several behaviors have been examined following middle cerebral artery ligation including shock induced aggression (Robinson et al., 1975), intracranial self-stimulation (Robinson and Bloom, 1978), and response to various schedules of reinforcement for water reward (Kubos et al., 1985). Most of the experimental attention, however, has been focused on spontaneous activity (Robinson et al., 1975; Robinson and Bloom, 1977; Robinson and Coyle, 1980). Measurement of activity has been done using 24 hour running wheel revolutions in most experiments (Robinson, 1979), although computerized photocell chambers (Robinson et al., 1975) and visual observations in an open field environment (Robinson, 1979) has also been used.

Following right middle cerebral artery ligation, rats are significantly more active than animals given either craniotomy or left middle cerebral artery ligation (Robinson and Coyle, 1980). Increased activity begins about 4 days after middle cerebral artery ligation and continues to rise until about day 12 when it is approximately 150% of preoperative baseline (Figure 5.5). Activity then tapers off to pre-operative levels by about 20 days postoperative. In

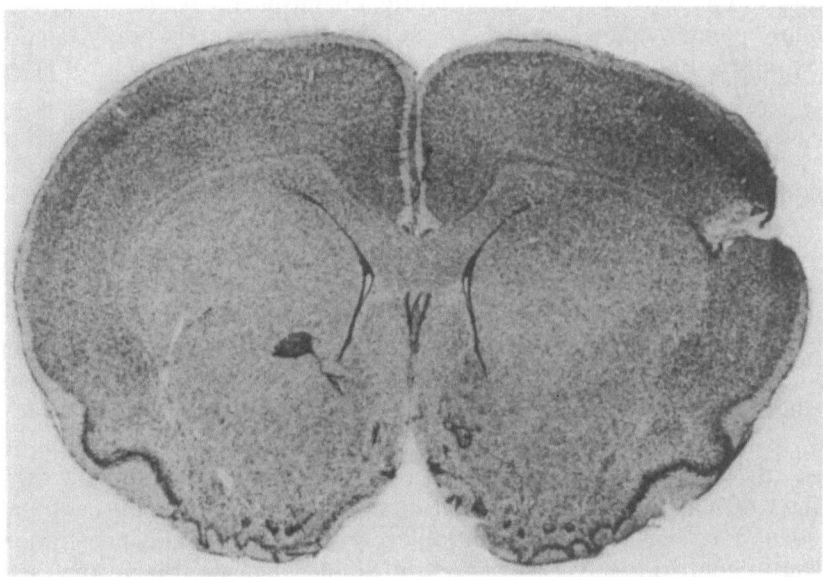

FIGURE 5.4. Photomicrograph of rat brain 2 weeks following middle cerebral artery ligation. The section is 25-m thick and is stained with toluidine blue. Note, the lesion is restricted to the cortex and does not involve the underlying striatum.

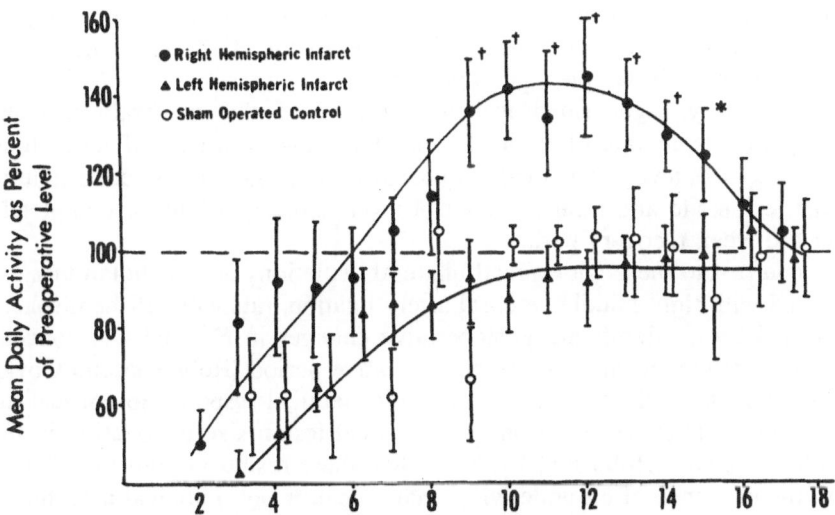

FIGURE 5.5. Mean 24-hour running wheel activity during the 3 week postoperative period. The daily activity of each animal is expressed as a percent of its mean preoperative value. Bars indicate SEM. From Robinson RG, *Science* 205:707–710:1979, © 1979 by the AAAS. Reprinted by permission.

contrast to this hyperactivity following right hemisphere lesions, the activity of sham-operated animals or those with left middle cerebral artery ligation slowly returns to baseline about one week after surgery and levels off at the preoperative baseline values (Figure 3).

Effect of Experimental Stroke on Biogenic Amines

By 12 hours following right middle cerebral artery ligation, norepinephrine (NE) concentrations in both the ipsilateral and contralateral cerebral cortex decrease about 75% compared with operated controls. During a 40-day postoperative period, NE concentrations return slowly towards control levels with complete recovery on the contralateral side and partial recovery on the ipsilateral side (Robinson et al., 1980).

By 12 hours following right middle artery ligation, there is bilateral decrease of NE concentration in the locus coeruleus of approximately 60%. During a 40-day postoperative period, concentrations of NE in the contralateral locus coeruleus recover to control levels while ipsilateral concentrations recover to approximately 75% of control. Similar depletions have been seen in dopamine (DA) concentrations of the substantia nigra and in the A-10 cell group. Right middle cerebral artery ligation reduces DA concentrations bilaterally between 60% and 70% by 12 hours after operation in the substantia nigra. During a 40-day postoperative period, there is relatively less recovery than with NE since DA concentrations return to only 45% to 50% of control levels during this time (Robinson and Coyle, 1980). DA concentrations in the A-10 cell groups are bilaterally depleted to 30% of control levels by 12 hours after operation and recover to only 45% to 50% of control levels during a 30-day postoperative period.

In summary, right sided ischemic lesions of the rat brain produce widespread depletions of both NE and DA concentrations of more than 50% of control levels. The depletion occur both ipsilateral and contralateral to the lesion site and return to control levels either partially or completely over a 6 week recovery period.

In contrast to these widespread bilateral depletions of catecholamines associated with right middle cerebral artery ligation, rats with left hemisphere lesions show no significant postoperative changes in NE and DA concentrations throughout the 30-days postoperative period (Robinson and Coyle, 1980). This lateralized behavioral and biochemical response to cortical injury is not explicable based on asymmetrical lesion size or location in the two hemispheres (Robinson, 1979). Rather, these results are most consistent with the existence of an underlying anatomical or physiological asymmetry in the brain.

Relationship Between Spontaneous Hyperactivity and Changes in Biogenic Amines

Although right but not left middle cerebral artery ligation produces both hyperactivity and depletions of norepinephrine and dopamine, it was also necessary to determine whether the biochemical changes were a consequence of the hyperactivity, a cause of the hyperactivity, or a parallel phenomenon that did not relate directly to the activity changes. In order to investigate the relationship between noradrenergic neurons in the lateralized phenomenon of spontaneous hyperactivity, several experiments have been conducted in which specific lesions of the noradrenergic pathways have been made.

Specific injury to the cortical noradrenergic pathways was produced by microinjections of either 6-hydroxydopmaine (Robinson and Stitt, 1981) or DSP-4 (N-2-chloroethyl-N-ethyl-bromobenzylamine hydrochloride) (Kubos et al., 1984) using stereotaxic cannula placement in the frontal lateral cortex (in the area where ischemic lesions are produced) 1 mm below the brain's surface. Either 2 or 4 μg of 6-OHDA or 10 μg of DSP-4 produced hyperactivity when injected into the right hemisphere. No change in activity occurred when injected into the left hemisphere.

In another experiment, designed to investigate the relationship between NE function and hyperactivity, whole brain catecholamine lesions were accomplished by intracisternal injections of 250 micrograms of 6-OHDA (Robinson and Bloom, 1977). Three weeks after intracisternal injection, ischemic lesions of the right hemisphere were produced by ligation of the right middle cerebral artery. The 6-OHDA produced a 65% reduction of cortical NE and prevented the development of post-middle cerebral artery (MCA) ligation hyperactivity. That is, although right MCA ligation is sufficient to produce hyperactivity, pre-ligation destruction of the catecholaminergic neurons prevented the development of post-ischemic lesion activity without affecting activity in non-ligated controls.

In summary, these experiments suggest that the noradrenergic pathways are important in the production of post-ischemic lesion hyperactivity and that partial lesions (i.e., without destruction of the entire pathway) of the noradrenergic cortical pathways in the right hemisphere are sufficient to produce lateralized hyperactivity.

The effect of enhancement of noradrenergic function was examined by postoperative administration of desipramine (DMI), the tricyclic antidepressant that blocks the reuptake of norepinephrine. A dose of 10mg/kg DMI administered intraperitoneally once per day beginning on the day of right MCA ligation, prevented the development of hyperactivity (Robinson and Bloom, 1977). This effect was specific to post-lesion hyperactivity since DMI did not alter the activity of nonligated animals.

Relationship of Hyperactivity to Non-Biogenic Amine Neurons

In an effort to study neurons which are postsynaptic to the cortical noradrenergic pathways, noncatecholaminergic injury was produced by destroying cell bodies within the cerebral cortex either by using microinjections of kainic acid, or by severing the connections between the cortex and underlying subcortical structures, or by injuring the nucleus accumbens with an electrolytic lesion.

Microinjections of 5nmol of kainic acid into the territory of the MCA 1 mm below the brain's surface, produced a significantly higher degree of spontaneous hyperactivity when injected into the right hemisphere as compared with symmetrical injections of the left hemisphere (Kubos et al., 1982). Kainate is frequently found to diffuse from the site of injection. In spite of possible diffusion, however, there was a differential sensitivity of the two hemispheres to this neurotoxin.

As a second experimental approach to destroying neurons which were postsynaptic to the noradrenergic cortical terminals, disc shaped lesions were produced in the right or left hemisphere white matter underlying the cortical area that produces post-lesion hyperactivity. The disc shaped lesions were 2 mm in diameter overlying the head of the caudate. In a finding similar to those with kainic acid lesions, cortical undercut lesions did not alter catecholamine concentrations but produced spontaneous hyperactivity when made in the right but not left hemisphere. These findings are consistent with the hypothesis that the anatomical or neurochemical basis for the asymmetrical effects of lesions in hyperactivity resides in the post-catecholaminergic projections from the cortex to subcortical structures or at some point further "downstream" in the neural pathway leading to hyperactivity.

One possible subcortical projection area that may mediate hyperactivity is the nucleus accumbens. In recent experiments, the authors and colleagues have demonstrated a lateralized behavioral response to injury of the nucleus accumbens. Partial destruction of the nucleus accumbens was produced by an electrolytic lesion without significant destruction outside the nucleus. This lesion produced spontaneous hyperactivity following right but not left hemisphere lesions. Biochemical measurement revealed a 35% reduction in dopamine concentrations within the accumbens regardless of whether left or right lesions were made (Kubos et al., 1987). These DA depletions, however, did not appear to be responsible for spontaneous hyperactivity since 6-OHDA lesions in the nucleus accumbens which produced 70% to 80% depletions of DA led to hypoactivity rather than hyperactivity (Kubos et al., 1987).

The Role of Interhemispheric Interaction in Hyperactivity

Previous investigators have suggested that behavioral changes following brain injury may result from the "release" of the opposite hemisphere (Denenberg, 1981). In order to investigate whether lateralized hyperactivity following MCA ligation is dependent upon an intact left hemisphere, the author and colleagues examined the effect of bilateral hemispheric lesions on spontaneous hyperactivity. Bilateral lesions were made either simultaneously or a left lesion was followed one week later by a right hemisphere lesion. Regardless of the time between lesions, bilateral lesions led to spontaneous hyperactivity and bilateral depletions of cortical norepinephrine concentrations (Dewberry et al., 1986). This finding that interhemispheric release did not play a major role in hyperactivity response was also supported by the finding that rats given corpus callosum sectioning as neonates and frontal cortical suction lesions as adults developed hyperactivity only if the right hemisphere was injured. The animals did, however, show distinct lateralization of spontaneous activity in response to cortical lesion. Animals with right hemisphere lesions or bilateral lesions became hyperactive while those with left hemisphere lesions alone did not.

These data suggest that lateralized spontaneous hyperactivity, elicited by middle cerebral artery ligation of the right hemisphere, does not depend on interhemispheric release or interaction and that the cortical component of the mechanisms is within the right hemisphere itself.

Hypothesized Mechanisms of Hyperactivity

The authors and colleagues have suggested a tentative and partial neural mechanism involved in the asymmetrical response to cortical ischemia in the rat (Figure 5.6) (Robinson and Justice, 1986). The noradrenergic pathways are not anatomically asymmetrical but rather constitute the first link in asymmetrical neural pathways which lead to hyperactivity when various elements of the system are injured. Neurons postsynaptic to the noradrenergic terminals within the cortex constitute the second link of the pathway, and these intracortical neurons project subcortically to the nucleus accumbens either directly or via cortico-cortical connections. The third neuron in these hypothesized pathways projects to other brain regions such as the ventral pallidal region (Swerdlow et al., 1984; Mogensen et al., 1985) or other areas important in the regulation of spontaneous activity. This third neuron is likely to be the point of anatomical asymmetry. Presumably, the lateralized response to brain injury implies that a neural anatomical and/or neurochemical asymmetry must exist at some point within the pathway leading

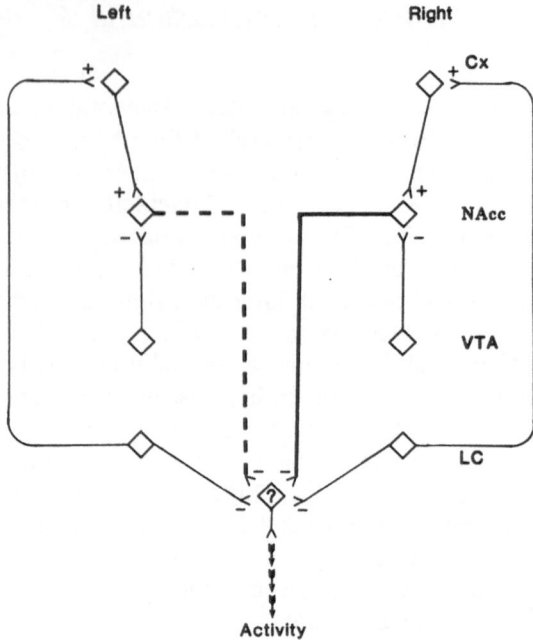

FIGURE 5.6. A schematic model for the hypothesized mechanisms of the lateralized hyperactivity response to brain damage. The three serial links in the system investigated have cell bodies in the LC, cortex, and nucleus accumbens in that order. The first two connections are depicted as facilitatory (+) to the next link because damage to either (by ischemia, suction, neurotoxins, or undercutting) has similar effects on activity. The resulting hyperactivity may be caused by removal of an inhibitory influence (−) that would normally affect spontaneous activity. The fact that right-sided damage produces greater hyperactivity than left-sided damage suggests an asymmetry within the accumbens or its projections or farther downstream. The direct inhibitory link from the LC to the activity area is proposed because dorsal bundle lesions do not have asymmetric effects on activity. The VTA influence is seen as inhibitory because its damage (by accumbal 6-OHDA injections) produces hyperactivity. (Abbreviations: cx, cortex; LC, locus coeruleus; NAcc, nucleus accumbens; VTA, ventral tegmental area.) Reprinted with permission from *Pharm Biochem Behav* 25:263–267:1986, Robinson RG, Justice A, Mechanisms of lateralized hyperactivity following focal brain injury in the rat, © 1986, Pergamon Journals, Ltd.

to hyperactivity. It is not known whether the neural asymmetry exists in the second neuron (from cortical to subcortical areas), or in the third neuron projecting from the accumbens to the ventral pallidum, or further downstream. These data, however, do argue for the existence of at least one lateralized subcortical structure or pathway and that the disruption or destruction of this asymmetrical pathway has important behavioral consequences.

Parallels Between Human and Animal Studies

The previously outlined studies have demonstrated some interesting parallels between post-stroke mood disorders in humans and hyperactivity in the rat. First, there is the lateralized response to injury. In humans, left hemisphere lesions lead to depression while right hemisphere injury leads to undue cheerfulness and apathy. Rats also have a differential behavioral response to injury depending upon whether the right or left hemisphere is injured. Another parallel between human and animal studies is the importance of anterior-posterior location of lesion. In humans, lesions closer to the left frontal pole lead to more severe depression (Robinson et al., 1984a). In rats, there is also an anterior-posterior gradient with more anterior lesions producing the greatest amount of hyperactivity (Pearlson et al., 1984). A third parallel between human and animal studies is the lack of important interhemispheric mechanisms. In humans, bilateral brain injury does not modify the expression of a single left frontal lesion (Lipsey et al., 1983). In rats, the phenomenon of spontaneous hyperactivity induced by a unilateral lesion is not modified by bilateral brain injury or severing the corpus callosum (Dewberry et al., 1986).

A major finding not parallel between humans and rats is that left hemisphere lesion leads to depression in humans, but left hemisphere lesions do not produce behavioral or biochemical changes. In addition, even if the right hemisphere lesions led to "depression" in rats, hypoactivity rather than hyperactivity might be expected as the behavioral manifestation of depression. These differences between humans and animal findings remain to be resolved through further investigation of the variables which influence these behavioral, emotional, and biochemical outcomes. It should also be emphasized, however, that changes in biogenic amine concentrations in animals do not necessarily reflect functional changes in these neurons and conversely changes in function (e.g., firing rate, transmitter release) may occur without demonstrable changes in brain transmitter concentrations. Thus, the parallels between human and animal experiments may involve other types of biochemical analyses such as receptor changes, electrophysiological changes and the like.

Proposed Mechanism of Post-Stroke Mood Disorders

Just as functional depression in nonbrain injured patients is likely to arise from several possible mechanisms, it is likely that there are multiple etiologies of post-stroke depression. The fact that post-stroke major and minor (dysthymic) depression have different two year outcomes (Robinson et al., 1987), different responses to dexamethasone administration (Lipsey et al., 1985), different relationships to anatomical lesion location (i.e., anterior lesions are associated with major depression and posterior lesions are associated with minor depression) (Robinson et al., 1984a) and different effects on intellectual function (Robinson et al., 1986) suggests that these two disorders have different etiologies.

The author and colleagues have proposed a role for catecholamine-containing neurons in major depressive disorder. The norepinephrine and serotonin containing pathways arise from the brainstem and project anteriorly into the frontal cortex and then pass anterior to posterior running through the deep layers of the cortex and arborizing throughout the cortex (Morrison et al., 1978). Focal injury such as occurs in stroke may cause partial damage to these catecholamine containing terminals in the cerebral cortex. Previous investigators have demonstrated that catecholamine concentrations are significantly altered following stroke in humans (Meyer et al., 1973). It has also been shown using animal models of stroke that cortical ischemic lesions can produce widespread depletions of biogenic amine neurotransmitters (Zervas et al., 1974; Brown et al., 1974; Kogrue et al., 1975). In addition, the neurochemical response may be lateralized (Robinson, 1979); that is, the degree of neurotransmitter depletions may depend upon which hemisphere is injured.

Reis and Ross (1973) have suggested that injured catecholaminergic neurons may switch from producing neurotransmitter for nerve transmission to synthesizing protein for nerve regeneration. This switch may lead to a decline in available transmitter throughout the uninjured as well as injured branches of the system.

The author and colleagues have suggested that this widespread depletion of norepinephrine throughout the brain following stroke may be expressed emotionally as major depression (Robinson and Szetela, 1981). In addition, differences in the emotional response to left anterior brain injury (major depression) as compared with right anterior brain injury (indifference and apathy) may be the result of differential biochemical responses to ischemia depending on which hemisphere is injured (Robinson and Chait, 1985).

Since noradrenergic axons arborize as they pass anteriorly to posteriorly through the deep layers of the cortex, a lesion in the anterior regions of the brain would interrupt these pathways closer to their source and thus cause greater disruption of transmitter concentrations than a posterior lesion which would be more distal. This anatomical organization of the noradrenergic pathways might lead to a "graded" effect of lesion location on norepinephrine concentrations (i.e., more anterior lesions produce greater depletion of NE concentrations and the effect gradually diminishes as the lesion is more posterior) and perhaps on depression or apathy. This anatomical relationship between anterior-posterior lesion location and depletion of biogenic amines might explain the clinical finding that the closer the lesion is to the frontal pole, the more severe the depression.

The neuronal circuit proposed to explain the mechanism of lateralized hyperactivity in the rat may be part of the anatomical pathways through which this asymmetry is expressed (Figure 6). This hypothesis that catecholamine containing neurons play an integral role in the etiology of post-stroke mood disorders is also consistent with the extensive literature implicating depletions of catecholamines in the etiology of functional depression (Schildkraut, 1978).

The mechanism of minor depression does not appear to be related to a family history of affective disorder or to a previous personal history of depressive disorder (defined as seeing a physician for treatment of depression). Premorbid personality vulnerabilities such as depressive or obsessional character traits or some other psychological or biological processes might play a role in the etiology of these disorders. As indicated previously, however, there are several lines of evidence which suggest that major and dysthymic depressions have different etiologies.

In summary, the author and colleagues have hypothesized that depletion of biogenic amines following injury to frontal brain regions may play a role in the production of depressive disorders and/or undue cheerfulness and apathy following stroke. The lateralized emotional reaction to stroke may reflect the outcome of neurophysiological responses to stroke within subcortical brain regions.

Summary, Conclusions, and Future Directions

Two types of depression have been recognized following stroke: major depression which has a natural course of approximately one year and minor depression which has a more chronic course and a relatively unfavorable prognosis without treatment. It is likely that at least some of these disorders have foundations in neuropathological processes initiated by stroke. The author and colleagues have developed an animal model of stroke to investigate the neural mechanisms of post-stroke mood disorder. Stroke in rats produces spontaneous hyperactivity and depletions of catecholamine neurotransmitters throughout uninjured as well as injured brain areas. The magnitude of the behavioral and neurochemical response to stroke in the rat depends upon which hemisphere is injured. The neural pathways involved in this lateralized response to injury have been traced, and both subcortical as well as cortical pathways appear to be involved.

Based on the integration of the clinical and laboratory findings, the authors and colleagues have hypothesized that lateralized neurotransmitter depletions initiated by stroke may be expressed in patients as depression or apathy depending upon which hemisphere is injured. The asymmetrical pathways which underlie these emotional responses may be the same ones that lead to lateralized hyperactivity in the rat model.

Although hyperactivity in rats may not be a model of post-stroke mood disorder, the fact that this "model" of depression is based on an underlying neuropathology rather than a nonspecific pharmacological treatment or environmental manipulation has some advantages over these other methods. For instance, the validation of the model is based on its neuropathology rather than the behavior of the animal. This method of validation is unusual for animal models of psychiatric disorders, but it may provide some additional insights into mechanisms or specific treatment modalities not obtained from traditional models. After all, there is no reason why the behavioral manifestation of similar neuropathological processes should be the same in both rats and humans.

The utility of this approach will ultimately be determined by the insights it may provide into the mechanism or treatment of post-stroke mood disorders. Future research may delineate which neuronal systems or neurotransmitter pathways may be involved in the emotional disorders associated with cerebrovascular disease and provide a more rational basis for treatment of these disorders. In addition, an increased understanding of the nature of neuronal regeneration may help investigators to elucidate the natural course and dynamic nature of recovery from these emotional disorders (Robinson et al., 1984b and 1985). Finally, it is hoped that the study of specific emotional response to cerebrovascular disease may help investigators to understand the production of emotional disorders in nonbrain injured patients and the neuronal mechanisms which underlie mood regulation in normal individuals.

Acknowledgments. This work was supported in part by the following NIH grants: Research Scientist Development Award MH00163, NS15178, NS15080 and MH40355.

References

American Psychiatric Association (1980). *Diagnosis and Statistical Manuals of Mental Disorders*, (3rd.). Washington D.C.: American Psychiatric Press

Bleuler EP (1951): *Textbook of Psychiatry*. New York: Dover Publications

Brown RM, Carlson A, Ljunggren B, Siesjo, Snider SR (1974): Effect of ischemia on monamine metabolism in the brain. *Acta Physiol Scand* 90:789–791

Denenberg VH (1981): Hemispheric laterality in animals and the effects of early experience. *Behav Brain Res* 4:1–49

Denny-Brown D, Meyer JS (1957): The cerebral collateral circulation. Production of cerebral infarction by ischemic anoxia and its reversibility in early stages. *Neurol* 7:567–579

Dewberry RG, Lipsey JR, Saad K, Moran TH, Robinson RG (1986): Lateralized response to cortical injury in the rat: Interhemispheric interaction. *Behav Neurosci* 100:556-562

Feibel JH, Springer CJ (1982): Depression and failure to resume social activities after stroke. *Arch Phys Med Rehabil* 63:276

Finkelstein S, Benowitz LI, Baldessarini RJ, Arana GW, Levine D, Woo E, Bear D, Moya K, Stoll AL (1982): Mood, vegetative disturbance, and dexamethasone suppression test after stroke. *Ann Neurol* 12:463-468

Fisher SH (1961): Psychiatric considerations of cerebral vascular disease. *Am J Cardiol* 7:379-385.

Folstein MF, Folstein SE, McHugh PR (1975): Mini-mental state: a practical method for grading the cognitive state of patients for the clinician. *J Psychiatr Res* 12:189-198

Folstein MF, Maiberger R, McHugh PR (1977): Mood disorders as a specific complication of stroke. *Neurol Neurosurg Psychiatr* 40:1018-1020

Harlow HE, Suomi RS (1974): Induced depression in monkeys. *Behav Biol* 12:273

Hecaen H (1962): Clinical symptomatology in right and left hemisphere lesions. In: *Interhemispheric Relations and Cerebral Dominance.* Mountcastle VB, ed. Baltimore: Johns Hopkins Press.

Katz RJ (1981): Animal models and human depressive disorders. *Neurosci Biobehav Rev* 5:231-246.

Kay DK (1962): Outcome and cause of death in mental disorders of old age: a long term follow-up of functional and organic psychoses. *Acta Psychiatr Scand* 38:249-267.

Kogure K, Scheinberg P, Matsumoto A, Busto R, Reinmuth OM (1975): Catecholamines in experimental brain ischemia. *Arch Neurol* 32:21-24.

Kraeplin E. (1921): *Manic depressive insanity and paranoia.* Edinburgh: E & S Livingston

Kubos KL, Brady JV, Moran TH, Smith CH, Robinson RG (1985): Asymmetrical effect of unilateral cortical lesions and amphetamine on DRL-20: a time loss analysis. *Pharmacol Biochem Behav* 22:1001-1006.

Kubos KL, Moran TH, Robinson RG (1987): Differential and asymmetrical behavioral effects of electrolytic or 6-OHDA lesions in the nucleus accumbens. *Brain Res* 401:147-151

Kubos KL, Moran TH, Saad KM, Robinson RG (1984): Asymmetrical locomotor responses to unilateral cortical injectons of DSP-4. *Pharmacol Biochem Behav* 21:163-167.

Kubos KL, Pearlson GD, Robinson RG (1982): Intracortical kainic acid induces an asymmetrical behavioral response in the rat. *Brain Res* 239:303-309

Kubos KL, Robinson RG (1984): Cortical undercuts in the rat produce asymmetrical behavioral response without altering catecholamine concentrations. *Exp Neurol* 83:646-653.

Kunitz SC, Gross CR, Heyman A, Kase CS, Mohr JP, Price TR, Wolf PA (1984): The pilot stroke data bank: definition, design and data. *Stroke* 15:740-746.

Levine S, Payan H (1966): Effect of ischemia and other procedures on the brain and retina of the gerbil. *Exp Neurol* 32:450-456.

Levy DE, Brierley JB, Silverman DG, Plum F (1975): Brief hypoxia ischemia initially damages cerebral neurons. *Arch Neurol* 32:450-456.

Lipsey JR, Robinson RG, Pearlson GD, Rao K, Price TR (1983): Mood changes following bilateral hemisphere brain injury. *Br J Psychiatry* 143:266-273.

Lipsey JR, Robinson RG, Pearlson GD, Rao K, Price TR (1984): Nortriptyline treatment of post-stroke depression: A double-blind study. *Lancet i:*297-300.

Lipsey JR, Robinson RG, Pearlson GD, Raok, Price TR (1985): Dexamethasone suppression test and mood following 'stroke. *Am J Psychiatry 142:*318-323

McKinney WT Jr (1974): Primate social isolation. *Arch Gen Psychiatry 31:*422-426.

Meyer JS, Stoica E, Pascu I, Shimazu K, Hartmann A (1973): Catecholamine concentrations in CSF and plasma of patients with cerebral infarction and hemorrhage. *Brain 96:*277-288

Mogensen GJ, Swanson LW, Wu M (1985): Evidence that projections form substantia innominata to zone incerta and mesencephelic locomotor region contribute to locomotor activity. *Brain Res 334:*65-76.

Moran TH, Zern KA, Pearlson GD, Kubos KL, Robinson RG (1986): Cold water stress abolishes hyperactivity produced by cortical suction lesions without altering noradrenergic depletions. *Behav Neurosci 100:*422-426

Morrison JH, Grzanna R, Molliver ME, Coyle JT (1978): The distribution and orientation of noradrenergic fibers in neocortex of the rat: an immunofluorescence study. *J Comp Neurol 181:*171-40

O'Brien MD, Waltz AG (1973): Transorbital approach for occluding the middle cerebral artery without craniectomy. *Stroke 4:*201-206

Parikh RJ, Robinson RG (1987): Mood and cognitive disorders following stroke. In: *Experimental Models of Dementing Disorders: A Synaptic Neurochemical Perspective*, Coyle JT, ed. New York: Alan R. Liss, Inc.

Pearlson GD, Kubos KL, Robinson RG (1984): Effect of anterior-posterior lesion location on the asymmetrical behavioral and biochemical response to cortical suction ablations in the rat. *Brain Res 293:*241-250.

Pearlson GD, Robinson RG (1981): Suction lesions of the frontal cerebral cortex in the rat induce asymmetrical behavioral and catecholaminergic responses. *Brain Res 218:*233-242

Peterson JN, Evans JP (1937): The anatomical end results of cerebral artery occlusion. *Trans Am Neurol Assoc 63:*88-93.

Porsolt RG, Anton G, Blavet N, Jalfre M (1978): Behavioral despair in rats: a new model sensitive to antidepressant treatments. *Eur J Pharmacol 47:*379-391.

Post F (1962): *The Significance of Affective Symptoms in Old Age* (Maudsley Monograph, no. 10). London: Oxford University Press

Reding MJ, Orto LA, Winter SW, Fortuna IM, Di Ponte PD, McDowell FH (1986): Antidepressant therapy after stroke: a double blind trial. *Arch Neurol 43:*763-765

Reis DJ, Ross RA (1973): Dynamic changes in brain dopamine B-hydroxylase activity during anterograde and retrograde reactions to injury of central noradrenergic axons. *Brain Res 57:*307-326

Robinson RG, Shoemaker WJ, Schlumpf M, Valk T, Bloom FE (1975): Effect of experimental cerebral infarction in rat brain on catecholamines and behavior. *Nature* (London) *295:*332-333

Robinson RG, Bloom FE (1977): Pharmacological treatment following experimental cerebral infarction: implication for understanding psychological symptoms of human stroke. *Biol Psychiatry 12:*669-680

Robinson RG, Bloom FE (1978): Changes in posterior hypothalamic self-stimulation following experimental cerebral infarction in the rat. *J Comp Physiol Psychol 92:*969-976

Robinson RG (1979): Differential behavioral and biochemical effect of right and left hemispheric cerebral infarction in the rat. *Science 205:*707-710.

Robinson RG, Coyle JT (1980): The differential effect of right versus left hemispheric cerebral infarction on catecholamines and behavior in the rat. *Brain Res 188*:63–78

Robinson RG, Shoemaker WH, Schlumpf M (1980): Time course of changes in catecholamines following right hemispheric cerebral infarction in the rat. *Brain Res 181*:202–208

Robinson RG, Stitt TG (1981): Intracortical 6-hydroxydopamine induces an asymmetrical behavioral response in the rat. *Brain Res 213*:387–395

Robinson RG, Szetela B (1981): Mood change following left hemispheric brain injury. *Ann Neurol 9*:447–453

Robinson RG, Price TR (1982): Post-stroke depressive disorders: a follow-up study of 103 outpatients. *Stroke 13*:635–641

Robinson RG, Starr LB, Kubos KL, Rao K, Price TR (1983): A two year longitudinal study of post-stroke mood disorders: findings during the initial evaluation. *Stroke 14*:736–741

Robinson RG, Kubos KL, Starr LB, Rao K, Price TR (1984a): Mood disorders in stroke patients: importance of location of lesion. *Brain 107*:81–93

Robinson RG, Starr LB, Price TR (1984b). A two year longitudinal study of post stroke mood disorders; prevalence and duration at six months follow-up. *Br J Psychiatry 144*:256–262

Robinson RG, Starr LB, Lipsey JR, Rao K, Price TR (1985): A two-year longitudinal study of post-stroke mood disorders: inhospital prognostic factors associated with six months outcome. *J Nerv Men Dis 173*:221–226

Robinson RG, Chait RM (1985): Emotional correlates of structural brain injury with particular emphasis on post-stroke mood disorder. *Crit Rev Clin Neurobiol 1*:285–318

Robinson RG, Bolla-Wilson K, Kaplan E, Lipsey JR, Price TR (1986a): Evidence for intellectual impairment related to depression in stroke patients. *Br J Psychiatry 148*:541–547

Robinson RG, Justice A (1986b): Mechanisms of lateralized hyperactivity following focal brain injury in the rat. *Pharmacol Biochem Behav 25*:1344–1354

Robinson RG, Bolduc PL, Price TR (1987): A two-year longitudinal study of post-stroke mood disorders: diagnosis and outcome at one and two year follow-up. *Stroke 18*:837–843

Ross ED, Rush AJ (1981): Diagnosis and neuroanatomical correlates of depression in brain damaged patients. *Arch Gen Psychiatry 38*:1344–1354

Schildkraut JJ (1978): Current status of the catecholamine hypothesis of affective disorders. In: *Psychopharmacology: A Generation of Progress*, Lipton MA, DiMascio A, Killam KF eds. New York: Raven Press, 1223–1234.

Seligman MEP, Maier SF (1967): Failure to escape traumatic shock. *J Exp Psychol 74*:1–9

Sinyor D, Jacques P, Kaloupek DG, Becker R, Gildenberg M, Coopersmith H (1986): Post-stroke depression and lesion location: an attempted replication. *Brain 109*:537–546

Symon L, Crockard HA, Dorsch NWC, Branston NM, Juhasz J (1975): Local cerebral blood flow and vascular reactivity in a chronic stable stroke in baboons. *Stroke 6*:482–492

Swerdlow NR, Swanson LW, Koob GF (1984): Electrolytic lesions of the substantia innominata and lateral preoptic area attenuate the 'supersensitive' locomotor response to apomorphine resulting from denervation of the nucleus accumbens. *Brain Res 306*:141–148.

Yamori, Okamoto K (1974): *Spontaneous hypertensionin the rat: a model of essential hypertension.* In: Proceedings of the 80th Congress of German Society for Internal Medicine. April 21–25. Wiesbaden: Springer Verlag, Berlin

Zervas NT, Hon H, Negora M, Wurtman RJ, Larin F, Lavyne MH (1974): Reduction of brain dopamine following experimental cerebral ischemia. *Nature* (London) *247*:283–284

6

Social Zeitgebers: A Peer Separation Model of Depression in Rats

CINDY L. EHLERS, TAMARA L. WALL,
STEPHAN P. WYSS, AND R. IAN CHAPLIN

Introduction

The concept that loss and/or separation might play a pivotal role in the generation of a depressive episode has been a key point in several theories seeking to explain the etiology of affective disorder. Freud's initial hypothesis suggested that a fixation (i.e., an arresting of psychological growth) at a highly dependent phase of development rendered a person much more vulnerable to the development of depression following a real or imagined loss in adult life (Freud, 1917). Elaborations of Freud's theory have broadened the concept of loss to include experiences of separation and social rejection.

Animal models have also been developed in order to study the role of separation and loss in the etiology of depression. Both primates and rodents have been utilized in various paradigms in an attempt to generate behaviors which either resemble human affective disorders or are modified by antidepressant drugs (McKinney et al., 1984). In the primate models, separation of infant monkeys (usually rhesus) from their mothers produces a series of behaviors described as the "protest and despair" reactions (Seay et al., 1962). In the protest stage, monkeys were found to display a generalized increase in activity accompanied by loud screeching, whereas animals in the despair stage were found to be in a general state of withdrawal consisting of decreased activity and vocalization. The response of rodents to maternal separation has also been described. Rat pups isolated at a critical stage of development have been shown to exhibit increased activity during wakefulness (Smith and Anderson, 1984; Garzon and Del Rio, 1981) and disorganized EEG during sleep (Hofer, 1976). While such studies have provided important information on mother-infant interactions, they have been challenged as to whether the observed behaviors truly model a major affective disorder. Although the protest-despair response in monkey infants has been suggested to resemble responses in human infants (Robertson and Robertson, 1971), this response may, in fact, represent a model of separation anxiety (Bowlby, 1960) or anaclitic depression (Spitz and Wolf, 1946).

Suomi et al. (1970) and Bowden and McKinney (1972) have suggested that separation of rhesus monkeys from their peers might represent a model which more closely simulates the social factors which may trigger human depressions. Studies by Suomi and collaborators (1975) have demonstrated that young adult monkeys will display severe and persistent signs of despair in response to being completely socially isolated. Response to peer separation has also been described in rats. Valzelli and Bernasconi (1976) have demonstrated that adult rats isolated a minimum of six weeks will display either highly "aggressive" behavior or "indifference" and have suggested that these abnormal behaviors reflect a series of emotional changes that can be altered by psychoactive drugs. It has also been shown that adult rats will exhibit hyperactivity in response to peer separation (Weinstock et al., 1976), although to a lesser degree than is observed in maternally isolated pups. One major criticism of all of these animal investigations and particularly of rodent studies has been that the limited repertoire of animal behaviors severely restricts the ability of animal studies to model the complexity of human psychopathology.

One set of measures, the study of biological rhythms, might represent a potential animal model of depression which may be less species specific. A large subgroup of patients with major depression display clinically-specific disturbances in their biological rhythms (Wehr and Wirz-Justice, 1982), particularly in their sleep/wake cycle (Kupfer, 1978) and in their neuroendocrine systems (Carroll et al., 1980). It has been postulated that the disturbance in biological rhythms that is seen during a depressive episode is the result of a disruption in the clock or clocks which synchronize the phase relationships between the various physiological functions (e.g., sleep/wake cycles, neuroendocrine rhythms) (Wehr and Goodwin, 1983). How the biological clock is perturbed in depression is not clear; however, several studies have suggested that various rhythms may become disentrained from their zeitgeber. A "zeitgeber" for a circadian rhythm is a person, or process, which is capable of synchronizing or entraining a rhythm, so that its cycle length is about 24 hours long. In many mammals, light is the most important zeitgeber; and thus, day length can entrain several biological rhythms. In the case of depression, the effect of light or day length on affective symptomatology has been studied in some human populations. In these studies, it has been hypothesized that light may play a role in the treatment of seasonal depressive disorders (James et al., 1985; Lewy et al., 1985). In the case of human major affective disorder, it is possible that social factors may be more potent synchronizing forces for neuroendocrine and sleep rhythms. Thus, the authors have postulated that separation or loss may trigger the biological rhythm disturbances seen in depression through the ability of a mother or peer to act as a "social zeitgeber" (Ehlers et al., 1988).

In order to test this theory in an animal model, the authors chose to socially isolate mature rats and measure several of their biological rhythms. Specifically, rhythms in eating, drinking, activity, sleep, and response to corticotropin-releasing factor (CRF) were measured following isolation. The authors have postulated that examination of the rhythm abnormalities associated with the lack of social interactions might aid in elucidation of the underlying physiological mechanisms and thus improve the understanding of the ways this model might be relevant to human studies of depression.

Description of the Model

Eight week old adult, male Wistar rats were surgically prepared for intracerebroventricular (ICV) injections and chronic EEG recordings as previously described (Ehlers et al., 1986). Following recovery from surgery, each animal was randomly assigned to one of three housing conditions aimed at producing various degrees of peer separation: isolated, partially isolated, or group housed. The isolated animals were housed singly in an opaque cage so that they were unable to see or interact with other animals. The partially isolated animals were also housed singly, but in transparent cages with small holes through which they could "nose poke" with another rat whose cage they were paired with. This arrangement allowed the two animals to see and smell each other, but they could not directly interact. The group housed animals were paired in cages that were twice the space of the isolated or partially isolated cages, giving all animals the same amount of living area. Environmental conditions were held constant and food and water were provided ad libitum. Rats were isolated in these housing conditions for a period of six weeks following which the animals were monitored for rhythms in food and water consumption, sleep EEG, locomotor activity rhythms, and response to ICV injection of CRF.

Food and Water Consumption

A disturbance in feeding behavior is a common biologic symptom in depression (Grant, 1979). Although loss of appetite and weight loss are typical features in affective disorder, increased food consumption can also be a prominent feature particularly in "atypical" presentations. In the authors' study, food and water consumption were measured over a 24-hour period in order to detect whether these rhythms were disturbed by social isolation in the previously described model. Evaluation of food and water consumption revealed that both of these rhythms were significantly modified in isolated and partially isolated animals as compared to control (group housed) rats. These rhythms appeared flattened in the case of food consumption and shifted in the case of water consumption due to the fact that isolated animals ate and drank significantly more at various time points over the 24-hour period, as seen in Figures 6.1 and 6.2. Based on these data one might suggest that this increase in food and water consumption may represent a

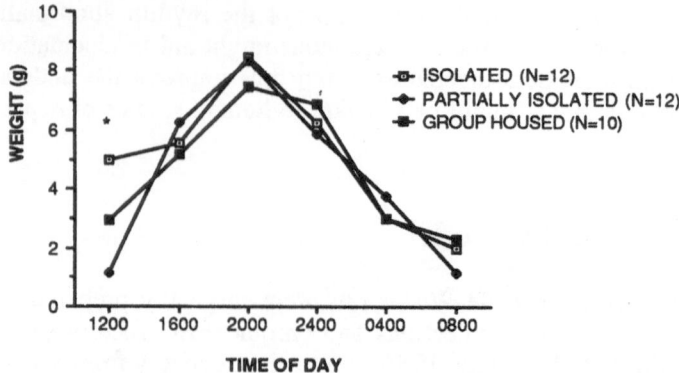

FIGURE 6.1. Food consumption (mean grams per 4-hour epoch) over 24-hour period. (ANOVA, group effect: F = 7.38, df = 2, 31, $p < .01$, Newman-Keuls a posteriori test indicated that the isolated animals ate significantly more over the total 24-hour period than both the partially isolated and group housed animals; group x time interaction: F = 2.05, df = 10,186, $p < .05$, * $p < .05$ difference for all groups, Newman-Keuls test following significant simple main effects.)

model of "atypical" depression. However, the reasons for changes in eating behaviors in these isolated rats are unknown and the observed increase may represent, for instance, a simple loss of competition for food or boredom due to isolation. In any case, the authors believe that the disturbance in the 24-hour rhythm of these behaviors due to social isolation to be an important feature in this model irrespective of the motivational variables.

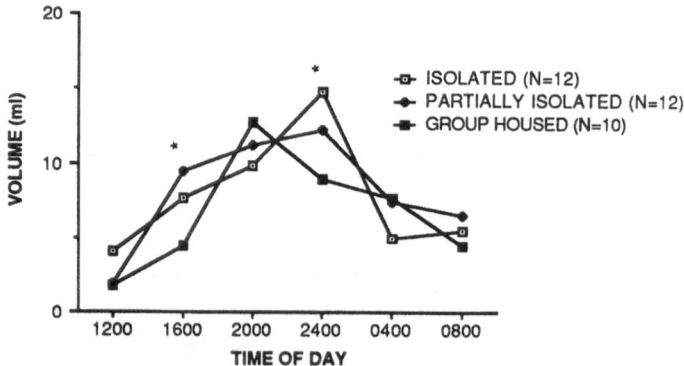

FIGURE 6.2. Water consumption (mean ml per 4-hour epoch) over 24-hour period. (ANOVA, group effect: F = 3.47, df = 2,31, $p < .001$, Newman-Keuls a posteriori test indicated that the partially isolated animals drank significantly more over the total 24-hour period than the group-housed animals did; group x time interaction: F = 4.15, df = 10,186, $p < .001$,* $p < .05$ difference between isolated and partially isolated animals, Newman-Keuls test following significant simple main effects).

Locomotor Behavior

A slowing down of spontaneous activity (psychomotor retardation) or an increase in certain behaviors (psychomotor agitation) is another symptom of depression that has been utilized as the basis for the development of affective disorder (see chapter by Teicher et al., this volume).

Several studies have demonstrated that locomotor behavior is disrupted in models in which animals are isolated from either maternal or peer contact. In monkeys, two complex series of behaviors are generated by isolation which usually culminate in reduced activity (see McKinney et al., 1984): whereas in rat studies, increased levels of locomotor activity are universally reported following both peer (Sahakian et al., 1974; Weinstock et al., 1976), and maternal (Garzon and Del Rio, 1981; Smith and Anderson, 1984) separation.

The authors also found increases in spontaneous activity following isolation when rats were initially placed in a novel environment. In addition, as seen in Figure 6.3, the degree of isolation determined the amount of increased activity as isolated animals displayed even greater increases in locomotion than partially isolated animals. It might be suggested that the increase in locomotion observed in rat studies may represent a model of the psychomotor agitation observed in some depressed patients. However, these results could also be interpreted as a possible increase in response to the novelty of the test apparatus produced by the prolonged sensory monotony which presumably occurs during social isolation. In addition, locomotor activity in rats varies with time of day and the technique of measurement. Therefore, a simple increase in activity observed over a short measuring period is not sufficient to fully characterize changes in locomotor activity.

In a study evaluating the effects of maternal separation in rats, Teicher

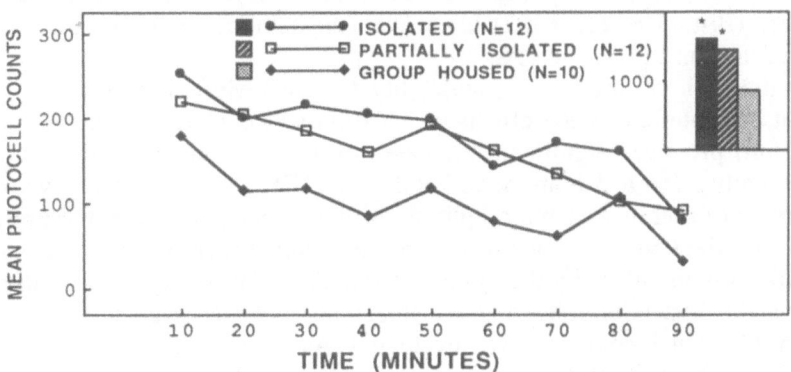

FIGURE 6.3. Locomotor activity (mean photocell counts per 10-minute epoch) following 6 weeks of housing condition. Increased activity was observed in the isolated and partially isolated animals. (ANOVA, group effect: $F = 5.79$, df $= 2,31, p < .01$, *$p < .05$ difference from group-housed animals, Newman-Keuls test.) Insert shows mean of total responses for 90-minute period for each rat.

et al. (this volume) found that the circadian entrainment of locomotion was weakened and that the rats ultradian rhythms were enhanced. The authors have evaluated ultradian rhythms in locomotion in rats isolated from their peers and have also found that the temporal organization of the isolated rats locomotor patterns were disturbed, particularly in response to administration of corticotropin-releasing factor.

Response to Corticotropin-Releasing Factor (CRF)

Evidence that the brain hypothalamic pituitary axis (HPA) may play a role in either the initiation or maintenance of a depressive episode has accumulated over the last two decades (Carroll et al., 1980). The discovery of the hypothalamic regulatory peptide corticotropin-releasing factor (CRF) has provided new strategies for the exploration of the HPA axis in depression, particularly in animal models. Recent clinical investigations have emphasized that the levels or activity of CRF may be modified in affective disorder. A blunting of the pituitary release of ACTH in response to intravenous CRF administration (Gold et al., 1984) and an increase in cerebrospinal fluid levels of CRF (Nemeroff et al., 1984) have been observed in depressed patients. Both findings are consistent with the hypothesis that CRF is dysregulated in the state of depression.

It has been demonstrated in animal studies that CRF can exert direct effects on behavior separate from its ability to cause the release of ACTH and β-endorphin from the pituitary. Acute intracerebroventricular (ICV) administration of CRF to rats has been found to produce a behavioral syndrome which resembles a spontaneous exaggerated "stress response." Dramatic increases in locomotion (Sutton et al., 1982), activation of the EEG (Ehlers et al., 1983a, 1986; Ehlers, 1986) and behavioral signs of increased anxiety (Britton et al., 1982) have all been demonstrated following ICV administration of CRF to rats.

In the authors' model, the possibility that behavioral response to CRF might be modified by the effects of isolation was explored. Although CRF (0.15nm) produced significant increases in mean locomotor activity in all three groups for a 3-hour period following ICV injections, there was no significant difference in the amplitude of the response between the groups. However, the temporal organization of the locomotor response to CRF was significantly modified by the effects of isolation. This effect was significant during the first hour following the injections, but was even more evident when all night locomotion was evaluated. As seen in Figure 6.4, animals who were group housed displayed clear ultradian rhythms in locomotion with broad peaks at 90 to 100 minute intervals. Animals who were partially isolated also displayed ultradian peaks; however, the peaks were narrower and occurred at about twice the frequency. The isolated animals' locomotor rhythms were found to be disrupted following CRF injections. This disruption was characterized by difficulty in identifying clear peaks of activity in the

FIGURE 6.4. Ultradian peaks in the locomotor time series produced in response to ICV administration of CRF. Using the Mann-Whitney U test, significant differences were found between all groups injected with CRF. Group housed versus partially isolated, U = 3.5, N1 = 5, N2 = 6, $p < .05$; group housed versus partially isolated, U = 1.0, N1 = 5, N2 = 6, $p < .01$; partially isolated versus isolated, U = 5.0, N1 − 6, N2 = 6, $p < .05$. No significant differences were found between the saline-injected groups.

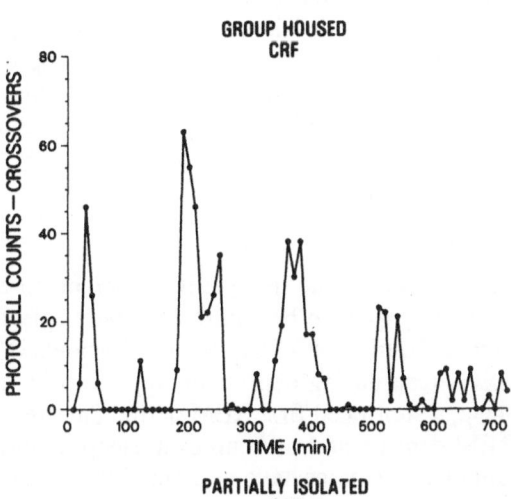

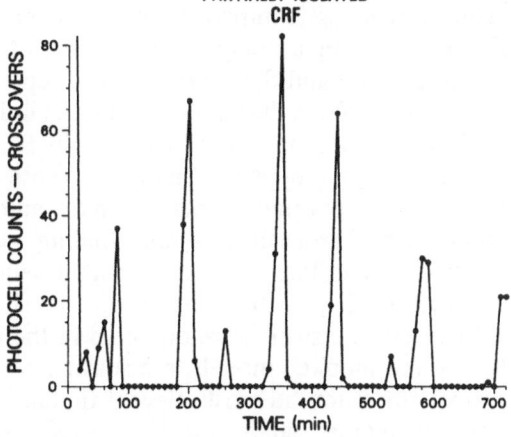

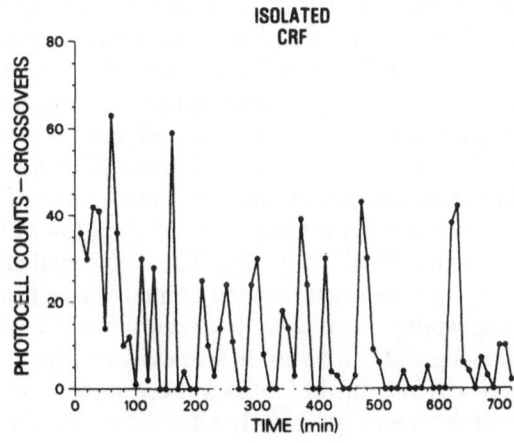

isolated group due to the appearance of low amplitude fast rhythms which appeared "chaotic" in nature. Thus, a very different pattern in locomotor response to CRF was observed depending on the degree of social isolation that the animal had been subjected to. This difference in the rhythmicity of the animals' behavior also appeared to be long-lasting since the effects were still measurable over a 12-hour period.

EEG Activity

The changes in the sleep-EEG which occur during depressive episodes have been suggested to be the most robust episode marker of a major affective illness (Gillin, 1983; Kupfer and Reynolds, 1983). The important features of EEG-sleep during depression appear to be a reduced amount of slow wave sleep, a shortened first REM period (shortened REM latency), an increase in REM density, and difficulties in sleep maintenance (increased awakenings). Only a few studies have attempted to determine whether sleep disturbance is an accompanying feature of the development of symptoms in animal models of depression. In a study of the effects of maternal separation on infant monkeys, Reite and Short (1977) have reported that sleep disturbances can occur in the infants during the time of isolation. Paradoxically, it was found that separation produced a decrease in REM density and an increase in REM latency, the opposite findings to those observed in depressed patients. However, they did find an increase in the amount of time spent awake and in the number of arousals. A similar finding was observed by Hofer (1976) in rats who were isolated from maternal influence. An increase in the frequency of state transitions with more frequent and shorter periods of slow wave sleep and paradoxical sleep were observed in the isolated rat pups. Thus, in both the rat and monkey models, a disturbance of sleep maintenance and sleep architecture were the most relevant findings.

In the authors' study, the sleep EEG in peer isolated adult rats were also evaluated. The EEG of the rats was quantitated by the use of spectral analysis, utilizing previously described methods (Ehlers and Havstad, 1982). An increase in mean EEG amplitude was observed in the sleep of the rats who were in total social isolation. This increase was significant for the frequencies between 4 to 6 and 6 to 8 Hz (ANOVA, group effect: $F = 5.16$, $df = 2,31$, $p < .05$) and represents an increase in theta wave activity. The isolated animals also displayed a loss of EEG stability as quantified by the increase in the coefficient of variation of the EEG in all frequency bands (Newman-Keuls, $p < .05$). This decreased stability can be visualized in the sleep spectra of an isolated animal where it can be seen that the isolated rat was rapidly fluctuating between various EEG states (Figure 6.5). This loss of sleep state structure is also evident when EEG power is viewed as a time series. Thus, in the authors' peer separation model, it was also found that isolation can significantly alter the architecture of the sleep EEG, producing an unstable rapid fluctuation between states.

Group Housed   Isolated

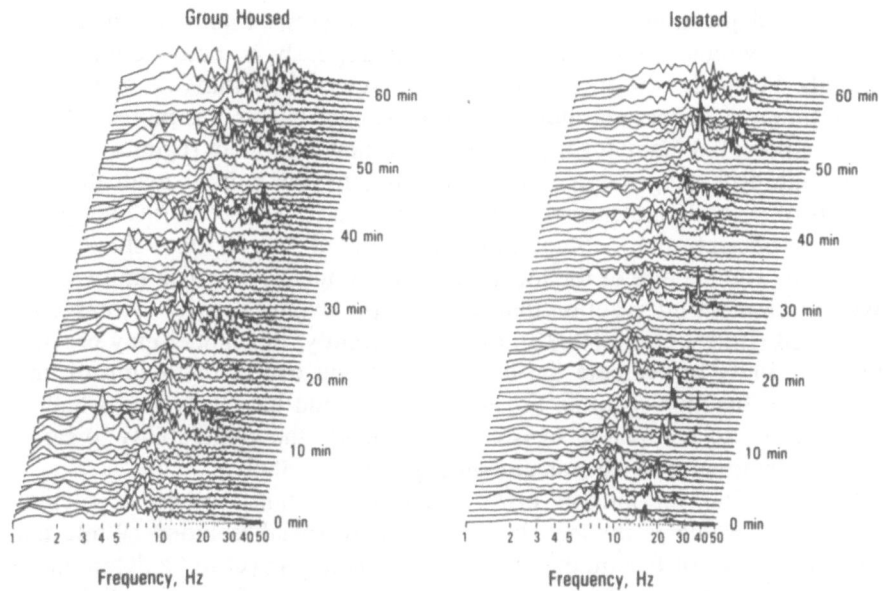

FIGURE 6.5. Two individual rats' EEG spectra following 6 weeks of housing conditions. In each graph, frequency log (Hz) is represented on the x-axis, power density (μv^2/octave) on the y-axis, and time (minutes) on the z-axis. Note in the group-housed animal, the presence of more slow waves (activity below 5 Hz) than the isolated subject. In addition, the isolated animal displayed a more "choppy" EEG record.

SOCIAL ZEITGEBERS AND DEPRESSION

Several studies have now demonstrated that human social interactions can act as "zeitgebers" through their ability to entrain circadian rhythms (Wever, 1985) In animal studies, the presence of the mother mouse has also been demonstrated to act as a synchronizing factor for several rhythms of mouse pups (Viswanathan and Chandrashekaran, 1985). In the study described here the authors investigated whether isolation from social contact in adult rats could produce biological rhythm disturbance. Measurements of eating, drinking, sleep, and activity all revealed that these rhythms were changed in isolated animals. In addition, the authors found that in the case of activity, it further appeared that the degree of isolation also influenced the severity of the dysynchronization. These studies, therefore, suggest that peer contact can act as a potent social zeitgeber in adult rats. Although the rhythms in these rats are altered during isolations, they are not altered in an identical way to rhythms in human depression. Thus, this model does not represent a homologous model of depression (see chapter by Kornetsky for definitions, this volume).

Nevertheless, the study of disturbances in biological rhythms in animal models of depression is important for several reasons. Firstly, the measurement of rhythms can be accomplished analogously in both humans and animals. Secondly, the fact that the disturbance in rhythms is induced by environmental factors and not by drug treatment allows this model to represent a potential assay for the evaluation of new treatment regimes without producing drug-drug interactions. In this regard, the authors evaluated the actions of several antidepressant treatments in prior studies on the temporal structure of nocturnal locomotion in non-isolated rats. In these studies, the pattern or architecture of the locomotion was found to be much more sensitive to treatment intervention than whether the overall activity increased or decreased (Ehlers et al., 1983b). Most importantly, this model may provide an opportunity to evaluate the development and course of rhythm disturbance as a function of time. For instance, it could be determined whether the exposure to one period of isolation increases the vulnerability to rapid rhythm disturbance when rats are subsequently isolated.

In conclusion, the authors have presented an animal model of rhythm disturbance which is induced in adult rats by social isolation. While the direct relevance of this model to human depression is yet to be determined, it may represent a new approach to the understanding of affective disorders.

References

Bowlby J (1960): Grief and mourning in infancy and early childhood. *Psychoanal Study Child* 15:9-52

Bowden D, McKinney W (1972): Behavioral effects of peer separation, isolation, and reunion on adolescent male rhesus monkeys. *Dev Psychobiol* 5:353-362

Britton D, Koob G, Rivier J, Vale W (1982): Intraventricular corticotropin-releasing factor enhances behavioral effects of novelty. *Life Sci* 31:363-367

Carroll B (1982): The dexamethasone suppression test for melancholia *Br J Psychiatry* 140:292-304

Carroll B, Greden J, Feinberg M (1980): Neuroendocrine disturbances and the diagnosis and aetiology of endogenous depression. *Lancet* i:321-322

Ehlers C (1986): EEG stability following corticotropin-releasing factors in rats. *Psychoneuroendocrinology* 11:121-125

Ehlers C, Frank E, Kupfer D (1988): Social zeitgebers and biological rhythms: A unified approach to understanding the etiology of depression. *Arch Gen Psychiatry* 45:948-952

Ehlers C, Havstad J (1982): Characterization of drug effects on the EEG by power spectral time band series analysis. *Psychopharmacol Bull* 18:43-47

Ehlers C, Henriksen S, Wang M, Rivier J, Vale W, Bloom F (1983a): Corticotropin-releasing factor increases brain excitability and convulsive seizures in the rat. *Brain Res* 278:332-336

Ehlers C, Reed T, Henriksen S (1986): Effects of corticotropin-releasing factor and growth hormone releasing factor on sleep and activity in rats. *Neuroendocrinology* 42:467-474

Ehlers C, Russo P, Mandell A, Bloom F (1983b): Architecture of rat nocturnal locomotion: a predictive descriptor of the effects of antidepressant and antimanic treatments. *Psychopharmacol Bull* 19:692-695

Freud S (1917): Mourning and melancholia. In: *Collected Papers of Sigmund Freud, Vol. IV* Jones E ed. (1957) London: Hogarth

Garzon J, Del Rio J (1981): Hyperactivity induced in rats by long-term isolation: further studies on a new animal model for the detection of antidepressants. *Eur J Pharmacol* 74:278-294

Gillin J (1983): Sleep studies in affective illness: Diagnostic, therapeutic, and patho-physiological implications. *Psychiatry Annals* 13:367-384

Gold P, Chrousos G, Kellner C, Post R, Roy A, Auqerinos P, Schulte H, Oldfield E, Loriaux L (1984): Psychiatric implications of basic and clinical studies with corticotropin-releasing factor. *Am J Psychiatry* 141:619-627

Grant I (1979): *Behavioral Disorders: Understanding Clinical Psychopathology.* New York: Spectrum Publications

Hofer M (1976): The organization of sleep and wakefulness after maternal separation in young rats. *Dev Psychobiol* 9:189-205

James S, Wehr T, Sack D, Parry B, Rosenthal N (1985): Treatment of seasonal affective disorder with light in the evening. *Br J Psychiatry* 147:424-428

Kupfer D (1978): Application of EEG sleep for the differential diagnosis and treatment of affective disorders. *Pharmakopsychiatrie-Neuro-Psychopharmakologie* 11:17-26

Kupfer D, Reynolds C (1983): Neurophysiological studies of depression: state of the art. In: *The Origins of Depression: Current Concepts and Approaches* (Dahlem Konferezen), Angst J ed. Berlin: Springer-Verlag

Lewy A, Sack R, Singer C (1985): Treating phase typed chronobiological sleep and mood disorders using appropriately timed bright artificial light. *Psychopharmacol Bull* 21:368-372

McKinney W, Moran E, Kramer G (1984): Separation in nonhuman primates as a model for human depression: neurobiological implications. In: *Neurobiology of Mood Disorders,* Post R, Ballanger J, eds. Baltimore: Williams and Wilkens

Nemeroff C, Widelov E, Bissette G, Walleus H, Karlsson I, Eklund K, Kilts C, Loosen P, Vale W (1984): Elevated concentrations of CSF corticotropin-releasing factor like immunoreactivity in depressed patients. *Science* 226:1342-1344

Reite M, Short R (1977): Nocturnal sleep in separated monkey infants. *Arch Gen Psychiatry* 35:1247-1253

Robertson J, Robertson J (1971): Young children in brief separation: a fresh look. *Psychoanal Study Child* 26:264-315

Sahakian B, Robbins T, Morgan M, Iverson S (1974): The effects of psychomotor stimulants on stereotypy and locomotor activity in socially-deprived and control rats. *Brain Res* 84:195-205

Seay B, Hanson E, Harlow H (1962): Mother-infant separation in monkeys. *Child Psychol Psychiatry* 3:123-132

Smith G, Anderson V (1984): Effects of maternal isolation on the development of activity rhythms in infant rats. *Physiol Behav* 33:751-756

Spitz R, Wolf K (1946): Anaclitic depression: an inquiry into the genesis of psychiatric conditions in early childhood, II. *Psychoanal Study Child* 2:313-342

Suomi S, Eisele C, Grady S, Harlow H (1975): Depressive behavior in adult monkeys following separation from family environment. *J Abnorm Psychol* 84:576-578

Suomi S, Harlow H, Domek E (1970): Effect of repetitive infant-infant separation of young monkeys. *J Abnorm Psychol* 76:161-172

Sutton R, Koob G, LeMoal M, Rivier J, Vale W (1982): Corticotropin-releasing factor (CRF) produces behavioural activation in rats *Nature* 297:331-333

Valzelli L, Bernassconi S (1976): Psychoactive drug effect on behavioral changes induced by prolonged socio-environmental deprivation in rats. *Phys Med* 6:271-276

Viswanathan N, Chandrashekaran M (1985): Cycles of presence and absence of mother mouse entrain the circadian clock of pups. *Nature* 317:530-531

Wehr T, Goodwin F (1983): Biological rhythms in manic-depressive illness. In: *Circadian Rhythms in Psychiatry* Wehr T, Goodwin F, eds. Pacific Grove: Boxwood Press.

Wehr T, Wirz-Justice A (1982): Circadian rhythm mechanism in affective illness and in antidepressant drug action. *Pharmacopsychiatria* 15:31-39

Weinstock M, Speiser Z, Ashkenazi R (1976): Biochemical and pharmacological studies on an animal model of hyperactivity states. In: *The Impact of Biology on Modern Psychiatry*, Gershon E, Belmaker R, Kety S, Rosenbaum M, eds. New York: Plenum

Wever R (1985): Man in temporal isolation: Basic principles of the circadian system. In: *Hours of Work: Temporal Factors in Work Scheduling*, Folkard S, Monk T, eds. New York: John Wiley and Sons

This work was supported by grants AA 00098 and AA06059 from NIAAA and MH 24642 from the John D. and Catherine T. MacArthur Foundation.

7

Electrophysiology of the Locus Coeruleus: Implications for Stress-Induced Depression

JAY M. WEISS AND PETER E. SIMSON

Stress-Induced Depression: An Animal Model

Exposing rats to uncontrollable shock produces behavioral and vegetative changes which bear considerable similarity to what is seen in human depression. This procedure represents the most widely studied animal model of psychopathology. As a result of this extensive study, similarity of the model to human depression has been shown with respect to etiology, symptomatology, and responsiveness to treatment. These similarities will now be reviewed briefly.

Etiology

One of the primary attributes of uncontrollable shock is that it is a highly stressful condition (e.g., Tsuda et al., 1983; Weiss, 1968; Weiss, 1971). In essence, the uncontrollable-shock model creates stress-induced depression. Stressful conditions have been widely proposed to increase vulnerability to depression in humans (reviewed by Anisman and Zacharko, 1982; see also Frank and Stewart, 1983; Leff et al., 1970; Lloyd, 1980). Furthermore, numerous studies of the animal model indicate that the shock required to produce depressive symptoms must be *uncontrollable shock* (Corum and Thurmond, 1977; Redmond et al., 1973; Seligman and Maier, 1967; Weiss, 1968; Weiss et al., 1981) since exposure of animals to equal amounts of controllable shock will not produce depressive symptomatology. Depressed individuals often report that they feel helpless, hopeless, or unable to control events, and consequently exposure to uncontrollable stressful events has been hypothesized to be quite significant in bringing about depression (Seligman, 1974; Seligman, 1975). Thus, in precipitating depressive symptomatology by using highly stressful conditions, and uncontrollable conditions in particular, the uncontrollable-shock model appears to share etiological factors with depression observed in humans, at least in certain cases.

Symptomatology

Animals exposed to uncontrollable shock show a number of behavioral and vegetative changes that can be related to what is seen in depression. These changes include:

1. weight loss and decreased intake of food and water (Brady et al., 1962; Pare, 1964, 1965; Weiss, 1968; Ritter et al., 1978);
2. decreased ability to produce active behavior (Overmier and Seligman, 1967; Seligman and Maier, 1967; Overmier, 1968; Weiss and Glazer, 1975; Weiss et al., 1975; Looney and Cohen, 1972; Glazer and Weiss, 1976a, 1976b; Seligman and Beagley, 1975; Seligman et al., 1975; Freda and Klein, 1976; Kelsey, 1977; Anisman et al., 1978; Lawry et al., 1978; Anisman and Sklar, 1979, 1981; Jackson et al., 1980; Sherman and Petty, 1980; Sutton et al., 1981; Altenor et al., 1977; Irwin et al., 1980; Weiss et al., 1980; Weiss et al., 1981);
3. decreased ability to compete with other animals and loss of normal aggressiveness (Peters and Finch, 1961; Maier et al., 1972; Corum and Thurmond, 1977);
4. decreased grooming and play activity (Redmond et al., 1973; Stone, 1978; Weiss et al., 1981);
5. decreased responding for appetitive rewards (Rosellini, 1978);
6. decreased responding for rewarding brain stimulation (Zacharko et al., 1983);
7. deficits in ability to make correct choices in an attentional situation (Jackson et al., 1978; Jackson et al., 1980); and
8. decreased sleep behavior marked especially by early morning waking (Weiss et al., 1985).

The effects of uncontrollable shock on the list above correspond directly with six symptoms of depression in the *Diagnostic and Statistical Manual* of the American Psychiatric Association *(DSM-III)*: poor appetite and significant weight loss, psychomotor alteration, loss of energy or fatigue, loss of interest in usual activities, sleep changes, and indecisiveness. The only symptoms found in the DSM-III not represented in the above list are two that require verbal reports from patients (i.e., feelings of worthlessness, recurrent thoughts of death and suicide), these being obviously inapplicable to an animal model. *DSM-III* requires the presence of four symptoms for a diagnosis of major depression.

Responsiveness to Treatment

Changes of behavior produced by exposure to uncontrollable shock (a) are reversed by electroconvulsive shock, MAO inhibitors, tricyclic antidepressants, and atypical antidepressants (Glazer et al., 1975; Leonard, 1984; Petty and Sherman, 1979; Sherman et al., 1982; Telner and Singhal, 1981) and (b) are not affected by single applications of drugs but instead require several days of drug administrations (Leonard, 1984; Petty and Sherman, 1979; Telner and Singhal, 1981). In these respects, the model parallels human depression, which responds to chronic but not acute administration of antidepressants. The most comprehensive pharmacological assessment was carried out by Sherman et al., (1982), who found that behavioral depression induced by uncontrollable shock was reversed by a variety of antidepressants (imipramine, desipramine, amitriptyline, nortriptylene, doxepin, iprincole, mianserin, iproniazid, and nialamide [all tricyclics except the last two]), and not reversed by anxiolytics, phenothiazines, certain stimulants, or tranquilizers.

Further Comments on the Model

The foregoing briefly describes how stress-induced behavioral depression corresponds to human depression with respect to etiology, symptomatology, and responsiveness to treatment. As stated earlier, this research presently represents the largest body of work relating human depression to an animal model. However, a few additional points need to be made concerning the model.

First, the summary presented above is synthesized from studies that have used different uncontrollable-shock procedures. These procedures are essentially of two types—those that employ (a) moderate-intensity shocks of long duration or (b) high-intensity shocks of brief duration. As the authors and their collaborators have pointed out previously (Glazer and Weiss, 1976a,1976b; Weiss and Simson, 1985b; Weiss et al., 1985), it is not yet clear that the two types of procedures produce all of the same effects. The studies carried out in our laboratory have used brief shocks of high intensity. All of the statements made above regarding etiology, symptomatology, and responsiveness to treatment apply to this procedure except for the fact that the extensive battery of tricyclic antidepressant drugs and atypical antidepressants described above have yet to be tested on this procedure. On the other hand, the list of symptoms produced by this shock procedure is presently longer than the list of symptoms produced by the other shock procedure; the brief, high-intensity shock procedure produces vegetative symptoms (weight loss, decreased food and water intake, sleep reduction marked by early morning awakening) not yet reported for the other type.

Second, many of the depression-like phenomena produced by uncontrollable shock are short-lived, most symptoms dissipating in 48 to 72 hours depending on the shock procedure used. The short-lived nature of the depression-like features is a drawback to the stress-induced depression model, as depression in humans clearly is longer lasting. It should be noted that certain behavioral effects, particularly deficits in learning ability, have been found to last longer than 48 to 72 hours; these effects have occurred in studies using longer shocks of moderate duration.

Neurochemical Basis of Behavioral Depression in the Uncontrollable Shock Model

As in the case of the behavioral study of the uncontrollable-shock model, much research has been carried out attempting to determine the neurochemical changes responsible for the behavioral depression produced by uncontrollable shock (Anisman, 1975; Anisman et al., 1978; Anisman et al. 1981; Anisman et al., 1979b; Anisman et al., 1983; Anisman et al. 1979c; Anisman et al. 1980b; Anisman et al. 1979a; Anisman and Sklar, 1979; Anisman and Sklar, 1981; Anisman et al. 1980a; Drugan et al., 1981; Drugan and Maier, 1983; Drugan et al., 1988; Drugan et al., 1982; Glazer et al. 1975; Goodman et al., 1983; Goodman et al., 1982; Hellhammer et al., 1983; Irwin et al. 1980; Johnson and Henn, 1980; Johnson et al., 1982; Katz, 1981; MacLennan et al., 1982; Maier et al., 1973; Maier et al., 1979; Maier et al., 1980; Maier et al., 1983; Maier et al., 1982; Maier and Jackson, 1977; Petty et al. 1982; Petty & Sherman, 1979; Sherman et al, 1979; Sherman & Petty, 1980; Sherman and Petty, 1982; Sherman et al., 1982; Sutton et al., 1981; Swenson and Vogel, 1983; Telner and Singhal, 1981; Telner and Singhal, 1984; Telner et al., 1980; Weiss et al., 1982; Weiss et al., 1985; Weiss et al., 1976; Weiss et al., 1979; Weiss et al, 1975.; Weiss et al., 1981; Weiss and Simson, 1985a; Weiss et al., 1985). Various investigators have, at one time or another, presented data linking behavioral depression following uncontrollable shock to changes in norepinephrine (Anisman, 1975; Anisman 1980a; Weiss et al., 1982; Weiss et al., 1976), dopamine (Anisman et al., 1980; Zacharko et al., 1983), serotonin (Petty and Sherman, 1980; 1983; Sherman and Petty, 1980), GABA (Petty and Sherman, 1981), and opioids (Maier et al., 1983; Maier et al., 1982) in the brain. From these studies, two hypotheses have emerged which are considerably more detailed and articulated than others. The first emphasizes changes in norepinephrine (NE) (Weiss et al., 1982; Weiss et al., 1981; Weiss et al., 1985), while the second, proposed by Sherman and Petty, originally focused on changes in serotonin (5-HT) and subsequently came to emphasize changes in GABA as well. In the most recent past, the first of these hypotheses has been the predominant focus of research; therefore, the remainder of this chapter will explain this hypothesis and describe some recent developments bearing on it. It should be mentioned, however, that this hypothesis and that offered by Sherman and Petty may not be mutually exclusive; rather, the neurochemical changes described in each may be part of the same neural sequence. To be specific, the first hypothesis proposes that stress-induced behavioral depression arises because of changes occurring in the locus coeruleus (LC) region of the brain which lead to increased release of NE in the forebrain regions to which the LC projects, while the second hypothesis proposes that such depression results from decreased release of 5-HT in the forebrain. Given the influence of NE on release of 5-HT (Frankhuyzen & Mulder, 1980; 1982; Jouvet, 1973), the noradrenergic changes emphasized by the first hypothesis may be responsible for the decreased release of 5-HT in the forebrain emphasized by the second hypothesis. Should this schema prove to be correct, the two hypotheses would simply have emphasized different steps in a common neural chain.

Noradrenergic Changes in the Locus Coeruleus

In 1981, after more than a decade of research indicating that a disturbance of NE in the brain played a major role in mediating stress-induced behavioral depression, the authors reported that stress-induced behavioral depression appeared to be highly correlated with a large fall in NE in the locus coeruleus region of the brain (Weiss et al., 1981). In that experiment, they observed that animals that had been exposed to uncontrollable shock evidenced behavioral depression when tested 1.5 or 48 hours after the shock and also showed a large depletion of NE in the locus coeruleus at those times. Changes in other amines and in NE in other brain regions were not nearly as pronounced as the change in NE in the LC. Moreover, animals were also tested 72 to 96 hours after exposure to the shock, and such animals showed neither behavioral depression nor depletion of NE in the LC region. (The loss of NE depletion in the LC at this time can be explained by induction of the enzyme tyrosine hydroxylase resulting from exposure to the shock [Weiss et al., 1981]). Thus, behavioral depression was accompanied by a large depletion of NE in the LC region, being present when NE was greatly depleted in the LC and being absent when large depletion of that amine in the LC was no longer present. Since that time, two additional laboratories have reported that behavioral depression produced by uncontrollable shock is accompanied by large depletion of NE in the LC region (Lenhart et al., 1984; Hughes et al., 1984).

To explore this finding, we attempted to establish the mechanism by which large-magnitude NE depletion in the LC might mediate stress-induced behavioral depression. As has been presented in detail in several previous publications (Weiss et al., 1982; Weiss et al., 1985; Weiss and Simson, 1986), it seemed most likely that large-magnitude depletion of NE would result in decreased NE available for release from the normal sites of release. But what was the consequence of this within the LC? Aghajanian and his colleagues established that alpha-2 adrenergic receptors were present within the LC and that activation of these receptors strongly inhibited the firing of LC cells (Svensson et al., 1975). Since these receptors respond to NE, it therefore seemed likely that the functional consequence of NE depletion leading to reduced release of NE in the LC region was a reduction in the normal stimulation of alpha-2 inhibitory receptors in the LC region.

Several experiments were carried out to test this formulation. First, it was found that normal (i.e., unshocked) rats could be made behaviorally depressed by pharmacological blockade of alpha-2 receptors in the LC region (Weiss et al., 1986). This effect was highly specific to the LC region, for application of pharmacological blocking agents to brain regions outside the LC did not produce this effect. Thus, by producing what was assumed to be the functional consequence of exposure to uncontrollable shock, depression of active behavior could be produced.

Next, it was found that the depressive effects of uncontrollable shock could be reversed in several ways consistent with the formulation described above. First, pharmacological stimulation of alpha-2 receptors in the LC was found to eliminate behavioral depression in animals that had been exposed to uncontrollable shock (Simson et al., 1986a). This result was predicted on the basis that pharmacological stimulation of alpha-2 receptors in the LC region should reverse the functional deficit resulting from reduced NE acting at these receptors. Second, reversal of the NE depletion that follows exposure to uncontrollable shock (reversal of depletion accomplished by pharmacological blockade of catabolism of NE in the LC region) was also shown to eliminate behavioral depression (Simson et al., 1986b). Finally, infusion of drugs into the LC region to block the reuptake of NE eliminated behavioral depression in animals that had been exposed to uncontrollable shock. This result was predicted on the basis that reuptake blockade in the LC region would increase the stimulation of alpha-2 receptors by NE even though the amount of NE released into synapses was reduced.

Influence of Alpha-2 Receptors on Activity in the LC

The function of alpha-2 receptors in the LC is critical to the hypothesis described above, since behavioral depression is said to result from decreased stimulation, or "functional blockade," of these receptors. As explained previously, Aghajanian and his colleagues made the initial discovery that alpha-2 receptors in the LC exhibit a strong inhibitory influence on LC firing. Moreover, these investigators proposed a rather elegant mechanism by which this occurs. Their schema was built on the observations that (a) clonidine suppressed LC firing, and (b) piperoxane, and alpha-2 antagonist, reduced the quiescence that followed bursts of LC activity after antidromic activation of LC neurons (Aghajanian et al., 1977). Taking these results together with anatomical data showing collateral branching of his LC axons (Shimizu and Imamoto, 1979; Swanson, 1976), Aghajanian and colleagues proposed that the LC is inhibited by transmitter released from LC collaterals onto alpha-2 receptors, the alpha-2 receptors being responsible for the post stimulation inhibition (PSI) (Aghajanian et al., 1977). Thus, the LC was thought to fire in a burst and then inhibit its own subsequent firing by releasing its transmitter (i.e.., NE) onto alpha-2 receptors in the LC. In support of this hypothesis, it was shown that blockade of alpha-2 receptors in the LC reduced the quiescent period that followed activation of the LC by orthodromic (sciatic nerve) stimulation (Cedarbaum and Aghajanian, 1978).

Shortly after this was proposed, however, Andrade and Aghajanian published a series of papers in which they cast doubt on the idea that alpha-2 receptors were responsible for the period of inhibition that followed a burst of firing in the LC. Recently, in fact, Andrade and Aghajanian have abandoned this idea and proposed a different mechanism to account for PSI of LC neurons. Based on intracellular *in vivo* and *in vitro* LC recordings, they have proposed that a calcium-dependent change in potassium conductance arising directly from depolarization is responsible for PSI (Aghajanian et al., 1983; Andrade and Aghajanian, 1984a; 1984b). In that membrane changes in ion conductance arising directly from depolarization are now thought to be responsible for the quiescent period following bursts of LC activity; the role of alpha-2 receptors in regulating LC activity remained to be determined.

The authors have recently been studying the way in which alpha-2 receptors in the LC region influence the activity of these cells. Two observations previously made by other investigators suggested that alpha-2 receptors might influence a different aspect of LC activity than had been studied heretofore. The first observation was that the magnitude of the neuronal response of LC neurons to both simple stimuli (tone, light flash, touch) and complex stimuli (food, novel objects) varies positively with the viligance of the animal (Foote et al., 1980; Aston-Jones and Bloom, 1981): The more vigilant the state of the animal, the greater the magnitude of the neuronal response. This suggested to the authors that the initial response of LC neurons to stimulation was a particularly important aspect of LC activity. The second observation was that the doses of the highly selective alpha-2 adrenergic antagonist idazoxan required to elevate LC firing were much higher than the doses of the same drug required to reverse (or block) the effect of the alpha-2 agonist, clonidine (Freedman and Aghajanian, 1984). Thus, a low dose of an adrenergic antagonist that appeared to be affecting alpha-2 receptors so as to block the action of an adrenergic agonist had no effect on the baseline rate of the LC, suggesting that the antagonist could be affecting the receptors in some way that was not apparent from examining the baseline rate of firing. Putting these two observations together, the authors considered, first, whether or not alpha-2 receptors might influence the initial response of the LC to sensory input rather than just the baseline firing rate of the LC and, second, whether or not alpha-2 blocking agents might affect this initial LC response at doses below that required to elevate the basal firing of unstimulated LC. They therefore tested whether alpha-2 receptors modulate the responsiveness of LC neurons to excitatory stimuli by blocking alpha-2 receptors pharmacologically with doses both above and below those required to increase spontaneous LC activity.

Effect of Alpha-2 Receptor Blockade on the Response of LC Neurons to Excitatory Stimulation

Standard electrophysiological techniques were employed to record extracellularly from LC neurons in male Sprague-Dawley rats. Locus coeruleus neurons fired spontaneously at a regular low rate (0.5 to 3.5 per sec) and were observed to respond to pressure applied to the contralateral hind paw with a burst of firing followed by a prolonged quiescent period. Pharmacological manipulations were achieved with intravenous drug administration through a lateral tail vein or through direct infusion of drug into the LC. In the latter method, the drug was infused into the LC via a cannula made of 31 gauge hypodermic tubing that was cemented to the recording electrode.

As stated above, compression of the contralateral hind paw produces a transient burst of LC neuronal activity that is followed by a quiescent period. Typically used in electrophysiological studies as a preliminary means of demonstrating that an electrode is recording from within the LC, this manipulation served as a stimulus to activate LC firing in three of the four experiments to be described. Paw compression was produced by a pair of surgical forceps, which made it possible to deliver repeated compression of similar magnitude.

The experiments utilizing contralateral hind paw compression (PC) shared a common paradigm. In these experiments, an LC cell was isolated and a baseline rate determined by recording spontaneous activity for a minimum of 5 min. While the baseline firing rate was being determined, spikes were integrated over 10 sec periods. The responsiveness of the cell to PC was then measured ten times, with each measure taken by applying PC for one second and allowing ten seconds between compressions. During these measurements of response to PC, the number of spikes occurring in each 1-sec period was recorded. Two minutes after completion of these trials, a drug manipulation was performed (by intravenous or direct LC administration). In the first study, groups of rats received intravenously administered idazoxan at varying doses; in the second study, groups received varying doses of idazoxan by infusion; and, in the third study, a group received intravenously administered yohimbine. Following administration of drug, a post-drug baseline rate was determined for 3 minutes, with spikes again being integrated over 10-sec periods. Finally, the responsiveness of the LC cell to PC in the presence of drug was measured ten times by giving 1-sec PCs with 10-sec intervals.

The fourth study employed intravenously administered idazoxan, but nicotine sulfate (25 mg I.V.) replaced PC as the means for transiently increasing LC activity. After determining the base rate of spontaneous activity, the responsiveness of the cell to nicotine was determined by intravenously administering the drug, then integrating spikes over 10-sec periods for 7 min. This amount of time allowed the cell to respond to 25 mg/kg nicotine and then return to its previous level of baseline activity. The procedure was then repeated: A second injection of nicotine was administered and the response recorded following the same procedure as with the first injection. (This second injection of nicotine tested for the possibility that the response to nicotine might increase simply from repeated administration of the drug.) After completion of these trials, the drug manipulation was performed: Idazoxan was intravenously administered and a post-drug baseline rate was again determined by integrating spikes over 10-sec periods for 3 min. After this, the responsiveness of the neuron to nicotine in the presence of idazoxan was de-

FIGURE 7.1. *(Top)* Typical positive-negative wave form of a spontaneously firing locus coeruleus (LC) neuron. Note the notch on the ascending limb. *(Bottom)* Typical response pattern of a spontaneously firing LC neuron to the brief application of a noxious stimulus in the form of contralateral hind paw compression (PC). All recorded cells fired at a slow, regular rate (0.5 to 3.5 Hz). Note the increase in firing rate with stimulus onset, followed by a prolonged quiescent period after stimulus offset.

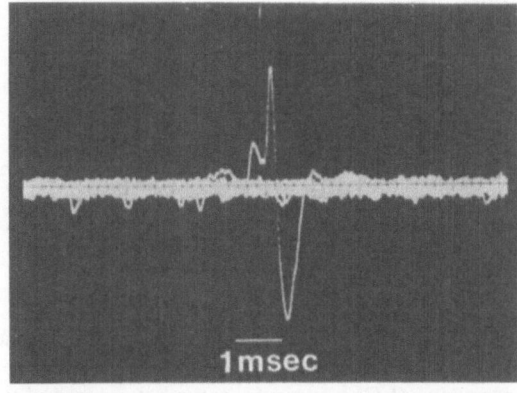

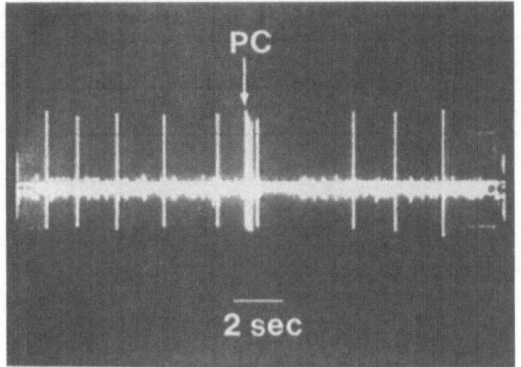

termined by repeating the nicotine procedure as described above, carrying out at least two nicotine injections.

Recordings were made from 51 LC neurons in an equal number of rats. All LC neurons displayed a similar wave form (Figure 7.1, top) and responded to PC with an increase in firing rate which was followed by a period of quiescence (Figure 7.1, bottom).

Intravenously Administered Idazoxan Increases the Responsiveness of LC Neurons to Noxious Stimulation

Twenty-eight LC neurons (one neuron in each of 28 rats) were recorded from in this phase of the study. Six groups of 4 rats received a single dose of idazoxan through a lateral tail vein. The doses, in ug of drug per kg body weight, were as follows: 10, 18, 25, 60, 140, and 640. A control group received 0.9% saline through a lateral tail vein. Idazoxan solutions were such that all animals were administered volumes ranging from 0.05 to 0.15 ml.

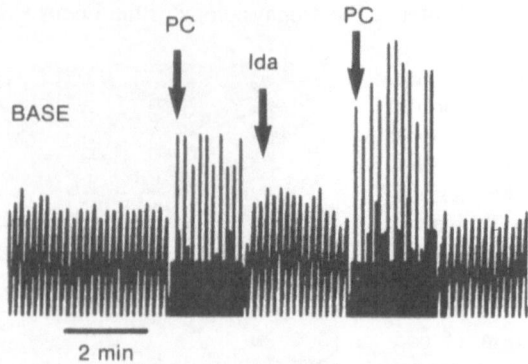

FIGURE 7.2 Strip chart record showing the augmentation by the alpha-2 adrenergic antagonist idazoxan (25 μg/kg i.v.) of the response to paw compression (PC) of a locus coeruleus (LC) neuron in a typical subject. After a baseline rate of activity was established, the magnitude of the response of the LC neuron to PC was recorded in 10 trials. Idazoxan was then administered, and the postdrug neuronal response to PC was tested. This procedure was repeated in four subjects at each of the doses of idazoxan employed (see Figure 7.3).

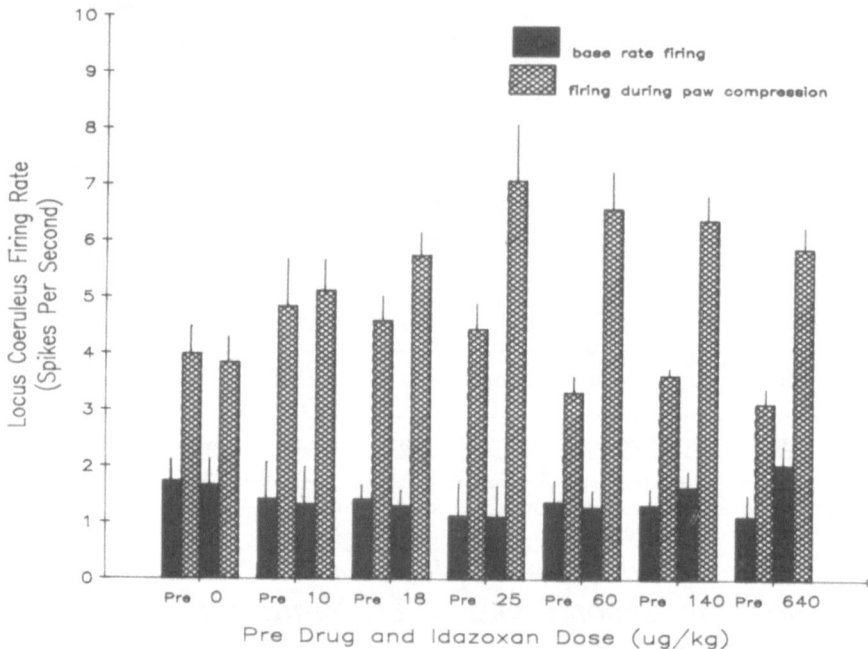

FIGURE 7.3. Increase in the responsiveness of locus coeruleus (LC) neurons to contralateral hind paw compression (PC) by idazoxan. Solid bars indicate spontaneous firing rates of locus coeruleus (LC) neurons, whereas hatched bars indicate firing rates during the application of PC. After a predrug baseline firing rate and predrug response to PC rate were determined (Pre), idazoxan was administered intravenously (dose indicated on graph), and baseline firing rate and response to PC were measured again. Each dosage of idazoxan was administered to four animals, and no animal received more than one dose. Note that although baseline activity was not increased until 640 μg/kg idazoxan was administered, there was a significantly larger response to PC following 18 μg/kg idazoxan.

As shown in Figure 7.2 and Figure 7.3, intravenous administration of idazoxan markedly increased the responsiveness of LC neurons to PC at doses far below those required to alter baseline activity. Although baseline firing rates were not significantly increased until doses of 640 uk/kg were reached, the response to PC was significantly increased following idazoxan administration beginning with doses as low as 18 ug/kg. The increased response to PC was seen in every animal given a dose above 18 ug/kg despite the fact that there was wide variation in the responsiveness of LC neurons to PC in the pre-drug condition. Interestingly, the magnitude of increase in LC firing produced by PC did not correlate with the baseline activity rate either pre-drug (r=0.08) or after drug was given (r=0.16).

LC firing rates in the first sec following the termination of PC are shown in Figure 7.4. Although a post-stimulation quiescent interval can be observed in both the pre-drug and idazoxan conditions, poststimulation activity was increased following idazoxan treatment beginning with doses of 25 ug/kg. By the second sec following PC termination, LC activity in both pre-drug and idazoxan conditions returned essentially to the baseline firing rates (Figure 7.5).

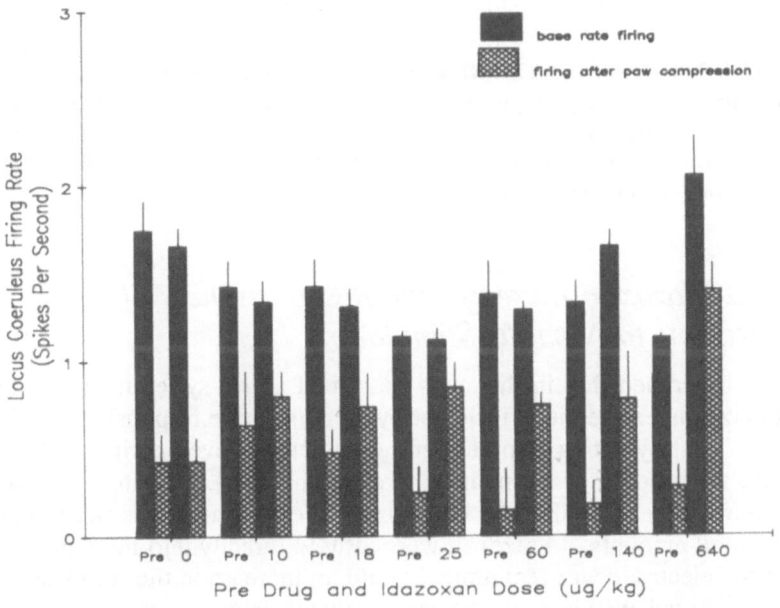

FIGURE 7.4. Locus coeruleus (LC) firing rates during the first second following the termination of paw compression (PC). Solid bars indicate spontaneous firing rates, whereas hatched bars indicate the firing of the LC neuron during the one-second period immediately following the offset of PC. Although a poststimulation quiescent interval can be observed in both the predrug and idazoxan conditions, there is significantly greater poststimulation activity following idazoxan treatment beginning with doses of 25 μg/kg.

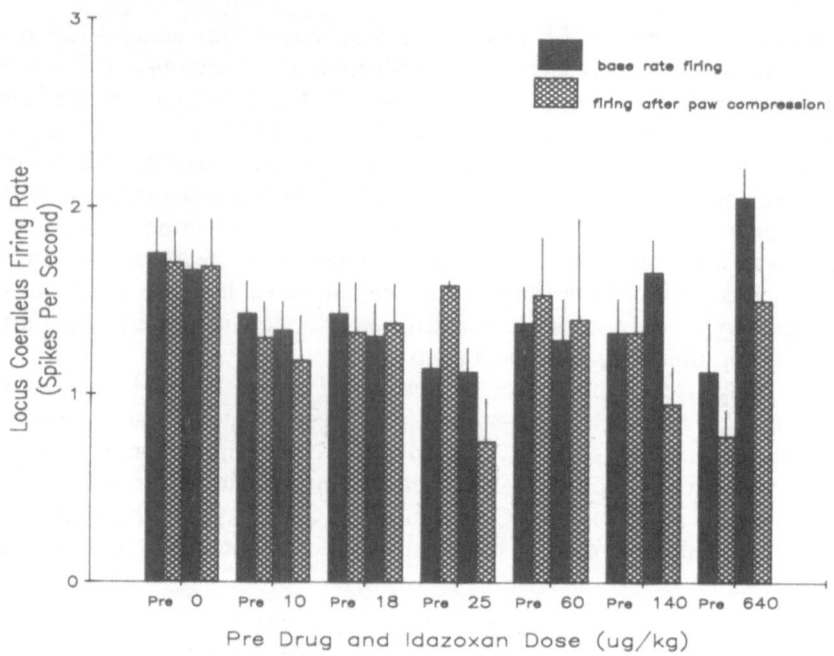

FIGURE 7.5. Locus coeruleus (LC) firing rates during the second second following the termination of paw compression (PC). Solid bars indicate spontaneous firing rates, whereas hatched bars indicate the firing of the LC neuron during the second second following PC offset. Both the predrug and postdrug conditions essentially recovered from poststimulation quiescence.

Infused Idazoxan Increases the Responsiveness of LC Neurons to Noxious Stimulation

It could be argued that in the study described above systemically administered idazoxan exerted its effects not by increasing the responsiveness of the LC to PC but by acting at other central or perhaps even peripheral sites to increase the amount of stimulation received by the LC during PC. To determine whether this might be occurring, an experiment was performed in which small amounts of idazoxan were infused directly into the region from which the electrode was recording. Should an increase in the responsiveness to PC be found in this case, the results would lend support to the notion that idazoxan is acting directly at the LC to increase LC responsiveness to noxious stimulation.

Twelve LC neurons (1 neuron in each of 12 rats) were recorded from in this study. Two groups of 4 rats received idazoxan through a cannula mounted adjacent to the recording electrode. The first group received 1.0 ng of idazoxan; while the second group received 22.0 ng. A third group received 0.9% saline via a cannula. All volumes were 1.0 ul delivered over

15 sec. In addition, 2 animals from the first group received increasing doses of idazoxan.

As shown in Figure 7.6, infusion of idazoxan increased the response to PC in both groups given drug, without having significant effects on baseline rates. This effect can also be seen in the two animals receiving multiple idazoxan injections (Figure 7.7), in which the rate of increase in responsiveness to PC can be seen to increase progressively with little increase in baseline firing rate.

Yohimbine Increases the Responsiveness of LC Neurons to Noxious Stimulation

To ensure that the effects presented in the two studies above were not unique to the particular alpha-2 antagonist employed (i.e., idazoxan), another alpha-2 antagonist, yohimbine, was intravenously administered to five animals at

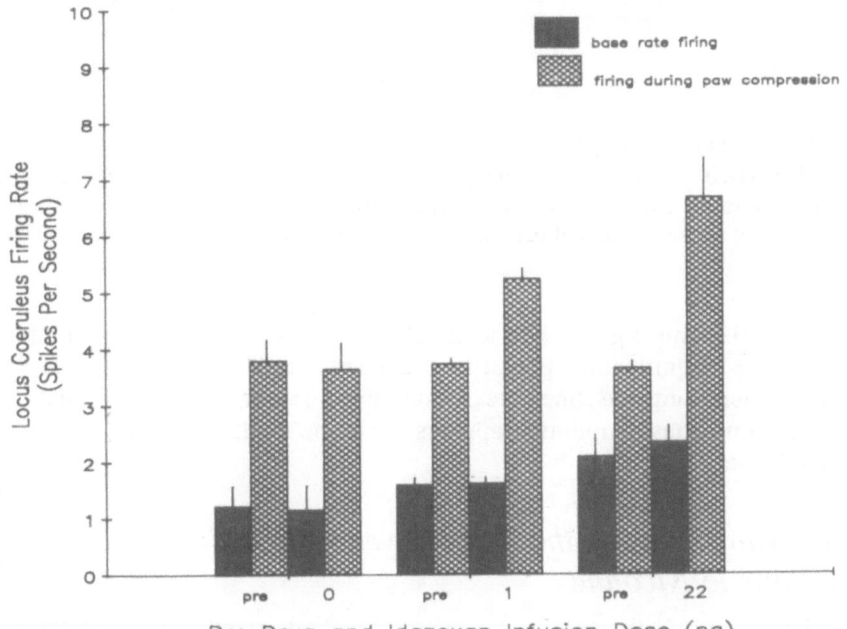

FIGURE 7.6. Augmentation of the responses of locus coeruleus (LC) neurons to paw compression (PC) resulting from direct infusion of idazoxan into the LC. Solid bars indicate spontaneous firing rates of LC neurons, whereas hatched bars indicate firing rates during the application of PC. Idazoxan was infused directly into the LC after a predrug baseline firing rate and predrug response to PC rate were determined. Each dosage of idazoxan was administered to four animals, and no animal received more than one dose. Note that there was an increase in LC responsiveness to PC following infusion of just 1.0 ng idazoxan, which represents approximately 1/10,000 of the smallest intravenous dose required for a similar effect (see Figure 7-3).

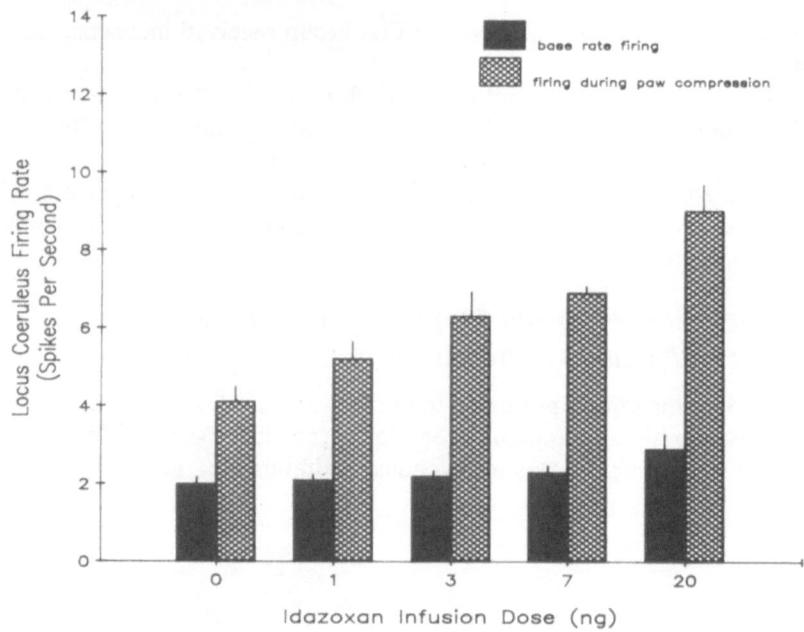

FIGURE 7.7. Dose-response curve showing the augmentation of the responses of locus coeruleus (LC) neurons to paw compression (PC) resulting from direct infusion of idazoxan into the LC in two animals. Solid bars indicate spontaneous firing rates of LC neurons, whereas hatched bars indicate firing rates during the application of PC.

a dose of 0.25 mg/kg. As with idazoxan, the alpha-2 antagonist yohimbine (Figure 7.8) significantly potentiated the response of LC neurons to PC without increasing baseline rates. Thus, the increase in responsiveness of the LC to noxious stimulation appears not to be confined to one particular alpha-2 antagonist.

Idazoxan Increases the Responsiveness of LC Neurons to Nicotine

To determine whether alpha-2 receptor blockade would augment the response of LC neurons to excitatory input other than that produced by noxious peripheral stimulation, nicotine was given instead of PC. Nicotine was substituted for PC because (1) it has been shown to transiently increase LC single unit firing rates (Engberg and Svensson, 1980) and (2) it offered a different modality of stimulation than noxious cutaneous stimulation.

Six LC neurons were recorded in six rats (i.e., 1 cell per rat). As shown in Figures 7.9 and 7.10, injections of nicotine (25 mg/kg I.V.) transiently increased the firing rate of LC neurons. This effect on LC neuronal activity typically reached a maximum within 1 minute of injection, with LC activity

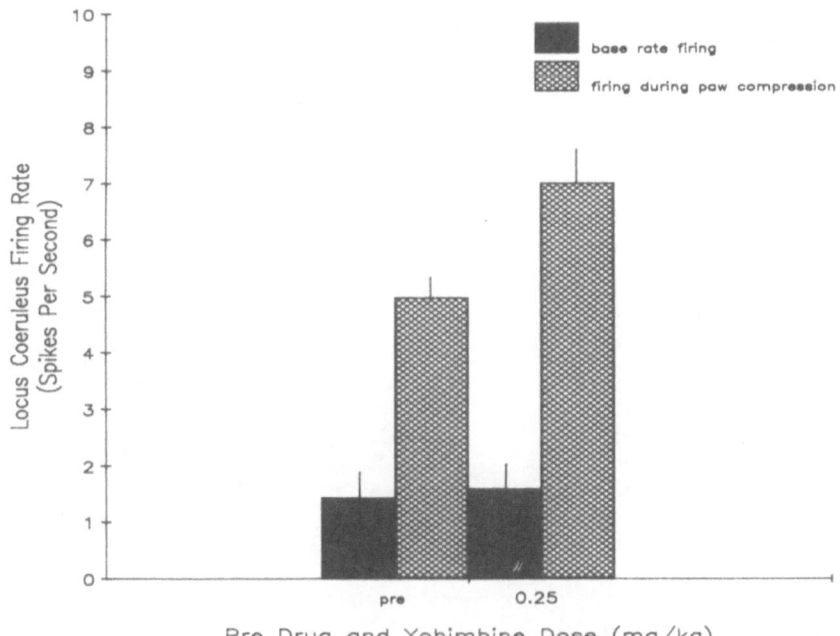

FIGURE 7.8. Increase in the responsiveness of locus coeruleus (LC) neurons to contralateral hind paw compression (PC) by the alpha-adrenergic antagonist yohimbine. Solid bars indicate spontaneous firing rates of LC neurons, whereas hatched bars indicate firing rates during the application of PC. Yohimbine was administered intravenously after a predrug baseline firing rate and predrug response to PC rate were determined. Yohimbine (0.25 mg/kg, i.v.) was administered to five animals. Although baseline activity was not increased by yohimbine at this dose, there was a significantly larger postdrug response to PC.

FIGURE 7.9. Strip chart record showing the augmentation by idazoxan (140 µg/kg i.v.) of the response to nicotine injection (25 mg/kg i.v.) of a locus coeruleus (LC) neuron in a typical subject. After a baseline rate of activity was established, the magnitude of the response of the LC neuron to nicotine injection was recorded for 7 minutes. Following idazoxan administration, the post-drug neuronal response to nicotine injection was tested.

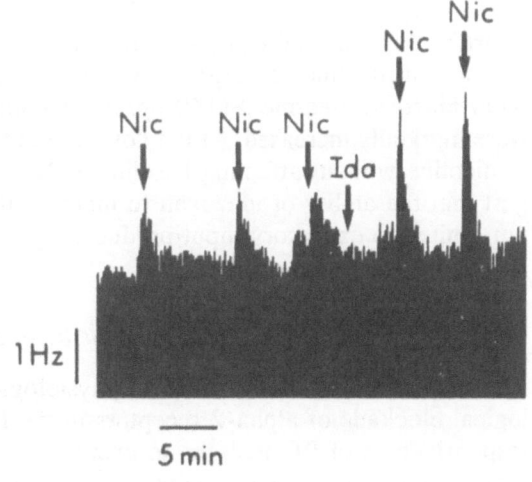

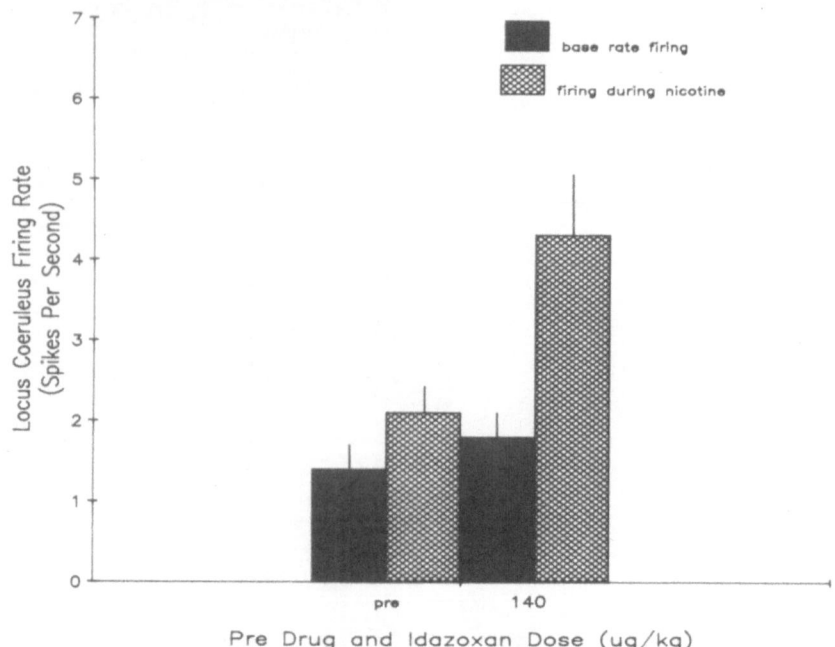

FIGURE 7.10. Augmentation of the responses of locus coeruleus (LC) neurons to nicotine injection (25 mg/kg i.v.) following intravenously administered idazoxan (140 μg/kg i.v.). Solid bars indicate spontaneous firing rates of LC neurons, whereas hatched bars indicate firing rates during the first minute following the nicotine injection. Idazoxan was administered intravenously after a predrug baseline firing rate and predrug response to nicotine injection were determined. Idazoxan (140 μg/kg i.v.) was administered to six animals. Note that although baseline activity was not increased by idazoxan, there was a significantly larger response to nicotine injection following idazoxan treatment.

returning to baseline (i.e., pre-nicotine) levels within 7 min. A second administration of nicotine (25 mg/kg I.V.) showed similar effects on LC activity. After idazoxan treatment (140 ug/kg I.V), nicotine's effects on LC activity were markedly increased. At the dose used, idazoxan increased the response to nicotine without affecting baseline activity of the LC. These results suggest that the ability of idazoxan to increase the reactivity of LC neurons is not limited to excitatory input produced by noxious cutaneous stimulation.

Comments on the Electrophysiological Results

The major finding of these electrophysiological studies is that pharmacological blockade of alpha-2 receptors in the LC can markedly increase the responsiveness of LC neurons to excitatory stimulation at doses far below those required to increase spontaneous activity. This finding suggests that alpha-2 receptors in the LC play an important role in modulating the excitability of LC neurons. The fact that idazoxan augmented the response of LC neurons to paw compression when administered directly into the LC as well as intravenously argues against the notion that idazoxan achieves its effects by increasing the amount of excitatory stimulation received by LC neurons. Indeed, increases in the responsiveness of LC neurons to paw compression by infusion of idazoxan occurred at doses as low as 1/10,000 of the smallest systemic dose needed to show a similar effect. That yohimbine as well as idazoxan was capable of increasing LC responsiveness without altering spontaneous activity indicates that the effects of alpha-2 receptor blockade on LC excitability are not unique to one particular alpha-2 adrenergic antagonist. Also, the fact that idazoxan potentiated the response of LC neurons to nicotine injection as well as noxious peripheral stimulation demonstrates that the effects of blockade of alpha-2 receptors on LC responsiveness are not limited to a specific modality of stimulation.

While recent evidence has indicated that PSI is mediated by a calcium-dependent potassium conductance increase independent of alpha-2 receptor involvement (Andrade and Aghajanian, 1984a; 1984b), it is interesting to note that the increase in LC activity observed during the period of poststimulation quiescence after idazoxan treatment parallels earlier findings by Aghajanian and colleagues in which the alpha-2 antagonist piperoxane reduced the length of the post-stimulation quiescent interval (Aghajanian et al., 1977; Cedarbaum and Aghajanian, 1978). The present studies demonstrate, however, that this decrease in the amount of postactivation inhibition by idazoxan is preceded by a large increase in LC responsiveness. Thus, an increase in LC responsiveness would explain the reduced quiescence (i.e., increased poststimulation activity) that Aghajanian and his colleagues observed following alpha-2 blockade. Such an explanation fits well with recent evidence suggesting that inhibitory recurrent collaterals are not involved in PSI (Andrade and Aghajanian, 1984a; 1984b).

The finding that low doses of alpha-2 adrenergic antagonists affect the initial excitability of LC neurons to stimulation without altering spontaneous activity fits well with data indicating that the LC appears to be particularly responsive to variations in the environment rather than steady states or repetitive events. That is, LC neurons exhibit biphasic responses to a variety of sensory stimuli (Foote et al., 1980; Aston-Jones and Bloom, 1981) and respond to continuous noxious input with a burst of activity that gradually returns to near baseline levels (Cedarbaum and Aghajanian, 1978). If, as these data suggest, part of the function of the LC is to signal environmental change, then examining the effects of pharmacological substances on LC activity might be most appropriate while those neurons are concurrently activated by sensory stimuli. As shown in these experiments, the application of paw compression revealed effects of one type of pharmacological manipulation (alpha-2 blockade) on LC activity at doses of the receptor blockers that would not have appeared to have had an effect had one limited oneself to an analysis of spontaneous activity.

Implications of the Electrophysiological Results for Stress-Induced Depression

The behavioral and neurochemical studies discussed earlier demonstrate that exposing animals to uncontrollable shock produces symptoms of depression and that these changes are associated with large-magnitude depletion of NE levels in the LC region of the brain. The functional consequence of this NE depletion is thought to be reduced release of NE in the LC region which leads to a reduction in the normal stimulation of alpha-2 inhibitory receptors in the LC region. Thus, stress-induced behavioral depression is said to be mediated by a functional blockade of alpha-2 receptors in the LC. The results described above show that pharmacological blockade of alpha-2 receptors in the LC causes a marked increase in the responsiveness of LC neurons to excitatory stimulation. Consequently, one would expect that exposure to an uncontrollable stressor, by producing a functional blockade of alpha-2 receptors, would result in increased responsiveness of LC neurons to excitatory stimulation and that this consequence plays a significant part in producing stress-induced behavioral depression. Experiments to determine whether uncontrollable shock indeed produces heightened responsiveness of LC neurons are presently underway.

Acknowledgments. This research was supported by MH-40406. Peter E. Simson was supported by NIMH Neurobehavioral Science Research Training Program Fellowship MH-15177. The authors would like to thank Reckitt and Colmann, Kingston-upon-Hull, England for their generous gift of idazoxan hydrochloride.

References

Aghajanian GK, Cedarbaum JM, Wang RY (1977): Evidence for norepinephrine-mediated inhibition of locus coeruleus neurons. *Brain Res* 136:570-577

Aghajanian GK, Vandermaelen CP, Andrade R (1983): Intracellular studies on the role of calcium in regulated activity and reactivity of locus coeruleus neurons *in vivo. Brain Res* 273:237-243

Altenor A, Kay E, Richter M (1977): The generality of learned helplessness in the rat. *Learn Motiv* 8:54-61

Andrade R, Aghajanian GK (1984a): Locus coeruleus activity in vitro: Intrinsic regulation by a calcium-dependent potassium conductance but not alpha-2 adrenoreceptors. *J Neurosci* 4:161-170

Andrade R, Aghajanian GK (1984b): Intrinsic regulation of locus coeruleus neurons: electrophysiological evidence indicating a predominant role for autoinhibition. *Brain Res* 310:401-406

Anisman H (1975): Time-dependent variations in aversively motivated behaviors: non-associative effects of cholinergic and catecholaminergic activity. *Psychol Rev* 82:359-385

Anisman H, Sklar LS (1979): Catecholamine depletion upon re-exposure to stress: mediation of the escape deficits produced by inescapable shock. *J Comp Physiol Psychol* 93:610-625

Anisman H, Sklar LS (1981): Social housing conditions influence escape deficits produced by uncontrollable stress: assessment of the contribution of norepinephrine. *Behav Neural Biol* 32:406-427

Anisman H, Zacharko RM (1982): Depression: the predisposing influence of stress. *Behav Brain Sci* 5:89-137

Anisman H, deCatanzaro D, Remington G (1978): Escape performance following exposure to inescapable shock: deficits in motor response maintenance. *J Exp Psychol [Anim Behav]* 4:197-218

Anisman H, Remington G, Sklar LS (1979a): Effects of inescapable shock on subsequent escape performance: Catecholaminergic and cholinergic mediation of response initiation and maintenance. *Psychopharmacology* 61:107-124

Anisman H, Irwin J, Sklar LS (1979c): Deficits of escape performance following catecholamine depletion: implications for behavioral deficits induced by uncontrollable stress. *Psychopharmacology* 64:163-170

Anisman H, Suissa A, Sklar LS (1980a): Escape deficits induced by uncontrollable stress: antagonism by dopamine and noradrenaline agonists. *Behav Neural Biol* 28:34-47

Anisman H, Pizzino A, Sklar LS (1980b): Coping for stress, norepinephrine depletion and escape performance. *Brain Res* 191:583-588

Anisman H, Glazier SJ, Sklar LS (1981): Cholinergic influences on escape deficits produced by uncontrollable stress. *Psychopharmacology* 74:81-87

Anisman H, Irwin J, Beauchamp C, Zacharko R (1983): Cross-stressor immunization against the behavioral deficits produced by uncontrollable shock. *Behav Neurosci* 97:452-461

Aston-Jones G, Bloom FE (1981): Activity of norepinephrine-containing locus coeruleus neurons in behaving rats anticipates fluctuations in the sleep-waking cycle. *J Neurosci* 1:876-886

Brady JP, Thornton DR, Fisher D (1962): Deleterious effects of anxiety elicited by conditioned pre-aversive stimuli in the rat. *Psychosom Med* 24:590-595

Cedarbaum JM, Aghajanian GK (1978): Activation of locus coeruleus neurons by peripheral stimuli: Modulation by collateral inhibitory mechanism. *Life Sci* 23:1383-1392

Corum CR, Thurmond JB (1977): Effects on acute exposure to stress on subsequent aggression and locomotion performance. *Psychosom Med* 39:436-443

Drugan RC, Maier SF (1983): Analgesic and opioid involvement in the shock-elicited activity and escape deficits produced by inescapable shock. *Learn Motiv* 14:30-48

Drugan RC, Grau JW, Maier SF, Madden J, Barchas JD (1981): Cross tolerance between morphine and the long-term analgesic reaction to inescapable shock. *Pharmacol Biochem Behav* 14:677-682

Drugan RC, Moye TB, Maier SF (1982): Opioid and nonopioid forms of stress-induced analgesia: some environmental determinants and characteristics. *Behav Neural Biol* 35:251-264

Drugan RD, Ryan SM, Minor TR, Maier SF (1988): Librium prevents the analgesia and shuttlebox escape deficit typically observed following inescapable shock. *Pharmacol Biochem Behav* 21:749-754

Engberg G, Svensson TH (1980): Pharmacological analysis of a cholinergic receptor mediated regulation of brain norepinephrine neurons. *J Neural Transm* 49:137-150

130   Jay M. Weiss and Peter E. Simson

Foote SL, Aston-Jones G, Bloom FE (1980): Impulse activity of locus coeruleus neurons in awake rats and monkeys is a function of sensory stimulation and arousal. *Proc Natl Acad Sci USA* 77:3033-3037

Frankhuyzen AL, Mulder AH (1980): Noradrenaline inhibits depolarization-induced 3-H-serotonin release from slices of rat hippocampus. *Eur J Pharmacol* 63:179-182

Frankhuyzen AL, Mulder AH (1982): Pharmacological characterization of presynaptic alpha-adrenoceptors modulating [³H]norepinephrine and [³H]hydroxytyptamine release from slices of hippocampus of the rat. *Eur J Pharmacol* 81:97-106

Frank E, Stewart BD (1983): Treatment of depressed rape victims: An approach to stress-induced symptomatology. In: *Treatment of Depression: Old Controversies and New Approaches*, Clayton PJ, Barrett JE, eds. New York, Raven Press

Freda JS, Klein SB (1976): Generality of the failure-to-escape (helplessness) phenomenon in rats. *Anim Learn Behav* 4:401-406

Freedman JE, Aghajanian GK (1984): Idazoxan (RX 781094) selectively antagonizes alpha-2 adrenoreceptors on rat central neurons. *Eur J Pharmacol* 105:265-272

Glazer HI, Weiss JM (1976a): Long-term interference effect: an alternative to "learned helplessness." *J Exp Psychol [Anim Behav]* 2:202-213

Glazer HI, Weiss JM (1976b): Long-term and transitory interference effects. *J Exp Psychol [Anim Behav]* 2:191-201

Glazer HI, Weiss JM, Pohorecky LA, Miller NE (1975): Monoamines as mediators of avoidance escape behavior. *Psychosom Med* 37:535-543

Goodman PA, Weiss JM, Hoffman LJ, Ambrose MJ, Bailey WH, Charry JM (1982): Reversal of behavioral depression by infusion of an alpha-2 adrenergic agonist into the locus coeruleus. *Soc Neurosci Abstr* 8:362

Goodman PA, Weiss JM, Ambrose MJ, Cardle KA, Bailey WH, Charry JM (1983): Infusion of a monoamine oxidase inhibitor into the locus coeruleus can protect against stress-induced depression. *Soc Neurosci Abst* 9:553

Hellhammer DH, Bell M, Ludwig M, Rea MA (1983): Learned helplessness: Effects on brain monoamines. *Soc Neurosci Abstr* 9:555

Hughes CW, Kent TA, Campbell J, Oke A, Croskell H, Preskorn SH (1984): Cerebral blood flow and cerebrovascular permeability in an inescapable shock (learned helplessness) animal model of depression. *Pharmacol Biochem Behav* 21:891-894

Irwin J, Suissa A, Anisman H (1980): Differential effects of inescapable shock on escape performance and discrimination learning in a water escape task. *J Exp Psychol [Anim Behav]* 6:21-40

Jackson RL, Maier SF, Rapaport PM (1978): Exposure to inescapable shock produces both activity and associative deficits in the rat. *Learn Motiv* 9:69-98

Jackson RL, Alexander JH, Maier SF (1980): Learned helplessness, inactivity, and associative deficits: the effects of inescapable shock on response choice escape learning. *J Exp Psychol [Anim Behav]* 6:1-20

Johnson JO, Henn FA (1980): Control of hippocampal beta-receptors in learned helplessness. *Psychology [Anim Behav]* 6:21-40

Johnson JO, Sherman A, Petty F, Taylor D, Henn F (1982): Receptor changes in learned helplessness. *Soc Neurosci Abstr* 8:392

Jouvet M (1973): Pharmacology of the future of man. In: *Brain, Nerves and Synapses, Vol 4*, Bloom FE, Acheson, GH eds. Basel: Karger

Katz RJ (1981): Animal models and human depressive disorders. *Neurosci Biobehav Rev* 5:231-246

Keisey JE (1977): Escape acquisition following inescapable shock in the rat. *Anim Learn Behav* 5:83-92

Lawry JA, Lupo V, Overmier JJ, Kochevar J, Hollis KL, Anderson DC (1978): Interference with avoidance behavior as a function of qualitative properties of inescapable shocks. *Anim Learn Behav* 6:147-154

Leff MJ, Roatch JF, Bunney WE (1970): Environmental factors preceding the onset of severe depression. *Psychiatry* 33:293-311

Lenhart H, Reinstein D, Strowbridge B, Wurtman R (1984): Neurochemical and behavioral consequences of acute, uncontrollable stress: Effects of dietary tyrosine. *Brain Res* 303:215-223

Leonard BE (1984): Pharmacology of new antidepressants. *Prog Neuropsychopharmacol Behav Psychiatry* 8:97-108

Lloyd C (1980): Life events and depressive disorder reviewed. *Arch Gen Psychiatry* 37:541-648

Looney TA, Cohen PS (1972): Retardation of jump-up escape responding in rats pre-treated with different frequencies of noncontingent electric shock. *J Comp Physiol Psychol* 78:317

MacLennan AJ, Drugan RC, Hyson RL, Maier SF, Madden JIV, Barchas JD (1982): Dissociation of long-term analgesia and the shuttle box escape deficit caused by inescapable shock. *J Comp Physiol Psychol* 96:904-912

Maier SF, Jackson RL (1977): The nature of the initial coping response and the learned helplessness effects. *Anim Learn Behav* 5:404-414

Maier SF, Anderson C, Lieberman DA (1972): Influence of control of shock on subsequent shock-elicited aggression. *J Comp Physiol Psychol* 81:94-100

Maier SF, Albin RW, Testa TJ (1973): Failure to learn to escape in rats previously exposed to inescapable shock depends on the nature of the escape response. *J Comp Physiol Psychol* 85:581-592

Maier SF, Coon DJ, McDaniel MA, Jackson RL, Grau J (1979): The time course of learned helplessness inactivity, and nociceptive deficits in rats. *Learn Motiv* 10:467-487

Maier SF, Davies S, Grau JW, Jackson RL, Morrison DH, Moye T, Madden IV, J Barchas JD (1980): Opiate antagonists and the long-term reaction induced by inescapable shock. *J Comp Physiol Psychol* 94:1172-1183

Maier SF, Drugan RC, Hyson RL, MacLennan AJ, Madden J, Barchas JD (1981): Opioid and nonopioid mechanisms of stress-induced analgesia. In: *Advances in Endogenous and Exogenous Opioids,* Takagi H, Simon EJ eds. Amsterdam, Elsevier Biomedical

Maier SF, Drugan R, Grau JW, Hyson R, MacLennan AJ, Moye T, Madden IV, Barchas JD (1983): Learned helplessness, pain inhibition, and the endogenous opiates. In: *Advances in Analysis of Behavior, Vol. 3*, Zeiter MD, Harzem P, eds. New York: John Wiley and Sons Ltd.

Overmier JB (1968): Interference with avoidance behavior: Failure to avoid traumatic shock. *J Exp Psychol* 78:340-343

Overmier JB, Seligman MEP (1967): Effects of inescapable shock on subsequent escape and avoidance learning. *J Comp Physiol Psychol* 63:23-33

Pare WP (1964): The effect of chronic environmental stress on stomach ulceration, adrenal function, and consummatory behavior in the rat. *J Psychol* 57:143-151

Pare WP (1965): Stress and consummatory behavior in the albino rat. *Psychol Rep* 16:399-405

Peters JE, Finch SB (1961): Short- and long-range effects on the rat of a fear-producing stimulus. *Psychosom Med* 23:138-152

Petty F, Sherman AD (1979): Reversal of learned helplessness by imipramine *Commun Psychopharmacol* 3:371-373

Petty F, Sherman AD (1980): Regional aspects of the prevention of learned helplessness by desipramine. *Life Sci* 26:1447-1452

Petty F, Sherman AD (1981): GABAergic modulation of learned helplessness. *Pharmacol Biochem Behav* 15:567-570

Petty F, Sherman AD (1983): Learned helplessness induction decreases *in vivo* cortical serotonin release. *Pharmacol Biochem Behav* 18:649-651

Petty F, Sacquitne JL, Sherman AD (1982): Tricyclic antidepressant drug action correlates with its tissue levels in anterior neocortex. *Neuropharmacology* 21:475-477

Redmond DE Jr, Maas JW, Dekirmanjian H, Schlemmer RE (1973): Changes in social behavior of monkeys after inescapable shock. *Psychosom Med* 35:448-449

Ritter S, Pelzer NL, Ritter RC (1978): Absence of glucoprivic feeding after stress suggests impairment of noradrenergic neuron function. *Brain Res* 149:399-411

Rosellini RA (1978): Inescapable shock interferes with the acquisition of an appetitive operant. *Anim Learn Behav* 6:155-159

Seligman MEP (1974): Depression and learned helplessness. In: *The Psychology of Depression: Contemporary Theory and Research*, Friedman RJ, Katz MM, eds. Washington D.C.: V.H. Winston pp. 83-125

Seligman MEP (1975): *Helplessness: On Depression, Development, and Death.*, San Francisco, W.H. Freeman

Seligman MEP, Beagley G (1975): Learned helplessness in the rat. *J Comp Physiol Psychol* 88:534-541

Seligman MEP, Maier SF (1967): Failure to escape traumatic shock. *J Exp Psychol* 74:1-9

Seligman MEP, Bosellini RA, Kozak MJ (1975): Learned helplessness in the rat: time course, immunization, and reversibility. *J Comp Physiol Psychol* 88:542-547

Sherman AD, Petty F (1980): Neurochemical basis of the action of antidepressants on learned helplessness. *Behav Neural Biol* 30:119-134

Sherman AD, Petty F (1982): Additivity of neurochemical changes in learned helplessness. *Behav Neural Biol* 35:344-353

Sherman AD, Allers GL, Petty F, Henn FA (1979): A neuropharmacologically relevant animal model of depression. *Neuropharmacology* 18:891-893

Sherman AD, Sacquitne JL, Petty F (1982): Selectivity of the learned helplessness model of depression. *Pharmacol Biochem Behav* 16:449-454

Shimizu N, Imamoto K (1979): Fine structure of the locus coeruleus in the rat. *Arch Histol Jpn* 31:229-246

Simson PG, Weiss JM, Ambrose MJ, Webster A (1986a): Infusion of a monoamine oxidase inhibitor into the locus coeruleus can prevent stress-induced behavioral depression. *Biol Psychiatry* 21:724-734

Simson PG, Weiss JM, Hoffman LJ, Ambrose MJ (1986b): Reversal of behavioral depression by infusion of an alpha-2 agonist into the locus coeruleus. *Neuropharmacology* 25:385-389

Stone EA (1978): Possible grooming deficit in stressed rats. *Res Commun Psychol Psychiatr Behav* 3:109-115

Sutton BR, Coover GD, Lints CE (1981): Motor debilitation, short- and long-term shuttlebox deficits, and brain monoamine changes following footshock pretreatment in rats. *Physiol Psychol* 9:127-134

Svensson TH, Bunney BS, Aghajanian GK (1975): Inhibition of both noradrenergic and serotonergic neurons in brain by the alpha-adrenergic agonist clonidine. *Brain Res* 92:291-306

Swanson LW (1976): The locus coeruleus; a cytoarchitectonic, Golgi, and immunocytochemical study in the albino rat. *Brain Res* 110:39-56

Swenson RM, Vogel WH (1983): Plasma catecholamine and corticosterone as well as brain catecholamine changes during coping in rats exposed to stressful footshock. *Pharmacol Biochem Behav* 18:689-693

Telner JI, Singhal RL (1981): Effects of nortriptyline treatment on learned helplessness in the rat. *Pharmacol Biochem Behav* 14:823-826

Telner JI, Singhal RL (1984): The learned helplessness model of depression. *J Psychiatr Res* 18:207-215

Telner JI, Singhal RL, Lapierre YD (1980): Brain monoaminergic systems and inescapable shock: support for the learned helplessness hypothesis. *Neuroendocrinol Lett* 2:285-290

Tsuda A, Tanaka M, Hirai H, Pare WP (1983): Effects of coping behavior on gastric lesions in rats as a function of predictability of shock. *Jap Psychol Res* 25:9-15

Weiss JM (1968): Effects of coping responses on stress. *J Comp Physiol Psychol* 65:251-260

Weiss JM (1971): Effects of coping behavior in different warning-signal conditions on stress pathology in rats. *J Comp Physiol Psychol* 77:1-13

Weiss JM, Glazer HI (1975): Effects of acute exposure to stressors on subsequent avoidance-escape behavior. *Psychosom Med* 37:499-521

Weiss JM, Stone EA, Harrell N (1970): Coping behavior and brain norepinephrine level in rats. *J Comp Physiol Psychol* 72:153-160

Weiss JM, Glazer HI, Pohorecky LA, Brick J, Miller NE (1975): Effects of chronic exposure to stressors on avoidance-escape behavior and on brain norepinephrine. *Psychosom Med* 37:522-533

Weiss JM, Glazer HI, Pohorecky LA (1976): Coping behavior and neurochemical changes: an alternative explanation for the original "learned helplessness" experiments. In: *Animal Models in Human Psychobiology*, Serban G, King A, eds. New York, London: Plenum Press.

Weiss JM, Glazer HI, Pohorecky LA, Bailey WH, Schneider LH (1979): Coping behavior and stress-induced behavioral depression: studies of the role of brain catecholamines. In: *The Psychobiology of the Depressive Disorders: Implications for the Effects of Stress*, Depue E, ed. New York: Academic Press

Weiss JM, Bailey WH, Pohorecky LA, Korzeniowski D, Grillone G (1980): Stress-induced depression of motor activity correlates with regional changes in brain norepinephrine but not in dopamine. *Neurochem Res* 5:9-22

Weiss JM, Goodman PA, Losito BG, Corrigan S, Charry JM, Bailey WH (1981): Behavioral depression produced by an uncontrollable stressor: relationship to norepinephrine, dopamine, and serotonin levels in various regions of rat brain. *Brain Res Rev* 3:167-205

Weiss JM, Bailey WH, Goodman PA, Hoffman LJ, Ambrose MJ, Salman S, Charry JM (1982): A model for neurochemical study of depression. In: Behavioral Models and the Analysis of Drug Action, Spiegelstein MY, Levy A, eds. Amsterdam, Elsevier Scientific pp. 195-223.

Weiss JM, Simson PG, Ambrose MJ, Webster A, Hoffman LJ (1985): Neurochemical basis of behavioral depression. In: *Advances in Behavioral Medicine*, Katkin E, Manuck S, eds. Greenwich, CT: JAI Press

Weiss JM, Simson PG, Hoffman LG, Ambrose MJ, Cooper S, Webster A (1986):

Infusion of adrenergic receptor agonists and antagonists into the locus coeruleus and ventricular system of the brain: effects on swim-motivated and spontaneous motor activity. *Neuropharmacology* 25:367-384

Zacharko RM, Bowers WJ, Kokkinidis L, Anisman H (1983): Region-specific reductions of intracranial self-stimulation after uncontrollable stress: possible effects on reward processes. *Behav Brain Res* 9:129-141

8

Motor Activity and Antidepressant Drugs: A Proposed Approach to Categorizing Depression Syndromes and Their Animal Models

MARTIN H. TEICHER, NATACHA I. BARBER,
JANET M. LAWRENCE, AND ROSS J. BALDESSARINI

Affective disorders are characterized by disturbances in mood, cognition, and neurovegetative processes. Among the latter, dysregulation of appetite and weight, sleep, energy, and libido have received particular clinical attention. Current standard diagnostic criteria for major depression in the American Psychiatric Association Diagnostic Manual (DSM-III) accommodate neurovegetative disturbances in either direction—deficits as well as excesses—and evaluation of these features is routine in the clinical assessment of depressed patients. These dysfunctions often serve as target symptoms for psychopharmacological treatment and as quantifiable behaviors that can be used to bridge the gap between animal models and relevant clinical disorders.

Attention to altered activity levels and cycles is implicit in queries to patients about insomnia or hypersomnia and psychomotor agitation or retardation; but aside from such general questions, biological rhythms are rarely assessed specifically and quantitatively in patients with mood disorders, outside of a few research centers. Before discussing the effects of drugs on behavioral arousal as expressed in activity rhythms, it is important to summarize how depression affects activity in man, and to emphasize the dichotomous (excess vs. deficit) nature of clinical symptom clusters in depression.

Depressive Subtypes and Locomotor Activity Levels

Given the scope of symptoms commonly affected by major affective disorders, it is not surprising that several relatively homogeneous clusters of depressive subsyndromes can be identified. DSM-III emphasizes a unipolar-bipolar division, and then subdivides each major division by a severity factor (dysthymic vs. major depression, cyclothymic vs. bipolar disorder), and by the presence of melancholic or psychotic features. Other researchers have emphasized syndromes such as anxious depression, involutional melan-

TABLE 8.1 Proposed subtypes of endogenous depression based on locomotor activity levels

Features	Clinical Clusters	
	Agitated	Retarded
Locomotor activity	Increased	Decreased
Sleep	Decreased	Increased
Appetite-weight	Decreased	Increased (Can be decreased if depression is very severe or associated with high stress).
Clinical polarity	Unipolar (can be confused with dysphoric mania)	Bipolar (Bipolar I or II, Seasonal affective disorder, pseudounipolar depression)
Typical onset age	Older (> 40) (involutional melancholia)	younger (< 30 years)
Family affective history	Sometimes	Usual
Preferred treatments	Sedating antidepressants amitriptyline imipramine doxepin trazodone amoxapine	Stimulating antidepressants nortriptyline desipramine protriptyline tranylcypromine
Additional treatments	Neuroleptics often added	Lithium salts, anticonvulsants; stimulants sometimes used

cholia, agitated depression, pseudo-unipolar disorder, hysteroid dysphoria, and neurotic depression (Paykel, 1971; Liebowitz and Klein, 1979; Nelson and Charney, 1981; Baldessarini, 1983; Akiskal et al., 1985; Akiskal, 1986). In their review of the literature on core symptom clusters in "autonomous" (endogenous) depression, Nelson and Charney (1981) found that psychomotor change was the symptom most clearly associated with such depression. Furthermore, autonomous depression could *not* be conceptualized as a single entity. Based on behavioral, neurochemical, and treatment response factors, at least two distinct autonomous states were described: a retarded anhedonic group and an agitated delusional group of depressed patients. Although sleep and appetite disturbances have not consistently distinguished endogenous from nonendogenous depression (Nelson and Charney 1981), there is a strong tendency for sleep and appetite changes to covary with activity levels in a meaningful way in major depression (Kupfer and Foster, 1973).

Table 8.1 presents our summary proposal of two distinct subtypes of endogenous depression, based on neurovegetative features and emphasizing alterations in psychomotor activity. Recognition of these and other subtypes may be particularly important in categorizing animal models or in searching for clinically useful biological markers associated with depression. This conceptualization also highlights the potential clinical usefulness of quantitative assessment of locomotor activity in patients with a major effective disorder.

Depression and Disturbances in Overall Activity Levels

In major depression, locomotor activity is disturbed in its mean 24 hour *levels,* and in the *temporal modulation* of its circadian rhythm. Differences in activity levels between depressive subtypes can be impressive. Using ambulatory activity monitors, it has been found that middle-aged, adult *bipolar* depressed patients are less active than younger normal subjects, and only half as active as middle-aged *unipolar* depressed patients (Kupfer et al., 1974; Wehr et al., 1980). These findings support the earlier empirical observations of Beigel and Murphy (1971) who found that pacing behavior in unipolar patients most clearly separated them from cases of bipolar depression: in 25 bipolar-unipolar pairs of patients matched for age and studied for 14 drug-free days on an inpatient research unit, not one bipolar patient had a higher pacing rate than his paired-matched unipolar control. Wolff et al. (1985) found middle-aged bipolar patients to have much lower activity monitor levels when depressed than when euthymic and to be less active than younger controls. However, since two of these activity monitor studies carried out at the NIMH appear to be confounded by age differences (control subjects averaged 21-24 years, and the bipolar depressed patients averaged 40-42 years [Wehr et al., 1980; Wolff et al., 1985]), further clinical studies of this kind involving close matching by age and polarity would be desirable.

Depression and Disturbances in the Timing of Activity

In recent years, chronobiological theories have emerged to explain several specific features of major affective illnesses, such as sleep disturbances, early morning awakening, and diurnal variation of symptom severity; long-term cyclic and seasonal recurrences; and responses to mood-altering (thymoleptic) drugs (Goodwin et al., 1982). These theories are based largely on observed disturbances in the frequency or phase of circadian rhythms in depressed patients. Foster and Kupfer (1975) reported that middle-aged depressed patients (mean age, 47 years) showed blunting of the amplitude, or "desynchronization", of their 24-hour activity cycle and emitted a higher than normal percentage of their daily activity during the night (8.5% vs. ca. 5%) than age-matched controls. Younger adult patients with primary depression (mean age 25 years) emitted a significantly lower percentage of total daily activity during the night (2.2%) than an age-matched control group (5.1%). Since this group of depressed young adults suffered from prominent anergic and hypersomnic symptoms, they may have been predominantly bipolar.

Using various methods to analyze circadian rhythms, Wehr and his colleagues (1980) found that the activity acrophase (peak time of activity determined by a *circadian cosinor analysis*) of middle aged *bipolar* depressed patients was 97 min earlier than in younger controls; a similar *phase-advance* held for body temperature and excretion of the norepinephrine metabolite

3-methoxy-4-hydroxyphenethyleneglycol (MHPG). Wolff et al. (1985) did not statistically replicate this result using a less sophisticated *centroid analysis* (time of day from 7:00 A.M. on, when half of total daily activity was produced). Nevertheless, there was a trend for *bipolar* patients to show a 20-min *phase-advance* and for *unipolar* depressed patients to show an 8-min *phase-delay* compared to younger normal controls. These trends may have been more prominent if phase differences had not been constricted by entraining both patients and controls to the same set ward environment with a fixed wake-up time (7:00 A.M.), fixed meal times, and a scheduled milieu (Wehr et al., 1980; Wolff et al., 1985).

Intriguing preliminary data also suggest that certain cycling affectively ill subjects may suffer from an internal "desynchronization" between different circadian neural oscillators ("clocks"). Theoretical schemes have proposed the existence of multiple coupled circadian oscillators such as an "X-oscillator" governing circadian rhythms in body temperature, cortisol secretion, and propensity for REM sleep and a "Y-oscillator" modulating rest-activity cycles, growth hormone release, and propensity for slow-wave sleep (Moore-Ede et al., 1983). Furthermore, Halberg (1968) and Kripke and coworkers (1978) have proposed that one such coupled oscillator might depart from its normal circadian frequency in some patients and that oscillators thus operating at disparate frequencies would drift into and out of phase with one another, producing episodic bouts of depression and mania. It has also been suggested that affectively ill patients with an aberrant circadian oscillator with an accelerated periodicity of less than 24 hours might respond to lithium treatment, which appears to slow certain circadian processes (Kripke and Wyborney, 1980), whereas patients with an aberrant oscillator running at periods greater than 24 hours would not respond to lithium (Kripke et al., 1978).

Recent Findings in Geriatric Patients with Unipolar Depression

Much of the available data on disturbances of activity rhythms in major depression has been derived from middle-aged or young adult bipolar patients, including some, with particularly rapid mood cycles, who may not be typical. Accordingly, we recently carried out activity measurements in geriatric patients with non-bipolar depression and in normal persons of similar age. We found that the elderly depressed patients were, on average, 29% *more active* than controls ($p < 0.02$). While the amplitude of their circadian rhythm (expressed as activity counts above and below the mean, or as a percentage of mean daily activity) was not different from controls, the depressed patients were markedly *phase delayed*. The acrophase (circadian peak activity) of the two groups was 1:50 P.M. in controls and 3:54 P.M. in depressed patients (2.04 hour phase delay; $p < 0.0001$).

Analysis of individual acrophase times indicated that the peak period of motor activity occurred between 1:00 P.M. and 6:30 P.M. in all of the elderly subjects. Control subjects invariably had their acrophase occur relatively early in the day, whereas depressed subjects had later activity peaks. There was no overlap between the two groups, though four of eight depressed patients had acrophase times near the extreme of the normal group. These patients were similar to the other four patients in Hamilton Depression Rating Scale total scores, but all of them suppressed cortisol normally at 4:00 P.M. on the day following 1.0 mg of dexamethasone at bedtime; three of the four patients with the most extreme degrees of phase-delay were cortisol nonsuppressors (> 5 μg/dl).

Summary: Locomotor Activity in Major Depression

Overall, activity levels in depressed patients can be either low or excessive compared to normal subjects. Patients with bipolar depressive disorders, and especially younger patients, tend to be psychomotorically retarded and to have diminished mean activity levels and increased sleep times. Their time of peak activity (acrophase), based on cosinor analysis, probably occurs earlier in the day than in nondepressed individuals and suggests the presence of a *phase-advance* of their circadian activity oscillator (internal clock). Older depressed patients, especially those with non-bipolar disorders, have elevated activity relative to most bipolar patients and to age-matched normal controls. Our findings in older, non-bipolar depressives suggest, in contrast, the intriguing possibility that the peak activity period (cosinor acrophase) may occur later in these patients and that their rest-activity oscillator may be *phase-delayed* rather than phase-advanced. The finding of hypomotility in many patients with recurring major depression with an early age of onset in one piece of evidence supporting the hypothesis that some apparently unipolar depressives may have biological characteristics of bipolar patients ("pseudo-unipolar" disorder: Akiskal et al., 1985; Akiskal, 1986).

Locomotor Activity and Animal Models of Depression

One question to be addressed when evaluating an animal model of a psychiatric disorder, is whether the behavioral disturbances produced resemble, in some meaningful way, the disturbances found in the clinical condition. In relation to major depression, it may be important to know whether the pattern of behavioral disturbances observed in an animal model are phenomenologically similar to disturbances occurring in a recognized clinical subtype of depression (Wilner 1984). Following the hypothesis in the preceding section that psychomotor activity levels may meaningfully subdivide depressive syn-

TABLE 8.2 Classification of animal models of depression based on differences in psychomotor activity

Descriptive name (ref.)	Appetite	Associated features
Retarded		
Amphetamine withdrawal (1)	↑	Elevated threshold for brain stimulation
Exhaustion stress (2)	?	Loss of cyclicity, constant diestrus
Learned helplessness (3,4)	↓	Decreased aggression, poor performance in appetitive tasks
Chronic unpredictable stress (5)	↓	Increased corticosteroids, decreased reactivity to acute stress
Separation-Despair (6)	↓	Decreased play and social interaction, sad expression
Agitated		
Olfactory bulbectomy (7)	↓	Irritable, decreased passive avoidance learning
Isolation-induced hyperactivity (8)	?	Not irritable
Focal brain injury (9,10)	↓	Increased corticosteroids
Separation-protest (6,11)	↓	Sleepless, distress calling, increased corticosteroids

References: (1) Simpson & Annau, 1977; (2) Wilner, 1984; (3) Anisman et al., 1979; (4) Weiss, 1968; (5) Katz et al., 1981; (6) McKinney & Bunney, 1969; (7) Craincross et al., 1981; (8) Garzon & Del Rio, 1981; (9) Pearlson & Robinson, 1981; (10) Finklestein et al., 1983a,b; (11) Higley et al., 1982. Appetite changes are indicated as increased (↑), decreased (↓), or uncertain (?).

dromes, some of the proposed animal models of depression have been categorized by activity criteria (Table 8.2).

Briefly, animal models that claim to replicate some of the clinical features of depression can be classified by changes in arousal or motor activity. Models which are characterized by psychomotor agitation appear to be fairly homogeneous. Those which have been studied extensively also have several associated features that correspond to the same clinical classification: decreased appetite, decreased sleep, and increased plasma corticosteroid levels. Models in which psychomotor retardation is prominent appear to be less homogeneous in their associated effects on appetite and sleep. Amphetamine withdrawal appears to correspond particularly well with the psychomotor-retarded depression seen in most bipolar patients and in many patients with seasonal affective disorders. Animals withdrawn from amphetamine are inactive, have increased appetite, and may be less able to perceive "pleasure" as inferred from an elevated stimulus intensity threshold for intracranial electrical self-stimulation. Most other models with psychomotor retardation are essentially based on stress-induced behavioral depression. In these models appetite and sleep are often diminished along with activity, and so they differ markedly from the characteristic psychomotor-retarded depression seen in bipolar patients (Kupfer and Foster 1973). Stress-related behavioral syndromes may represent a distinct group of models separable on behavioral and neurochemical parameters, and they may more closely resemble the form of depression observed in patients with stress-related reactive depressions or adjustment disorders with depressed mood rather than the depression of patients with endogenous major affective disorders.

Methodological and Statistical Issues in the Assessment of Motor Activity

Most studies examining locomotor activity disturbances in animal behavioral models provide relatively little information regarding the complicated but probably important matters of quantitation and statistical analysis. In some studies, activity has been quantified using a brief (ca. 5 minute) "open field" motility and exploration test, or by summating interruptions of a light beam detected by photocells in a cage during a very limited (typically, 1-4 hour) sample of daily activity. Locomotor activity varies enormously with the time of day, according to the setting in which it is assessed, and is strongly affected by the technique of measurement. Thus, for a full description of activity, one can not simply indicate its *magnitude* during an arbitrary time period, but should also attend to the *temporal organization* of the activity and possibly to the detailed forms and features of the behaviors emitted.

We have been collecting data in animals and man that permit relatively detailed, quantitative time-series analyses in which activity levels are summated over short (1-15 min) intervals during monitoring over periods from one to 14 days. For rats, activity is detected as perturbations in a radio-frequency field using a six-channel Stoelting Electronic Activity Monitor, which has been interfaced to an Apple II microcomputer. Human activity typically is monitored as accelerations of the nondominant arm measured by use of a wrist-worn ambulatory activity monitor with solid-state memory and a bilaminar pizoelectric acceleration detector. Initial studies employed a Colburn-NIMH device of this type with a 256 byte memory and a 15-min epoch interval (Colburn et al., 1976). Currently, an improved device is used. Developed by Precision Control Design, it has a 16 kilobyte memory, built-in microprocessor control, and a programmable epoch interval. Both devices interface to an Apple IIe computer.

By using these devices it is possible to obtain comparable activity profiles in laboratory animals and man, although with an expected phase difference of several hours when a nocturnal species such as the rat is compared with man. Figure 8.1 displays the activity profiles of a representative young male adult rat and a healthy young man. These profiles are strikingly similar—in fact, about as similar as profiles encountered between individual subjects of the same species.

These profiles contain an enormous amount of data and powerful analytical techniques are needed to provide meaningful descriptive summary statistics. The first technique used is a modified form of cosinor analysis. Traditionally, this technique provides a linear least-squares fit of a cosine function to raw time series data (Halberg et al., 1972; Nelson et al., 1979). The function is made linear by selecting a set frequency (1 cycle per day [cpd] for

142   Martin H. Teicher, et al.

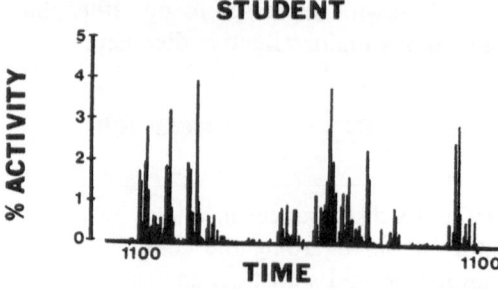

STUDENT

% ACTIVITY

TIME

FIGURE 8.1. Activity time-series for a representative 25-year-old male graduate student and a normal 60-day-old male Sprague-Dawley rat. Activity is expressed as the percentage of total 24-hour activity counts per 15-minute interval over 24 hours. Start times for each profile are indicated in the lower left-hand corner of the record.

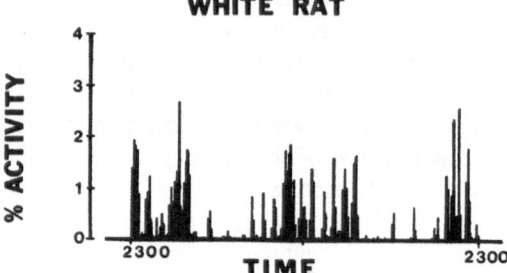

WHITE RAT

% ACTIVITY

TIME

circadian rhythms) and deriving least-squares estimations of mesor (mean activity level), amplitude, and acrophase (time of activity peak). A nonlinear least-squares fit enables frequency to be fit as an additional variable and makes it possible to fit raw data simultaneously to oscillators operating at several frequencies (Rummel et al. 1974). With a microcomputer program prepared by Teicher and Barber 1988, time-series for different groups can be analyzed simultaneously, parameters can be shared or constrained, and differences between parameter fits can be assessed by ANOVA methods.

Cosinor analysis of the data from rat and man shown in Figure 1 (using a fixed, 1 cpd frequency) indicates that these subjects had a comparable mesor and circadian amplitude but, as expected, very different acrophases separated by nearly 9 hours (see Table 8.3). Nonlinear analysis of these data to determine the optimal frequency indicated that the activity of both subjects was highly circadian with a frequency of very close to 1.0 cpd in both subjects.

Although it is well known that both normal *adult* rat and man display a significant circadian rhythm, additional rhythmic components in their activity pattern may also merit scrutiny. In particular, we have been interested in higher frequency or *ultradian* (more than 1 cpd) rhythms in locomotor activity. The analysis of such rhythms is complex and unsettled, and most researchers in this area have developed their own unique statistical approach to the problem. We have followed the recommendation

TABLE 8.3 Cosinor properties of young male rat and young man

	Mesor	Amplitude	Acrophase
Rat	1.64 ± .21	1.37 ± .30	24:48:12 ± 0:49:37
Man	1.71 ± .26	1.68 ± .36	15:57:07 ± 0:49:14
F-test	F = 0.04	F = 0.44	F = 16.62
df probability	1,90; p = 0.84	1,90; p = 0.51	1,90; p < 0.001

Mesor and amplitude are expressed as percent of total activity per 15-minute interval. Acrophase is expressed as actual clock time in hours:minutes:seconds. All data are means ± SEM.

F-test ratio is determined by comparing the residual variance for the two curves simultaneously fit with independent parameters (mesor, amplitude, phase) to the residual variance for an optimized simultaneous fit with one of the parameters constrained to be equal for the two curves. The degradation in goodness-of-fit resulting from constrained parameter sharing, balanced by the increase in degrees of freedom (df), determines the probability that the selected parameter is the same for the two curves (Teicher and Barber, 1988).

of Kripke (1974) and use a harmonic technique known as *variance spectral analysis* (Black and Tukey, 1958; Halberg and Panofsky, 1961). This approach represents one way of contending with the critical problem of the *nonstationary* character of the dominant ultradian rhythm. This problem occurs because ultradian activity rhythms are not entrained to any known regularly occurring external timing cue ("zeitgeber"); and, even under highly controlled conditions, they drift in and out of phase. Analysis of spectral variance uses a fairly narrow sampling window to scan over the entire time-series and sacrifices some degree of frequency resolution to obtain a more accurate estimation of spectral magnitude.

Figure 8.2 contains spectral plots for the same white rat and young man whose activity is depicted in Figure 1. Graphed is *percent of total spectral variance* as a function of frequency from 0-48 cpd, at 1 cpd increments. For convenience, this broad range of frequencies was subdivided into a circadian band (1 cpd), as well as "slow" (3-8 cpd), "medium" (9-16 cpd), and "fast" (17-24 cpd) ultradian bandwidths. These profiles are similar and are consistent with the observation (see Table 8.3) that normal young rats and humans are strongly circadian in early adult life. Ultradian rhythms at this age are rarely prominent and usually occur across a diffuse broad range of frequencies. They are much more prominent in immature subjects and in the aged, and they also become prominent under certain pathological conditions (Teicher et al., 1986; 1988a,b).

Locomotor Activity Rhythm Disturbances in Amphetamine-Withdrawal "Depression"

Given the technique for assessing activity rhythms just described, we are studying selected animal models of depression to ascertain whether characteristic rhythm disturbances result that might be sought in patients with major depression or other clinical disorders. We were particularly interested in

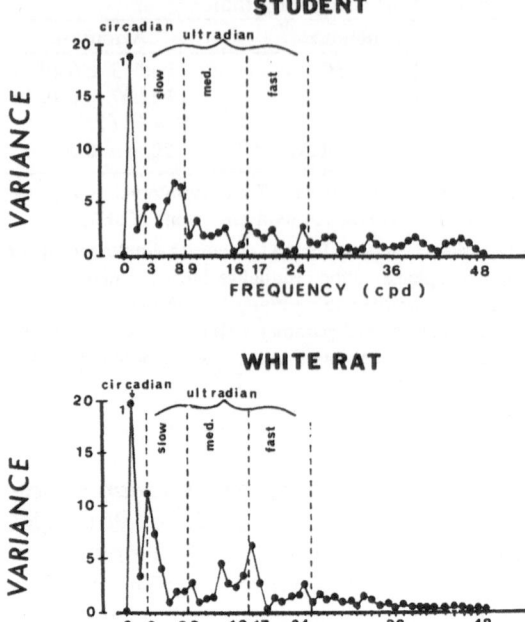

FIGURE 8.2. Variance spectral analysis of a male graduate student and young adult male rat. Results of spectral analysis based on the data of Figure 8.1 are expressed as the percentage of total variance present in the activity time-series that can be accounted for by rhythms at frequencies ranging from 0 to 48 cycles per day (cpd) at 1-cpd intervals. The spectral plots are divided into circadian (1 cpd), slow (3 to 8 cpd), medium (9 to 16), and fast (17 to 24 cpd) ultradian bandwidths, ignoring rarer ultrafast rhythms at 25 to 48 cpd.

determining whether animal models characterized by psychomotor retardation and psychomotor agitation might display similar or dissimilar disturbances in biological rhythms.

As a model of psychomotor retardation, stimulant withdrawal in the rat was selected. This model has a number of at least superficial parallels with bipolar depression and is a commonly encountered clinical entity in itself. Preliminary data arising from this model are intriguing. Adult male Sprague-Dawley rats weighting 400-440 grams, received 3 mg/kg of (+)amphetamine sulfate per 24 hours by continuous subcutaneous infusion by Alzet osmotic minipump over a 5 day period. Activity was monitored for 3 days following surgical removal of the infusion device. In humans, it is known that depression ratings peak at 48-72 hours after a last dose of amphetamine or cocaine, and that this "depression" coincides with a maximum decrease in excretion of MHPG (Watson et al., 1972).

A sample of preliminary data is shown is Figure 8.3 as successive 24-hour activity plots from one male Sprague-Dawley rat on days 1 through 3 following withdrawal of amphetamine. Testing took place under the usual 12-hour light/dark cycle of these nocturnal subjects, with illumination occuring from 7:00 A.M. to 7:00 P.M. The first point of interest is that this rat became progressively less active than controls (not shown) during the three days of testing, and was only 40% as active as controls on day

FIGURE 8.3. Activity time-series for an adult male rat following withdrawal from 5 days of continuously administered (+) amphetamine sulfate (3 mg/kg/day by Alzet minipump, s.c.). Data are expressed as activity counts $(\times 100^{-1})$ per 10-minute interval over 24-hour periods at 1, 2, and 3 days following stimulant pretreatment. The minipump was removed 3 hours prior to the start of testing.

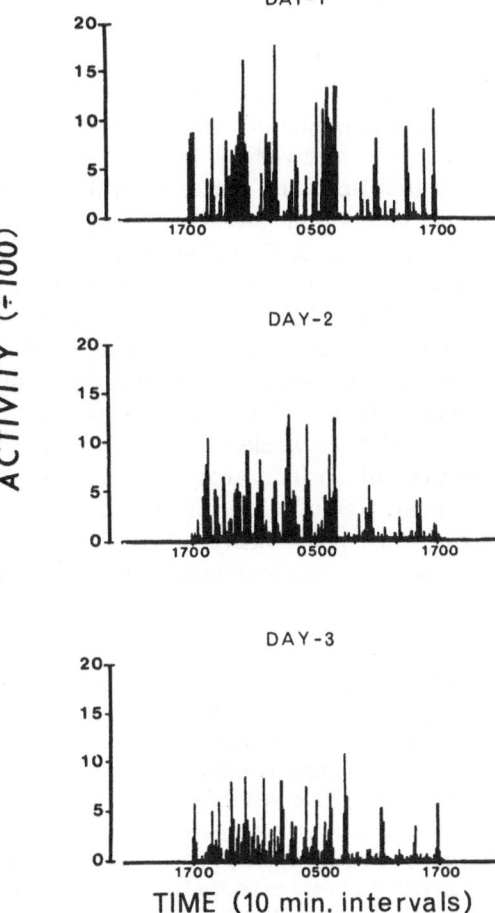

3. Second, the rat became active at about 5:00 P.M. on each test day. It is unusual for a normal adult rat to be so active prior to the onset of the dark-phase, and cosinor analysis revealed a phase-advance of about 68 min relative to controls that held throughout the 3-day test period. The third point to emphasize is that this rat had a prominent ultradian rhythm, with consistent episodic bursts of activity. Such a profile is also unusual for a normal adult rat (see Figure 8.1). Furthermore, the dominant ultradian rhythm was not stable and it appears to have accelerated each day (from 5 to 15 cpd) as overall activity levels declined. These observations suggest that, in this stimulant-withdrawal model of depression, there may be a quantifiable disturbance in locomotor activity rhythms that can also be sought in comparable human patients withdrawing from amphetamines or cocaine, or in bipolar patients with retarded depression.

Activity Rhythm Disturbances in a Focal Brain Injury Model of Depression

Finklestein and colleagues (1983b) reported on activity rhythms disturbances in an animal model of clinical post-stroke depression. Adult male Sprague-Dawley rats received bilateral aspiration lesions of 12 x 4 mm strips of ventrolateral cerebral cortex with the inferior border just above the rhinal fissure. Such lesions are accompanied by widespread alterations in cerebral monoamine metabolism (Finklestein et al., 1983a).

Activity was monitored prior to surgery and at one and three weeks postoperatively. Sham-operated controls and blood-loss controls displayed no significant differences in activity rhythm parameters between the times of testing. The overall 24-hour activity of rats receiving bilateral lesions increased about 80% at one week postoperatively and were still about 30% elevated at three weeks.

Results of cosinor analysis were unanticipated in that they were not consistent: 40% of bilaterally lesioned rats phase-advanced by an average of 54 min; 60% phase-delayed by a mean of 67 min. It may not be correct to conclude that there was no overall effect. Sham-operated control rats did not

FIGURE 8.4. Activity time-series of a representative adult male rat during a baseline period 3 days prior to bilateral aspiration lesions of 12 x 4 mm strips of frontal-parietal cerebral cortex, and at 1 and 3 weeks postoperatively. Data are expressed as activity counts ($\times 100^{-1}$) per 10-minute interval during the 36-hour period illustrated. These three plotted time-series started at 15:36, 22:15, and 21:16 hours, respectively.

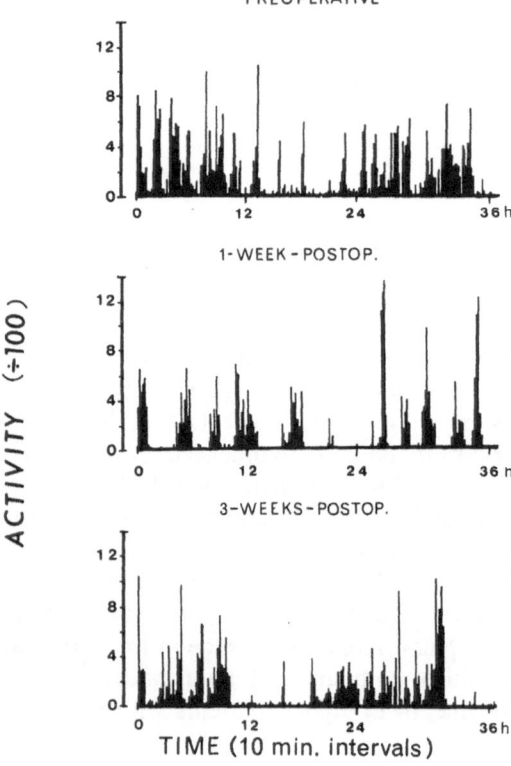

phase shift more than 12 min from test period to test period, but all of the lesioned rats displayed a disruption of 20 min or more. No covariate (such as lesion size, degree of activity enhancement, or body weight) has been identified that independently discriminates phase-advancing from phase-delaying lesioned rats.

However, it was found that the lesions produced very consistent short-term effects on ultradian activity rhythms. In Figure 8.4 a clear and distinct ultradian rhythm pattern can be seen at one week postoperatively. At 3 weeks there was still some ultradian rhythmicity, but less. Aggregate spectra from 18 rats pre- and post-surgery demonstrated a robust and consistent elevation of slow ultradian rhythms at 7 days postoperatively (from 15.1 ± 1.0 to 29.7 ± 2.5 percent total variance) with a peak frequency of 4-5 cpd. Circadian rhythmicity was not strongly affected (8.4 ± 0.6 vs 9.7 ± 0.9 percent total variance). What these observations mean is unclear, but, again, there appears to be a disruption in the rhythmic regulation of rest-activity states of the adult rats, and ultradian rhythms may be most significantly affected.

Activity Rhythm Disruptions During Protest Phase of Infant Separation

Several animal models of depression have focused on the effects of early maternal separation. Such models were the focus of an important review by McKinney and Bunney (1969), and they are particularly important in studies of the consequences of early experience. Often, primates have been used in this research, although some lower mammals (including rat pups) and imprintable birds can show equally dramatic separation phenomena (Willner 1984). Infant monkeys respond to maternal separation by an initial stage of *protest*, characterized by agitation, sleeplessness, and distress calls followed a few days later in some subjects by a stage of *despair*, characterized by decreased levels of activity, appetite, play, and social interactions (McKinney and Bunney 1969). The degree of plasma cortisol elevation during the protest phase may, to some degree, predict the severity of the depressive response during the despair phase (Higley et al., 1982).

Developing rats can also display a dramatic response to separation from their mother and littermates. This response is most evident at 15 days of age, coincident with eye-opening but before weaning. The activity level of isolated rats of this age is about 10-fold greater than at 12 or 20 days of age, or in adulthood (Campbell et al., 1969). This reaction appears to be a true separation-fear response and is not related to suckling deprivation or hunger because the protest response can be attenuated by housing a pup alone in its home cage, or by placing the pup with littermates or a nonlactating anesthetized dam (Campbell and Raskin, 1978; Randall and Campbell, 1976).

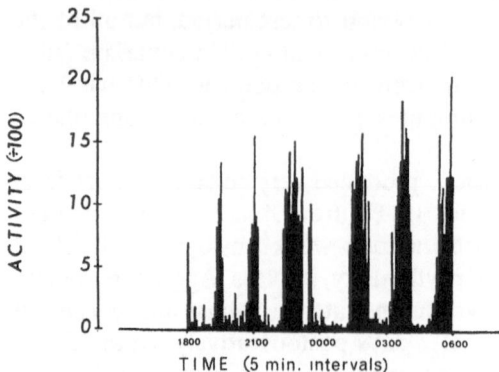

FIGURE 8.5. Activity time-series of an isolated 15-day-old rat tested in darkness. Data are expressed as activity counts ($\times$ 100^{-1}) per 5-minute interval during a 12-hour test period. Actual clock time is indicated on a 2400-hour basis (0000 = midnight; 1200 = noon).

Maternal separation and isolation have profound effects on rest-activity patterns in the immature rat (Teicher and Flaum, 1979). Figure 8.5 displays the activity profile of a representative 15-day-old rat studied on a vibrational activity monitor (Teicher and Green, 1977) for a 12-hour test period. Normal rats at this age display particularly robust and reliable ultradian rhythms (Teicher and Flaum, 1979). The dominant ultradian frequency varies from subject to subject, but usually involves the medium bandwidth of ultradian frequencies (9-16 cpd), with a maximum at about 10-12 cpd (and contrasts with the slower ultradian rhythms expressed in cortically lesioned adult rats.

Circadian disturbances are not easy to determine, as the circadian acrophase of preweanling rats is influenced strongly by the conditions of testing; and their acrophase activity peak may occur at any time of the day or night, depending on the time when the test period was started. In short, such young animals are not well-entrained to day/night lighting cycles (or to the mother's activity rhythm) and are remarkably ultradian until after weaning at 21 days of age.

Summary: Rest-Activity Rhythms in Animal Models of Depression

Data collected on amphetamine-withdrawal, focal cortical injury, and maternal separation models of depression in the rat all reveal prominent ultradian rest-activity rhythms. The dominant ultradian activity rhythm in rats given a focal cerebral cortical lesion is about 4-6 cpd. Preliminary evaluation of the effects of amphetamine withdrawal suggests that ultradian rhythms may increase in frequency from 5 to 15 cpd during the first three days of stimulant withdrawal. Normal immature rats have a dominant ultradian rhythm occurring between about 7 and 16 cpd when separated from their mother and littermates.

Disturbances in circadian rhythmicity have been harder to demonstrate in these animal models. Some of our preliminary data suggest that animals undergoing amphetamine withdrawal may be phase-advanced, as in young patients with bipolar major depression (Wehr et al., 1980) but this observation requires further evaluation. Rats with focal cerebral cortical injuries appear to have altered circadian acrophase times, but the direction of the disturbance is not consistent between subjects and may be affected by unknown and uncontrolled extraneous factors. A similar phenomenon has also been described by Kittrell and Satinoff (1988), who found that female rats segregate into phase-advancing and phase-delaying subgroups during lactation. Although the direction and magnitude of the phase shift is consistent within a given subject through reproductive cycles, independent factors determining the direction of the phase shift have not been identified.

Overall, the findings just summarized represent a set of prominent behavioral observations which still lack clear teleological significance or direct clinical relevance. Ultradian activity rhythms, which are evidently more prominent in rats, are not usually prominent in normal human adults; circadian disturbances are more accessible and may be more informative clinically. In pursuing these findings, it is particularly important to understand how drugs with known effects on human mood and behavior affect circadian and ultradian activity rhythms in animals and man.

Effects of Acute Drug Challenge on Ultradian Activity Rhythms

Initially, we attempted to define the effects of acute administration of representative mood-altering drugs on the modulation of ultradian activity rhythms. For convenience, we focused on the maternal separation protest phenomenon in 15-day-old preweanling rats. Littermates were injected with various drug doses or a vehicle control solution; and activity was monitored for 24 hours with each subject kept in isolation, without food and water, and in constant darkness to emphasize endogenous rhythms. Drugs were administered at various times to provide initial data on the relationship between injection time and effect on acrophase. Of particular interest were the effects of antidepressant drugs on these activity profiles.

Acute Effects of Drugs that Inhibit Monoamine Reuptake, Stimulate Monoamine Release, or Inhibit Monoamine Degradation on the Modulation of Activity Rhythms

Initially, acute injection paradigms were relied on even though they are limited in their comparability to the pharmacological treatment of clinical depression, where days or weeks of drug exposure elapse as beneficial clinical responses emerge. In one study we monitored changes in the activity pattern of 15 to 16 day-old rats acutely separated from their mothers and following single doses of desipramine, zimelidine, or GBR-13069, which are selective inhibitors of the neuronal uptake of norepinephrine (NE), serotonin (5-hydroxytryptamine, 5HT), and dopamine (DA), respectively (Heikkila and Manzino, 1984). Desipramine hydrochloride was administered at 4 and 20 mg/kg; zimelidine dihydrochloride was administered at 5.25 and 26.4 mg/kg (equimolar to the doses of desipramine); and GBR-13069 was given at 5 and 10 mg/kg i.p. At 10 min after drug administration, rats were placed in isolation chambers for 24 hours of monitoring.

Representative results of desipramine treatment are shown in Figure 8.6. The low dose of desipramine (4 mg/kg) dramatically altered the rest-activity pattern, appearing to eliminate the dominant 7–15 cpd ultradian rhythm observed in littermate controls. High doses of this NE-enhancing drug attenuated the dominant ultradian pattern but also introduced a slow, quasi-circadian modulation to the overall activity pattern. Results of aggregate

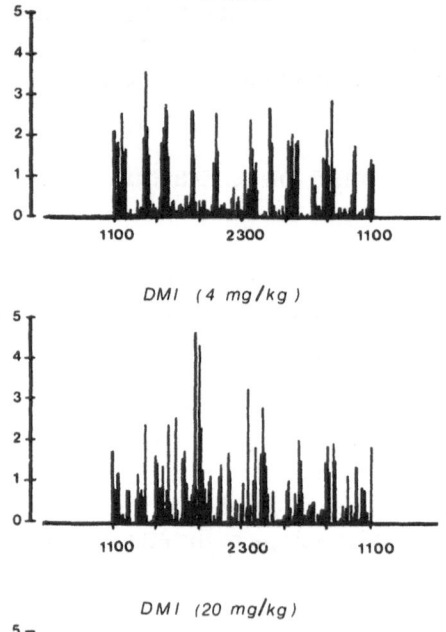

FIGURE 8.6. Activity time-series of isolated 2-week-old rats receiving acute intraperitoneal injections of isotonic saline or desipramine-HCl (4 or 20 mg/kg) at 1050 hours. Data are expressed as the percentage of total activity per 10-minute epoch during the 24-hour test period, which commenced 10 minutes after injection.

analysis of spectral variance from several subjects indicated that control pups had a broad ultradian peak extending from 7 to 15 cpd (31% of total spectal variance) and a moderate 1 cpd (circadian) peak (3.4% of variance) (Figure 8.7). The low dose of desipramine diminished ultradian rhythmicity in the 9-16 cpd band (35% mean reduction, $p < 0.001$) and enhanced the circadian peak by 39% ($p < 0.05$) Barber et al., 1988). At the higher dose (20 mg/kg), the ultradian band was further diminished (mean reduction of 48%, $p < 0.001$) while the circadian peak was increased enormously (180%, $p < 0.001$) (Barber et al., 1988).

In contrast, zimelidine had only small and inconsistent effects on the circadian component of activity, although both doses (5.25 and 26.4 mg/kg) tended to slow the peak frequency of the broad ultradian activity band from 12 cpd to 9-10 cpd (Barber et al., 1988). The effects of GBR-13069 on ultradian activity were complicated by an enormous acute perturbation in baseline activity. The low dose (5 mg/kg) produced an intense 5-8 hour period of continuous activity following drug administration, whereas the higher dose (10 mg/kg) produced a comparable burst of activity commencing about 3-4 hours after injection, a delay which may represent postsynaptic overstimulation. These periods were followed by a sharp depression of overall activity accompanied by the emergence of a slow ultradian rhythm (Barber et al., 1988).

(+)Amphetamine (1 mg/kg, i.p., of the sulfate salt), like GBR-13069, produced a dramatic effect on basal activity in 2-week-old rats and on the underlying ultradian rhythms (see Figure 8.8). During the first 4–6 hours following administration of this stimulant, basal activity was markedly elevated. During the rising phase of this expected stimulatory effect, the ultradian rhythm appeared to accelerate. As the arousal effect waned the ultradian rhythm appeared to slow, followed by a period of marked overall depression of activity that was particularly severe in immature rats (Teicher

FIGURE 8.7. Group variance spectra of 2-week-old rats (N=nine per group) given acute intraperitoneal injections of isotonic saline or desipramine-HCl (4 or 20 mg/kg). Results of the analysis are expressed as percentage of total spectral variance accounted for by rhythms at harmonic frequency bands from 0 to 36 cpd, at 1-cpd intervals.

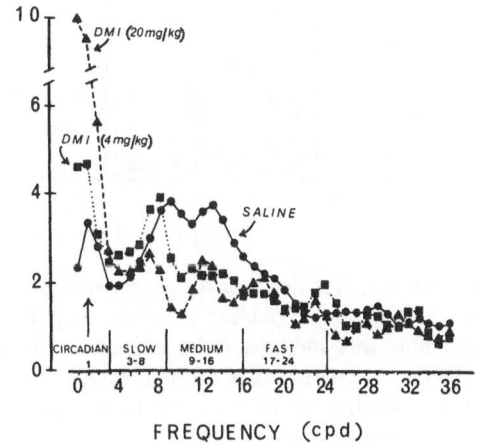

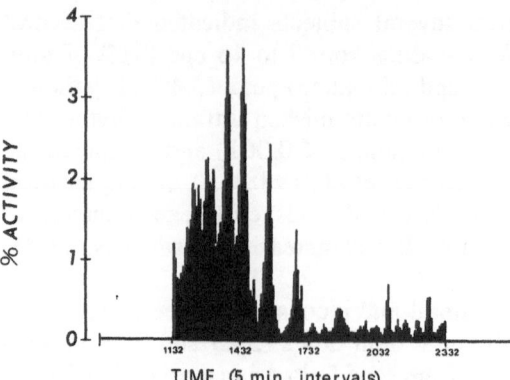

FIGURE 8.8. Activity time-series for a representative 16-day-old male rat given an acute intraperitoneal injection of (+)-amphetamine sulfate (1 mg/kg of the salt) 15 minutes prior to activity monitoring. Data are expressed as percentage of total activity per 5-minute interval during a 12-hour test period. Compare this profile with that of a normal control in Figure 8.5.

et al., 1988). During this depressed period, the evaluation of ultradian rhythms was obscured by a poor signal-to-noise ratio. If an ultradian rhythm was present, it was probably very slow (possibly similar to the 5 cpd rhythm seen initially after stimulant withdrawal in adult rats: see Figure 8.3).

The monoamine oxidase (MAO) inhibitor pargyline hydrochloride (40 mg/kg of the salt, i.p.), administered at 24 and 2 hours prior to testing, also dramatically disrupted the underlying ultradian rhythm of young rats (see Figure 8.9). Both the episodic ultradian bursts of activity and the usual circadian amplitude fluctuation appear to be replaced by a high-frequency diffuse (noisy) rhythm. In contrast, the neuroleptic drug haloperidol (0.01–0.3 mg/kg) had little discernable effect on these rhythms (data not shown).

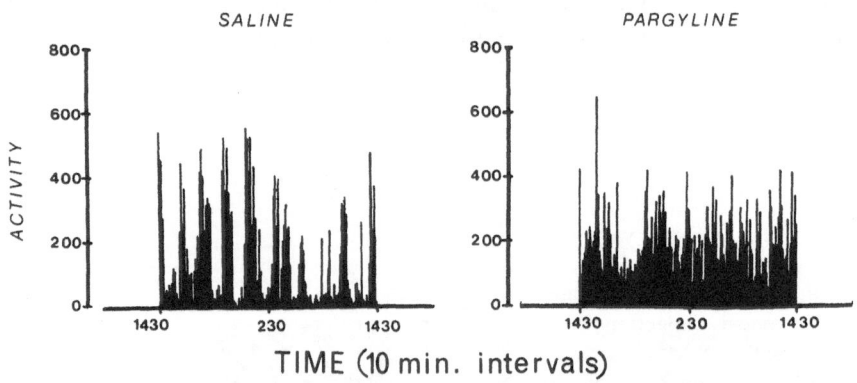

FIGURE 8.9. Activity time-series for littermate pups receiving treatment with saline or the monoamine oxidase inhibitor, pargyline hydrochloride (40 mg/kg). Saline or pargyline was administered intraperitoneally, 24 hours prior to testing and again at 12:30 hours on the day of testing. Data are expressed as activity counts per 10-minute interval during the 24-hour test period, which commenced two hours after the last drug administration.

Acute doses of haloperidol (1.0 mg/kg, i.p.) significantly accentuated the slow ultradian cycle, possibly by virtue of, an artifactual acute effect on baseline activity.

Overall, these data suggest that drugs which acutely enhance certain central monoaminergic functions may prominently alter activity rhythms in the rat. The data also suggest that the acute administration of the selective NE-uptake inhibitor desipramine, or the MAO inhibitor pargyline, can eliminate the expression of normally highly prominent ultradian rhythms of aroused, maternally-separated, preweanling rats.

Effects of Continuous Administration of Monoamine Uptake Inhibitors on Activity Rhythms

The previous studies suggest that some antidepressants and other drugs which acutely enhance central monoaminergic function may attenuate or slow ultradian rhythms in the immature rat pup. However, the data are partially obscured by the presence of acute arousing effects of drug administration. For this reason, continuous administration of agents was employed to help eliminate this potential source of error.

Drugs were administered continuously by Alzet osmotic minipumps, which infuse solutions at a constant, slow rate for up to two weeks. Pumps containing desipramine (0.5 mg/kg/hr), zimelidine (0.66 mg/kg/hr), or another DA-uptake inhibitor GBR-12909 (0.2 mg/kg/hr) were surgically implanted subcutaneously in 12-day-old rat pups anesthetized briefly with ether (Barber et al., 1988). Eight litters were used, with one pup per litter receiving a given drug. Control animals received either no treatment or saline-filled minipumps. Following surgery, rats were returned to their mothers for three days to recover and for the pump output and tissue drug levels to stabilize. At 15 days of age, rats were removed from their mothers, weighed, and placed in dark isolation chambers for 24 hours of activity monitoring.

Both untreated and saline control animals displayed normal, highly regular, ultradian patterns of rest and activity, although the circadian rhythm of the saline-treated pups was greater in amplitude and accounted for slightly more of their total spectral variance, evidently due to an artifactual effect of the pump itself. Thus, group comparisons were made only between saline-treated and drug-treated pups.

The continuous effects of these drugs on ultradian rhythms were remarkably similar to those found following acute administration. The activity profile of desipramine-treated pups (Figure 8.10) displayed a rapid, highly irregular, oscillation between activity and rest with no clear ultradian pattern as seen in the control animals. Variance spectral analysis showed that continuous desipramine infusion markedly diminished the entire 8-16 cpd band of ultradian activity (14% vs. 32% in controls), which is normally prominent at this age in isolated rat pups (Barber et al., 1988). This an-

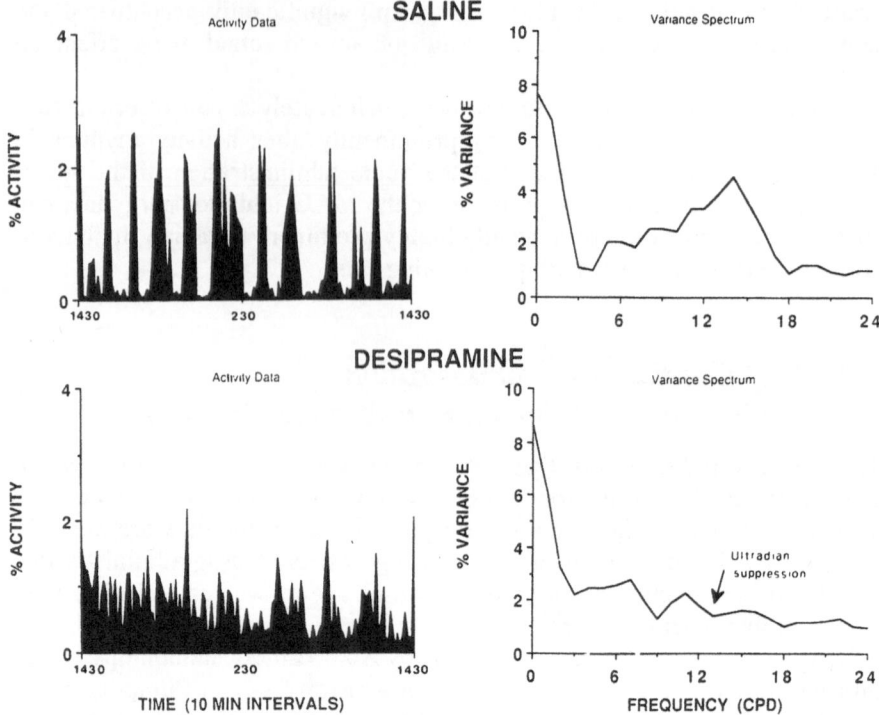

FIGURE 8.10. Activity time series and group variance spectrum for 2-week-old male and female littermate pups (N=eight per group) receiving continuous infusions of isotonic saline or desipramine-HCl (0.5 mg/kg/hour) via Alzet osmotic minipumps implanted subcutaneously 3 days prior to testing. Activity data are expressed as the percentage of total 24-hour activity per 10-minute interval during the 24-hour test period. Spectral data are expressed as the percentage of total variance accounted for by a rhythm with frequency ranging from 0 to 72 cpd at 1-cpd increments (only 0 to 24 cpd shown, as there was little variance accounted for by frequencies at 25 to 72 cpd).

tidepressant also significantly increased the overall circadian mesor, but did not significantly increase circadian amplitude or the percentage of spectral variability accounted for by this rhythm. There was also a trend for all three amine-uptake inhibitors to shorten circadian periodicity by about 2 hours, but this observation is not secure, based on only a 24-hour data collection period, despite the curve fitting powers of iterative non-linear least-squares procedures.

Zimelidine broadened the activity peaks and slowed the overall rhythms (see Figure 8.11). Spectral analysis revealed an enhancement in the very slow (3-4 cpd) ultradian band and suggested some slowing of the dominant ultradian rhythm from 8-15 cpd in saline-controls to about 6-11 cpd in the zimelidine-treated pups (Barber et al., 1988). Neither circadian amplitude

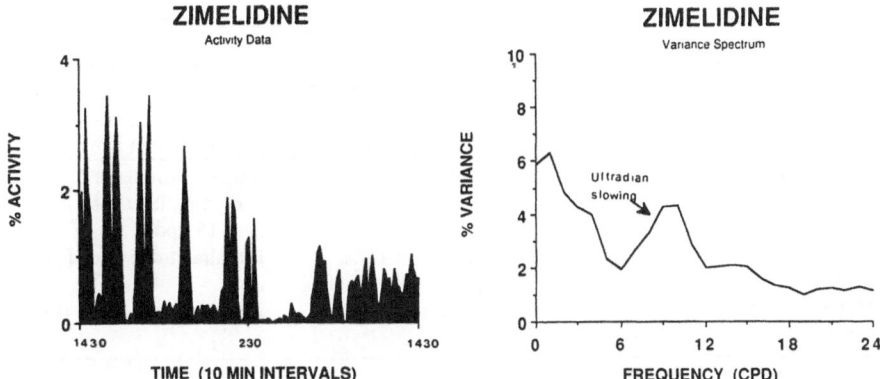

FIGURE 8.11. Activity time series and group variance spectrum for two-week-old male and female pups (N=8) receiving continuous infusions of zimelidine (0.7 mg/kg/hr) by means of an Alzet osmotic minipump implanted subcutaneously three days prior to testing. Representative activity data (left side) are expressed as the percentage of total 24-hour activity per 10-minute interval during the 24-hour test period. Spectral data (right side) are expressed as the percentage of total variance accounted for by a rhythm with frequency ranging from 0 to 72 cpd at 1-cpd increments (only 0 to 24 cpd shown, as there was little variance accounted for by frequencies at 25 to 72 cpd).

nor mean activity levels were affected by this treatment. These findings suggest that serotonin may play a role in the timing of slow ultradian rhythms in young rats.

Continuous infusion of GBR-12909, which is one of the most selective DA reuptake inhibitors available (Heikkila and Manzino 1984), had no discernable effects on ultradian rhythmicity (Barber et al., 1988). However, a relatively low dose of this new compound was used due to its limited solubility; further tests using higher doses are needed.

Discussion

The observations just presented suggest that ultradian activity rhythms can be influenced by conditions arising in several proposed animal models of major depression and shifted in opposite directions by mood-altering drugs, particularly antidepressants, MAO inhibitors, and stimulants that enhance the CNS activity of norepinephrine. Table 8.4 summarizes the effects of three potential animal models of depression on the locomotor activity rhythms of the rat. Both cortical lesions and amphetamine withdrawal acutely produced a prominent, 4-5 cpd slow ultradian rhythm in *adult* rats that is not found in normal animals of that age. More prolonged amphetamine withdrawal and recovery from stroke result in an acceleration of the dominant ultradian

TABLE 8.4. Summary of locomotor activity changes produced by potential rodent models of depression

| | Locomotor Activity Effects | | |
Depression Model	Mean Activity	Circadian Rhythm	Ultradian Rhythm
Amphetamine withdrawal	Decrease	Possible phase-advance	Initially prominent 5 cpd rhythm; later increasing to 15 cpd
Focal cortical injury	Increase	Phase-delay 60%, phase-advance 40%	Prominent 4–5 cpd rhythm
Maternal separation: protest	Increase	Indeterminant	Prominent 8–16 cpd rhythm

rhythm into the medium frequency ultradian range (9-16 cpd, which is a prominent rhythm in normal immature rats) and a slow decline in the magnitude of these ultradian rhythms.

The protest phase of maternal separation in the 15-day-old rat is characterized by the presence of a very strong 7-16 cpd (medium frequency) ultradian activity rhythm. This prominent ultradian rhythm was greatly attenuated by acute or continuous administration of the norepinephrine-reuptake inhibitor desipramine and by acute treatment with amphetamine or pargyline (Table 8.5). All of these treatments share a common locus of action on the central noradrenergic system, although amphetamine and pargyline affect other monoaminergic systems as well. However, more selective reuptake inhibitors of dopamine (GBR-12909) and serotonin (zimelidine) had little effect on this ultradian rhythm, underscoring the probable importance of norepinephrine. This conclusion is consistent with previous studies indicating a role for norepinephrine in such ultradian rhythmic phenomena as REM sleep (Hobson et al., 1975).

Further studies will need to focus on the specificity and sensitivity of drug effects and the effects of prolonged drug administration. Investigators should also examine the generalizability of these ultradian rhythms effects to other animal models and to clinical depression.

These observations are congruent with a model of depression recently proposed by Siever and Davis (1985), in which they postulated that depression may be linked to the stress-induced emergence of a deregulated, or "noisy," neurotransmitter system which generates false signals to biological oscillators and thus disturbs their normal periodicities. They proposed that clinically effective antidepressants may promote and restore normal rhythmicity by enhancing the normal regulation of synaptic neurotransmission. Maternal separation of preweanling rats produces dramatic large-amplitude noncircadian (ultradian) oscillations in rest-activity rhythms unrelated to feeding activity. Desipramine attenuated this effect, and its action is consistent with the model proposed by Siever and Davis (1985).

Ultradian processes are difficult to study and to analyze, but merit further

TABLE 8.5. Locomotor activity effects of antidepressant and stimulant drugs on 15-day-old maternally-separated rats

Drug	Dosage/Route	Mean Activity	Circadian Rhythm	Ultradian Rhythm
Desipramine	4 mg/kg, ip	Increased	Increased percent variance	Decreased 10–16 cpd band
	20 mg/kg, ip	Slight increase	Increased percent variance	Decreased 8–16 cpd band
	0.5 mg/kg/hr (mini-pump)	Increased	No effect	Decreased 8–16 cpd band
Zimelidine	5.25 mg/kg, ip	No effect	No effect	Slowed ultradian peak frequency
	26.4 mg/kg, ip	Decreased	No effect	Slowed ultradian peak frequency
	0.66 mg/kg/hr (mini-pump)	No effect	No effect	Slowed peak, increased 3–4 cpd
GBR-13069	5 mg/kg, ip	Initial activity peak	Indeterminant	Possibly slowed peak frequency
	10 mg/kg, ip	Delayed activity peak	Indeterminant	Possibly slowed peak frequency
GBR-12909	0.2 mg/kg/hr (mini-pump)	No effect	No effect	No significant effect
(+)Amphetamine	1 mg/kg, ip	Initial activity peak	Indeterminant	Decreased 8–16 cpd, slowed peak
Pargyline	40 mg/kg, ip	Increased	Decreased circadian variance	Decreased 8–16 cpd

scrutiny by chronobiologists and research psychiatrists since many physiological and endocrine functions of clinical psychiatric interest, such as REM sleep rhythms and cortisol secretory episodes, are essentially ultradian rhythms with circadian amplitude modulation and possibly frequency modulation. Psychomotor activity falls into the same category and requires fine-grained, *within-subject* analysis that may make these disturbances clearer, rather than obscuring ultradian effects by averaging data across subjects, or summating activity into large blocks of time. In the latter commonly employed approaches, investigators may not be filtering out noise but, rather, eliminating an important signal in clinical studies of chronobiological changes in depressed patients.

The present observations also emphasize the importance of separating psychomotor-retarded and psychomotor-agitated forms of depression clinically as well as in their animal models. These depressive subtypes are often treated clinically with different types of thymoleptic agents (e.g., stimulating versus sedating antidepressants), which are characteristically administered at different times of the day (e.g., upon arising versus at bedtime). It is not reasonable to presuppose that a single, all-purpose animal model of clinical depression exists. Depression is almost certainly a heterogeneous syndrome and may represent pathophysiologically different disorders. It is equally unreasonable to expect any single animal model of depression to respond coherently to every clinically effective antidepressant. Many patients with major depression seem to respond only to some antidepressant drugs. Progress is being made in identifying discrete depressive subtypes and in characterizing their respective biological features and therapeutic response profiles. Animal models of depression may be more useful in facilitating evaluation of *specific subtypes*, and in cataloging the range of their therapeutic responses; these models may be less useful in providing a universal model of depression, or an all encompassing screening strategy for potential treatments. We suggest that the study of motor activity levels and rhythms may provide a particularly valuable and noninvasive window on brain function, with a great degree of human and animal homology, that may help in the delineation of depressive subtypes and in the development of rational, innovative therapeutic strategies.

Acknowledgments. Supported, in part, by National Institute of Mental Health award MH-47370 and grants MH-31154, MH-36224, and MH-43743; by BRSG grant RR-05484 from the National Institute of Health, and from awards by the Marion Ireland Benton Trust Fund and the Milton Foundation. Donation of drug substances by the following manufacturers is greatfully acknowledged: Abbott Laboratories, North Chicago, IL (pargyline hydrochloride); Astra, Sodertalse, Sweden (zimelidine dihydrochloride); Gist-Brocades Laboratories, Haarlem, The Netherlands (GBR-12909 and GBR-13069); McNeil Pharmaceuticals, Spring House, PA (haloperidol); Merck, Sharp and Dohme Co, Philadelphia, PA ([+]amphetamine sulfate); USV Pharmaceuticals, Tuchahoe, NY (desipramine hydrochloride). We thank Seth Finklestein, M.D. for allowing us to use his data.

References

Akiskal HS (1986): The clinical significance of the "soft" bipolar spectrum. *Psychiatric Annals* 16:667-671

Akiskal HS, Downs J, Jordan P, Watson S, Daugherty D, Pruitt DB (1985): Affective disorders in referred children and younger siblings of manic-depressives. *Arch Gen Psychiatry* 42:996-1003

Anisman H, Remington G, Sklar LS (1979): Effects of inescapable shock on subsequent escape performance: catecholaminergic and cholinergic mediation of response initiation and maintenance. *Psychopharmacology* 61:107-124

Baldessarini RJ (1983): *Biomedical Aspects of Depression and Its Treatment.* Washington DC: APA Press

Barber NI, Teicher MH, Baldessarini RJ (1988): Effects of selective monoaminergic reuptake blockade on activity rhythms in developing rats. *Psychopharmacology* (in press)

Blackman RB, Tukey JW (1958): *The Measurement of Power Spectra.* New York: Dover

Campbell BA, Lyttle LD, Fibiger HC (1969): Ontogeny of adrenergic arousal and cholinergic inhibitory mechanisms in the rat. *Science* 166:635-637

Campbell BA, Raskin LA (1978): Ontogeny of behavioral arousal: the role of environmental stimuli. *J Comp Physiol Psychol* 92:176-184

Colburn TR, Smith BM, Guarnini JJ, Simmons NN (1976): An ambulatory activity monitor with solid state memory. *Instrument Soc Amer* 12:117-122

Craincross KD, Cox B, Forster C, Wren AF (1979): Olfactory projection systems, drugs and behaviour: a review. *Psychoneuroendocrinology* 4:253-272

Finklestein SP, Campbell A, Stoll AL, Baldessarini RJ, Stinus L, Paskevitch PA, Domesick VB (1983a): Changes in cortical and subcortical levels of monoamine metabolites following unilateral ventrolateral cortical lesions in rats. *Brain Res* 271:279-288

Finklestein SP, Teicher MH, Campbell A, Baldessarini RJ (1983b): Bilateral ventrolateral cortical lesions slow and accentuate ultradian activity rhythms in rats. *Soc Neurosci Abstr* 9:626

Foster FG, Kupfer DJ (1975): Psychomotor activity as a correlate of depression and sleep in acutely disturbed psychiatric inpatients. *Am J Psychiatry* 132:928-931

Garzon J, Del Rio J (1981): Hyperactivity induced in rats by long-term isolation: Further studies on a new animal model for the detection of antidepressants. *Eur J Pharmacol* 59:293-296

Goodwin FK, Wirz-Justice A, Wehr TA (1982): Evidence that the pathophysiology of depression and the mechanism of antidepressant drugs both involve alterations in circadian rhythms. *Adv Biochem Pharmacol* 32:1-11

Halberg F (1968): Physiological conditions underlying rhythmometry with special reference to emotional illness. In: *Cycles Biologiques et Psychiatrie,* DeAjuriaguerra J, ed. Paris: Masson et Cie

Halberg F, Johnson EA, Nelson W, Runge W, Sothern R (1972): Autorhythmometry procedures for physiological self-measurements and their analysis. *Physiol Teacher* 1:1-11

Halberg F, Panofsky H (1961): Thermovariance spectra: method and clinical illustrations. *Exp Med Surg* 19:284-309

Heikilla RE, Manzino L (1984): Behavioral properties of GBR 12909, GBR 13069, and GBR 13098: specific inhibitors of dopamine uptake. *Eur J Pharmacol* 103:241-248

Higley JD, Suomi SJ, Scanlon JM, McKinney WT (1982): Plasma cortisol as a predictor of individual depressive behavior in rhesus monkeys *(Macaca mulatta)*. *Soc Neurosci Abstr* 8:461

Hobson JA, McCarley RW, Wyzinsky TW: Sleep cycle oscillation: Reciprocal discharge by two brainstem neuronal groups. *Science* 189:55-58

Katz RJ, Roth KA, Carroll BJ (1981): Acute and chronic stress effects on open field activity in the rat: implications for a model of depression. *Neurosci Biobehav Rev* 5:247-251

Killrell EM, Satinoff E (1988): Diurnal rhythms of body temperature, drinking and activity over reproductive cycles. *Physiol Behav* 42:477-484

Kripke DF (1974): Ultradian rhythms in sleep and wakefulness. In *Advances in Sleep Research, Vol. 1*, Weitzmen ED, ed. New York: Spectrum Publication

Kripke DF, Mullaney DJ, Atkinson M, Wolf S (1978): Circadian rhythm disorders in manic-depressives. *Biol Psychiatry* 13:335-351

Kripke DF, Wyborney VG (1980): Lithium slows rat circadian activity rhythms. *Life Sci* 26:1319-1321

Kupfer DJ, Foster FG (1973): Sleep and activity in psychotic depression. *J Nerv Ment Dis* 156:341-348

Kupfer DJ, Weiss BL, Foster FG, Detre TP, Delgado J, McPartland R (1974): Psychomotor activity in affective states. *Arch Gen Psychiatry* 30:765-768

Liebowitz MR, Klein DF (1979): Hysteroid dysphoria. *Psychiatr Clin North Am* 2:555-575

McKinney WT, Bunney WE (1969): Animal model of depression: review of evidence and implications for research. *Arch Gen Psychiatry* 21:240-248

Moore-Ede MC, Czeisler CA, Richardson GS (1983): Circadian time keeping in health and disease. Part 1. Basic properties of circadian pacemakers. *New Engl J Med* 309:469-476

Nelson JC, Charney DS (1981): The symptoms of major depression. *Am J Psychiatry* 138:1-13

Nelson W, Tong YL, Lee JK, Halberg F (1979): Methods for cosinor-rhythmometry. *Chronobiologia* 6:305-323

Paykel ES (1971): Classification of depressed patients: A cluster analysis derived grouping. *Br J Psychiatry* 118:275-288

Pearlson GD, Robinson RG (1981): Suction lesions of the frontal cerebral cortex induce asymmetrical behavioral and catecholaminergic responses. *Brain Res* 218:233-242

Randall PK, Campbell BA (1976): Ontogeny of behavioral arousal in rats: effect of maternal and sibling presence. *J Comp Physiol Psychol* 90:453-459

Rummel J, Lee JK, Halberg F (1974): Combined linear-nonlinear chronobiologic windows by least squares resolve neighboring components in a physiologic rhythm spectrum. In: *Biorhythms and Human Reproduction*, Ferin M, Halberg F, Richart RM, Vande Wiele RL, eds. New York: John Wiley Sons

Siever LJ, Davis KL (1985): Overview: towards a dysregulation hypothesis of depression. *Am J Psychiatry* 142:1017-1031

Simpson DM, Annau Z (1977): Behavioural withdrawal following several psychoactive drugs. *Pharmacol Biochem Behav* 7:59-64

Teicher MH, Barber NI (1988): COSIFIT: Interactive program for simultaneous nonlinear least-squares multioscillator cosinor rhythm analysis of biological time-series data. (Submitted for publication)

Teicher MH, Barber NI, Baldessarini RJ, Shaywitz BA (1988a): Amphetamine accelerates and attenuates ultradian activity rhythms in preweanling rats. *Pharmacol Biochem Behav* 29:517-523

Teicher MH, Flaum LE (1979): The ontogeny of ultradian and nocturnal activity rhythms in the isolated albino rat. *Devel Psychobiol* 12:441-454

Teicher MH, Green WT (1977): A digital readout vibrational activity monitor for neonatal animals. *Physiol Behav* 18:747-750

Teicher MH, Lawrence JM, Barber NI, Finklestein SP, Lieberman H, Baldessarini RJ (1986): Altered locomotor activity in neuropsychiatric patients. *Prog Neuropharm Biol Psychiatry* 10:755-761

Teicher MH, Lawrence JM, Barber NI, Finklestein SP, Lieberman H, Baldessarini RJ (1988a): Circadian activity rhythm disruptions in geriatric major depression. *Arch Gen Psychiatry* 45:913-918

Teicher MH, Sussman AJ, Baldessarini RJ, Lieberman H (1988b): Motility rhythms in psychotic inpatients. *Am Psychiatric Assoc Abstr* 141:104

Watson R, Hartman E, Shildkraut JJ (1972): Amphetamine withdrawal: Affective state, sleep patterns, and MHPG excretion. *Am J Psychiatry* 129:39-45

Wehr TA, Muscettola G, Goodwin FK (1980): Urinary 3-methoxy-4-hydroxy-phenyl-glycol circadian rhythms: early timing (phase-advance) in manicdepressives compared with normal subjects. *Arch Gen Psychiatry* 37:257-263

Weiss JM (1968): Effects of coping responses on stress. *J Comp Physiol Psychol* 65:251-260

Willner P (1984): The validity of animal models of depression. *Psychopharmacology* 83:1-16

Wolff EA III, Putnam FW, Post RM (1985): Motor activity in affective illness: the relationship of amplitude and temporal distribution to changes in affective state. *Arch Gen Psychiatry* 42:288-294

9

Anhedonia as an Animal Model of Depression

GEORGE F. KOOB

Introduction

The clinical syndrome of depression is characterized by significant and long-term changes in mood and cognitive function. One of the most salient symptoms of a major depressive episode is a failure to obtain pleasure from activities that previously brought enjoyment (Klein, 1974). Recreational activities, sex, and eating now provide little or no gratification. The mood of the individual is characterized as depressed, sad, hopeless, and discouraged.

Indeed, this dysphoric mood and loss of interest or pleasure in all or most usual activities are considered essential features of a major depressive episode, providing it is persistent. The *Diagnostic and Statistical Manual of Mental Disorders (DSM-III)* of the American Psychiatric Association (1980) lists the following criteria for a major depressive episode:

Dysphoric mood or loss of interest or pleasure in all or most all usual activities and pasttimes. The dysphoric mood is characterized by symptoms such as the following: depressed, sad, blue, hopeless, low, down in the dumps, irritable. The mood disturbance must be prominent and relatively persistent, but not necessarily the most dominant symptom, and does not include momentary shifts from one dysphoric mood to another dysphoric mood, e.g., anxiety to depression to anger, such as are seen in states of acute psychotic turmoil. (For children under six, dysphoric mood may have to be inferred from a persistently sad facial expression.)

The subclassification criteria of a major depressive episode "with melancholia" places even more emphasis on this mood disturbance:

1. Loss of pleasure in all or almost all activities.
2. Lack of reactivity to usually pleasurable stimuli (doesn't feel much better, even temporarily, when something good happens).

Thus, *anhedonia*, which means "without *(an)* pleasure *(hedonia)*" and can be defined as "the diminished capacity to experience pleasure of any sort" (Fawcett et al., 1983) forms a cardinal symptom of a major depressive episode in man.

The purpose of this chapter will be to discuss how anhedonia may be used as a possible animal model of depression. First, anhedonia will be defined operationally. Second, theoretical and practical means to assess anhedonia and its neuropharmacological basis in animals will be explored. Finally, pharmacological experiments producing "anhedonia" measures will be described that may be isomorphic or even homologous with certain human clinical conditions. These studies give some promise to the possibility that behavioral measures of anhedonia may be possibly useful animal models for some aspects of affective disorders.

Measures of Anhedonia (or Measures of Reward)

Theory

Motivation can be simply defined as the "tendency of the whole animal to produce organized activity" (Hebb, 1972). Empirically, the term motivation can be reduced to the conditions under which a reinforcer is effective in initiating a response. For example, a rat will not perform an organized trained (operant) response to obtain food unless it is food deprived. Under these conditions, a *reinforcer* is operationally defined as "any event that increases the probability of a response." The term reward has been used in much the same way as reinforcer but usually connotes some additional emotional valence such as pleasure and "there does not seem to be any overwhelmingly compelling reason to choose one word over the other" (Kling and Schrier, 1971). For the purpose of this chapter both terms will be used interchangeably.

Another concept, incentive-motivation, has been used to explain how reinforcers are effective in the absence of energy deficits. Here, the reinforcers carry with them the motivation (or the conditions under which a reinforcer is effective). For example, nonhungry animals will perform tasks to obtain sweet solutions such as saccharin, and animals will perform a response to stimulate certain areas of their brains, that is, intracranial self-stimulation. However, regardless of the means by which an event acquires reinforcing properties, a major behavioral prediction is that increasing the size of the reinforcer increases its motivational or energizing properties, and thus the response that results.

A precise formulation of this concept of "quantitative hedonism" was originally developed from studies where choice was used to quantify the relative reward value of different conditions of reinforcement (Herrnstein, 1970). In these studies, an animal was tested in a continuous choice procedure in which two or more alternative schedules of reinforcement were simultaneously available on two levers. The animal could continually choose either alternative and the number of responses that the animal directed to each alternative was considered a measure of preference. The amount of behavior generated on a particular lever was a function of the frequency of reinforcement on that lever (Herrnstein, 1970).

Thus, the *relative* frequency of a' response on a given lever "matched" the relative frequency of reinforcement on that lever. With this "matching law," Herrnstein and associates have explained the relative allocation of behavior to the possible alternatives, even in single-response situations. Herrnstein (1971) suggested that absolute response rates in single-response situations (no choice procedures—one lever, one schedule) should vary in proportion to the density of reinforcement. He reasoned that every situation is basically a choice between alternatives of reinforcement. Even in a simple single-response situation, several alternative sources of reinforcement are available other than the contingencies arranged by the experimenter, e.g., in the rat sniffing, grooming, and exploration. Thus, the absolute response rate on single and concurrent schedules can be considered a function of the frequency of reinforcement for that response relative to all other sources of reinforcement for competing responses (De Villiers, 1977, Williams, 1988).

Intracranial Self-Stimulation (ICSS)

The discovery that electrical stimulation of particular areas of the brain could reinforce behavior provided a unique means to assess the motivation to respond to hedonic stimuli (Olds and Milner, 1954), and has been hypothesized to represent a "short-circuiting" of the reinforcement process (Gallistel, 1983).

ICSS is typically obtained from most regions of the limbic system, and can generate very high rates of responding (over 100 lever presses per minute at some sites). Although electrodes placed in a midbrain-forebrain system which courses through the lateral hypothalamus, produce the highest rates of responding, ICSS has also been produced in regions as far removed from the classical limbic system as the cerebellum and nucleus solitarius. In a typical ICSS preparation, a bipolar or monopolar electrode is lowered into the brain and permanently anchored to the skull using dental cement. Rats are then trained to perform an operant task to deliver a pulse train of current with each response. Typically, the operant is a lever press and the electrical current is 0.2 to 0.5 seconds pulse train of 60 Hz AC in the 15 to 75 μA range, RMS or a 0.2- to 0.5-seconds train of biphasic rectangular pulses of 0.1-mseconds interpulse interval between the positive and negative pulses at a frequency of 100 pulses/sec. For this biphasic stimulation, current usually ranges between 0.10 and 1.00 mA, peak to peak.

The potency of ICSS as a reinforcer led to its rapid use as a tool to measure activity in the brain "reward" systems. Increases in behavior to obtain ICSS were assumed to reflect increases in the activity of brain reinforcement circuits. This followed from the above principle, that the actual value of the reinforccer is directly related to the output of the organism to obtain the reinforcer.

Thus, the relative rate of responding in a situation is taken to indicate the degree (relative amount) of reward or hedonism (pleasure/pain) involved. While this is an arbitrary assumption, it is actually a small step from discussing response strength in terms of the relative value of it to all other sources of reinforcement.

Rate Measures of ICSS

One of the first measures of brain stimulation reward was that of *absolute* rate of responding. Here an experiment consisted of placing an animal in the test situation for a specified period of time and at a specified current that was definitely suprathreshold. Testing continued until response rates had stabilized from session to session. Any treatment that increased the rate of responding was thought to reflect a facilitation of reward or a lowering of threshold for reward, and any treatment that decreased responding was thought to reflect an inhibition of reward or a raising of threshold for reward. Unfortunately, measures of absolute response rates were subject to many problems of measurement and interpretation. Treatments in these situations were confounded by ceiling or floor effects, rate dependency, and nonspecific performance effects.

Rate Intensity/Frequency Measures of ICSS

Another means of evaluating threshold for brain stimulation or by extrapolation measuring "hedonia" is to establish within each test session the relationship between the intensity or frequency of the electrical stimulus and the response rate. Not unlike psychophysical tests, rats are systematically subjected to an ascending, descending, or random series of rate-intensity or rate-frequency functions. What typically results is a sigmoidal function whose slope increases the closer one approaches the low threshold areas of midbrain and lateral hypothalamus, as can be seen in Figure 9.1. Shifts of this function to the right or left without changes in the maximum rate of responding presumably reflect changes in threshold for reward that are less compromised by problems of performance variables while changes in the maximal response rate are assumed to reflect a nonspecific performance effect of the manipulation.

Comparison to Other Reinforcers

One of the more peculiar properties of ICSS as a reinforcer, determined in early work, was its rapid extinction and thus poor performance on schedules requiring intermittent reinforcement. Rats would not maintain responding beyond a variable interval of 16 seconds (VI-16sec) or a fixed ratio=7 (FR-7), (Sidman et al., 1955), limits well below those reached by food-deprived rats. However, this problem with schedules was easily resolved when steps

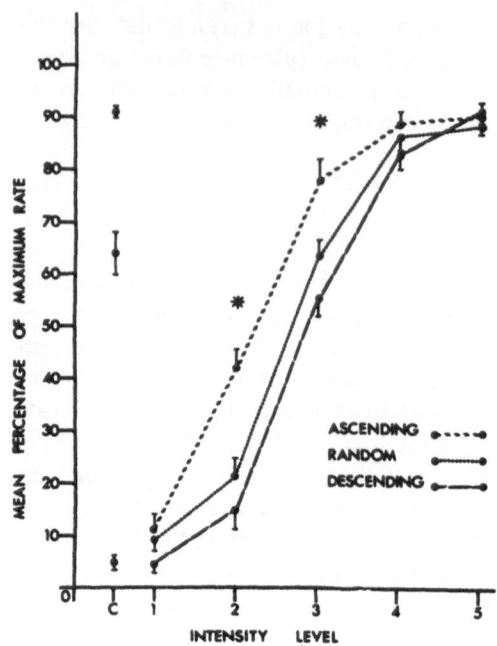

FIGURE 9.1. Intracranial stimulation rates for a series of stimulus intensity changes expressed as a percentage of maximal rate. Data are pooled for three replications and represent mean percent ± SEM for five rats. C refers to the rates observed during the constant intensity control sessions averaged across trials.* Significantly different from random or descending series at this intensity level. Figure reconstructed from Koob (1977).

were taken to lengthen the effective time of reinforcement. For example, Pliskoff, et al (1965) established schedules of reinforcement with ICSS equivalent to those obtained with food reward by chaining (Hawkins and Pliskoff, 1964). In these experiments, a permanent lever programmed on intermittent schedules produced a retractable lever where ICSS was available on a continuous reinforcement schedule. Similar effects have been observed using a nose poke as the terminal link (Robbins and Koob, 1978).

Rate-Independent Measures of Reward

More recently, Kornetsky and associates have exploited the high-incentive, low-drive properties of ICSS to develop a more rate-independent measure of reward (Kornetsky and Esposito, 1981). Rate measures per se can lead to misleading results when attempting to measure the relative value of ICSS. For example, in an early study Valenstein showed that rats allowed a choice between brain stimulation in two different sites demonstrated a preference for the site that maintained the lower rate of responding (Valenstein, 1964). Thus, threshold measures for a given reinforcer may be a more valid measure of the motivation to respond for a reinforcer and, by extrapolation, a more valid measure of a state of "hedonia."

In the Kornetsky threshold procedure, trials begin with the delivery of a noncontingent 0.5-second stimulus. If the animal responds by turning a wheel manipulandum one-quarter turn within 7.5 seconds of the onset of the "priming stimulus," a second stimulus is delivered that is identical to the first noncontingent stimulus. The current intensity of the stimulus is varied according to a psychophysical procedure where stimuli are presented in an alternating descending and ascending series, with five trails presented at each intensity level. The threshold value for each series is defined as the midpoint in current between the level at which the animal made three or more correct responses out of five stimulus presentations and the level where fewer than three correct responses were made (Marcus and Kornetsky, 1974). This "rate-free" measure of reward has proved sensitive to pharmacological manipulations with drugs that are self-administered by humans which is consistent with the hypothesis that it reliably measures "hedonic state."

Neuroanatomy of Reward

Other important variables needed to understand the measurement of brain stimulation reward include differences caused by the location of the intracerebral electrode and qualitative differences in the type of response required. For example, tail movement and running in a runway for ICSS had higher thresholds than lever pressing (White, 1976). Large differences in threshold can be obtained from different electrode placements, as well as large differences in the maximum rate of responding. As mentioned above, the highest rates for the lowest amount of current can be obtained in a midbrain-hypothalamic system that forms a U in the horizontal plane with the closed end of the U forming around the ventral tegmental area, and the legs of the U forming the medial forebrain bundle. Rate-intensity functions for electrodes located outside this system typically have lower slopes (Figure 9.2). Whether these differences in electrode location are more a reflection of a quantitative or qualitative difference in reward remains to be determined.

Pharmacological Manipulation of Reward

Catecholamine Hypothesis of Reward

Early work showed that ICSS was most easily obtained from a region in the midbrain-hypothalamus traversed by the medial forebrain bundle, and Stein suggested that the important neurochemical substrate supporting this behavior was norepinephrine (Stein, 1968). This work later led to a more general hypothesis that the catecholamines norepinephrine and dopamine have a critical role in reward (German and Bowden, 1974). Two major developments led to this hypothesis. First, the sites in the brain that supported ICSS corresponded well with the sites in the brain that contained either norepinephrine or dopamine. For example, ICSS can be obtained from the region surrounding the locus coeruleus and its dorsal projection to the forebrain (Crow et al., 1972), and ICSS can also be obtained from cell bodies of origin and terminal regions of the two major dopamine projections, the mesolimbic

168   George F. Koob

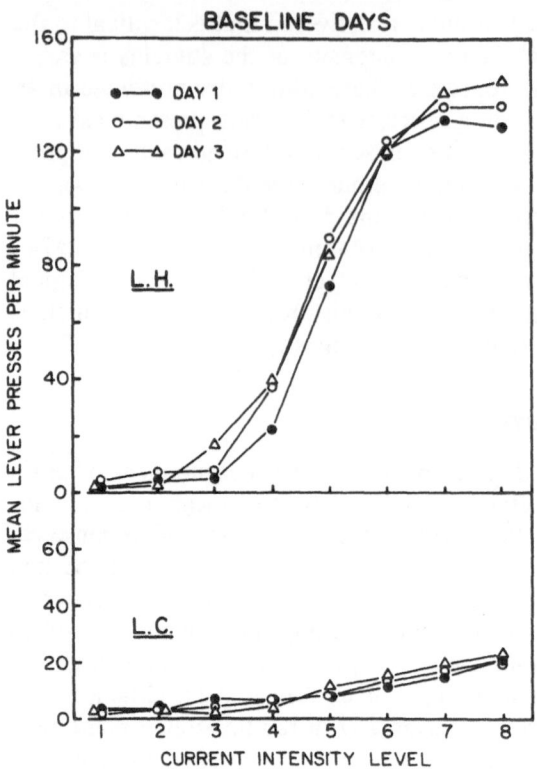

FIGURE 9.2. Rate versus current intensity functions for intracranial self-stimulation for eight rats with electrodes in the lateral hypothalamus (LH) and seven rats with electrodes aimed at the locus coeruleus (LC). Each curve represents one session with a descending sequence. Currents ranged from 5 to 40 μA (60 Hz AC RMS) for the LH and 15 to 50 μA for the LC.

and nigrostriatal systems (Phillips et al., 1975; 1976). However, both ICSS and catecholamines can be found in most regions of the brain leaving some doubt as to the importance of this correlation.

Second, drugs that influence catecholamine metabolism have pronounced effects on ICSS. In general, those drugs that increase the availability of catecholamines at the synapse facilitate ICSS, and those agents that deplete the brain of catecholamines or block catecholamine transmission inhibit ICSS. This facilitation of reward by various pharmacological agents that increase catecholamine neurotransmission is evident in studies using ICSS as the reinforcer (Cassens and Mills, 1973: Goodall and Carey, 1975; Koob et al., 1977; Phillips and Fibiger, 1973; Stein, 1964; Stein and Ray, 1960; Wauquier and Niemegeers, 1974). Studies have shown that amphetamine increases lever-pressing rates for ICSS and decreases ICSS thresholds in self-titration procedures. In these procedures, the current delivered decreases systematically with each lever press but the animal can reset the current back to its original level by pressing another lever (Stein and Ray, 1960; Zarevics and Setler, 1979). Functions relating rate of responding to current level, that is, rate/intensity (R/I) functions, shift to the left with administration of amphetamine (Steiner and Stokely, 1973; Phillips and LePiane, 1986), and can be seen in figure 9.3. Furthermore, amphetamine also shifts to the left rate/frequency functions (Gallistel and Karras, 1984). Other tests have been developed to minimize the contribution of nonspecific activating effects of amphetamine on the facilitation of ICSS. Amphetamine also enhances reward in these "rate-free" measures of ICSS threshold (Liebman and Butcher, 1974; Kornetsky and Esposito, 1979). Other psychomotor stimulants such as cocaine, pipradol and methylphenidate also facilitate ICSS (Crow, 1970; Sahakian and Koob, 1978; Tyce, 1968).

In contrast, drugs that interfere with central catecholamine mechanisms such as reserpine (Olds et al., 1956) and alpha methyl-p-tyrosine (Cooper et al., 1971) decrease ICSS (Table 9.1). These effects were originally interpreted as evidence for a role for norepinephrine in ICSS, but it must be noted that dopaminergic mechanisms can be equally affected by these treatments.

Norepinephrine vs. Dopamine

The strongest pharmacological evidence in support of the original norepinephrine hypothesis was the observation that the dopamine-β-hydroxylase (DBH) inhibitors, disulfiram and diethyldithiocarbamate (DDS) decrease

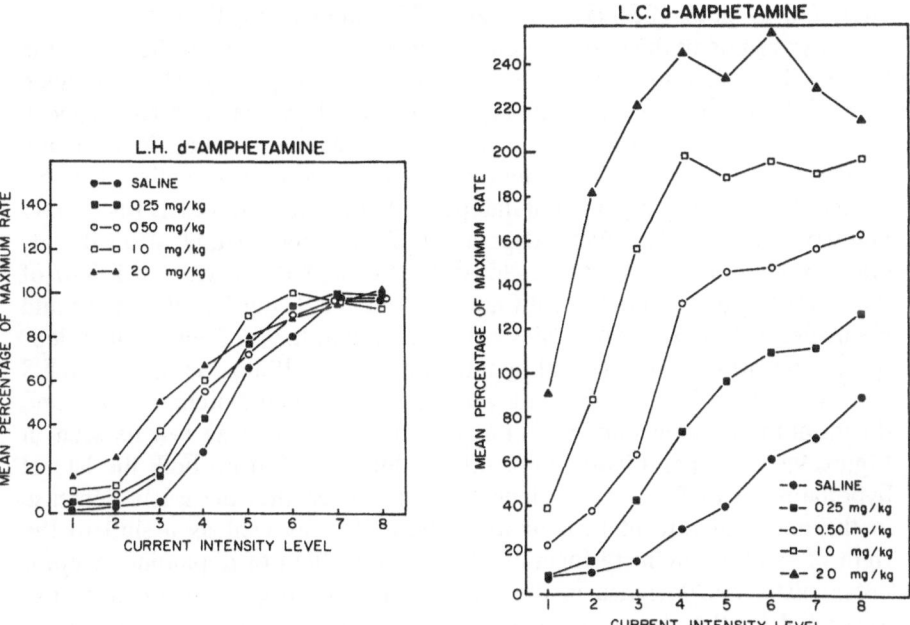

FIGURE 9.3. *Left:* Effects of d-amphetamine on intracranial self-stimulation in eight rats with LH electrodes using descending rate/intensity functions. Data are expressed as the mean percentage of maximal rate for all the rats. Maximal rate was determined by averaging all the predrug maximal rates. *Right:* The same as *left,* except it represents seven rats with electrodes aimed at the locus coeruleus.

TABLE 9.1. Summary of effects of drugs that interact with catecholamines on intracranial self-stimulation

Catecholamine synthesis inhibition	
Alpha methyl-p-tyrosine	Inhibits responding
Depletion of amine stores	
Reserpine	Inhibits responding
Tetrabenzazine	Inhibits responding
Enhanced amine release	
Amphetamines	Facilitates responding
Cocaine	Facilitates responding
Denervation of central catecholamine-containing neurons	
6-hydroxydopamine	Decreases response rates

ICSS and this effect could be reversed by intraventricular injection of norepinephrine (Wise and Stein, 1969). However, others have shown that these particular DBH inhibitors have sedative effects and that norepinephrine simply reversed these sedative effects (Roll, 1970; Rolls et al., 1974). Supporting this hypothesis are studies showing that different DBH inhibitors such as FLA-63 and U-14,624 without the sedative side effects of disulfiram and DDC fail to block ICSS (Lippa et al. 1973; Stinus et al., 1976).

Perhaps some of the strongest pharmacological evidence for the dopamine hypothesis is that the more specific dopamine receptor blockers produce an attenuation of ICSS at relatively low doses (Wauquier and Niemegeers, 1972). Both the phenothiazines and butyrophenones blocked ICSS in the posterior hypothalamus with chlorpromazine having an effective dose 50 percent (ED50) of 2 mg/kg, but haloperidol and spiroperidol ED50s of 0.05 and 0.01 mg/kg, respectively (Dresse, 1966). In another study, also with lateral hypothalamic electrodes, haloperidol blocked ICSS with an ED50 of 0.055 mg/kg and pimozide with an ED50 of 0.220 mg/kg (Wauquier and Niemegeers, 1972). Other studies have shown that these effects of neuroleptics are not restricted to lateral hypothalamic ICSS. Pimozide blocked ICSS from electrodes in the substantia nigra (Liebman and Butcher, 1974) and dorsal noradrenergic bundle (Phillips et al., 1976), and as can be seen in Figure 9.4, α-flupenthixol significantly reduces ICSS from both the lateral hypothalamus and locus coeruleus. Note, however, that there is a decrease in the maximal rate of responding at both sites as well as a shift to the right of the rate-intensity function. This motor effect of dopamine receptor antagonists (the decrease in the maximal rate of responding) can be exaggerated by the choice of operant (Ettenberg et al., 1981). Others have seen shifts to the right of ICSS rate-intensity and rate-frequency functions with pimozide at low doses (Phillips and LePiane, 1986; Gallistel and Karras, 1984), but most reports also show simultaneous decreases in the maximal rate of responding at higher doses (Ettenberg et al., 1981; Phillips and LePiane, 1986; Stellar et al., 1983). Although still controversial, a compromise interpretation of these effects is that dopamine receptor antagonists produce both motor and reward effects (Hamilton et al., 1985; Phillips and LePiane, 1986; Stellar et al., 1983).

In addition, dopamine receptor antagonists raise the threshold for ICSS in rate-independent measures. With the auto-titration procedure where the subject had access to two levers, both haloperidol (Schaefer and Holtzman, 1979) and pimozide (Zarevics and Setler, 1979) increased the average current at which the reset occurred. Consistent with a major dopamine role as opposed to norepinephrine role in ICSS are the findings that imipramine-like compounds (norepinephrine reuptake inhibitors) fail to alter ICSS except at very high doses that produce nonspecific effects (Stark et al., 1969), although a more recent study with *chronic* desmethylimipramine showed a facilitation of ICSS (Fibiger and Phillips, 1981). Based on pharmacological evidence alone, the catecholamine systems, particularly the dopamine systems appear to play an integral role in mediating ICSS. Nevertheless, the functional significance of the different dopamine systems in ICSS is still under study.

Other Transmitters and Reward

Work with pharmacological agents that act on other neurotransmitter systems has implicated serotonergic, opioid, and cholinergic systems as pos-

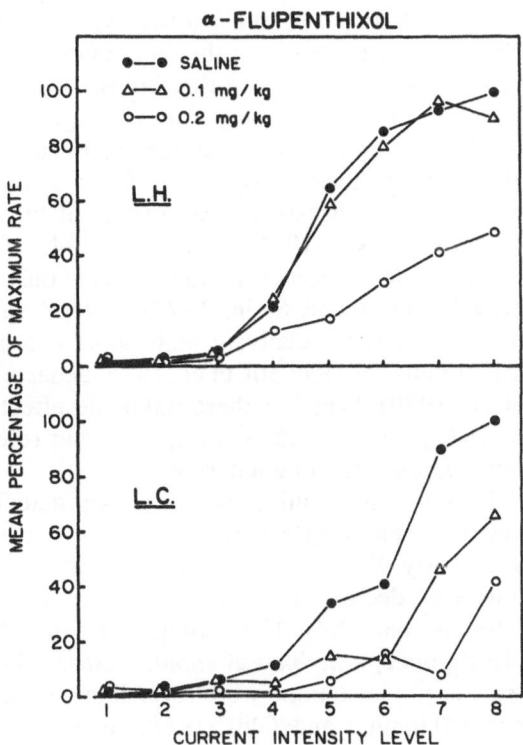

FIGURE 9.4. Effects of α-flupenthixol, a dopamine receptor blocker, on intracranial self-stimulation in the eight LH and seven LC rats. Data are expressed as in Figure 9.3.

sibly being involved in mediating ICSS. Electrodes aimed at the midbrain raphe will support ICSS (Miliaressis et al., 1975; Simon et al., 1976; Van der Kooy et al., 1978), and such behavior is attenuated by treatment with para-chorophenylalanine (PCPA), a tryptophan hydroxylase inhibitor, at least in some studies (Miliaressis et al., 1975; Van der Kooy et al., 1978, but see Simon et al., 1976 for opposite results). Indeed, acutely (24-96 hours postinjection) PCPA will inhibit ICSS from a variety of neuroanatomical sites (Gibson et al., 1970; Stark et al., 1970). Thus, serotonergic systems may support ICSS, but much more work needs to be done to establish a necessary role for serotonergic systems in ICSS.

Early studies examining the effects of opiates showed that morphine inhibited ICSS (Olds and Travis, 1960). Tests of longer duration with repeated doses of morphine, however, reveal a facilitative effect on ICSS that followed the initial suppressant effect (Adams et al., 1972; Lorens and Mitchell, 1973). This has also been observed with heroin (Koob et al., 1975), and there is at least one report of low doses of morphine producing a facilitation of ICSS without initial suppression (Glick and Rapaport, 1974). Results with "rate-free" measures of reward show an increase in "on" times with morphine or etorphine in a shuttlebox where subjects were trained to turn ICSS on and off by crossing back and forth (Baltzer et al., 1977; Levitt et al., 1978). Consistent with the view that morphine facilitates reward are studies using rate-intensity procedures (Maroli et al., 1978) and other threshold-measure procedures (Esposito and Kornetsky, 1977; Marcus and Kornetsky, 1974). These results suggest that the facilitation of ICSS by opiates results from an actual lowering of reward thresholds.

Several studies have shown that changes induced by morphine in ICSS can be reversed by naloxone (Holtzman, 1976; Koob et al., 1975; Wauquier et al., 1974). Although early studies failed to show any alteration in ICSS with naloxone or naltrexone alone except at high doses (Van der Kooy et al., 1977; Holtzman, 1976; Wauquier et al., 1974), and although one study showed that naloxone can block ICSS at only certain loci such as the central gray (Belluzzi and Stein, 1977), several subsequent studies have shown that naloxone at relatively high doses can attenuate lateral hypothalamic self-stimulation (Ichitani, et al., 1985; Schaefer and Michael, 1981; Stapleton et al., 1979). Whether these naloxone effects can be related to an action or endogenous opioid systems, and what role these opioid systems have in reward, remains to be determined.

Finally, a substantial amount of pharmacological evidence exists to suggest that central cholinergic mechanisms may have a role in ICSS (Table 9.2). Early studies established that ICSS from electrodes in the hypothalamus was decreased by physostigmine, a central cholinesterase inhibitor (Domino and Olds, 1968; Jung and Boyd, 1966; Stark and Boyd, 1963). Neostigmine, a peripheral cholinesterase inhibitor, had no effect. These effects of physostigmine were reversible with atropine, a muscarinic blocking agent (Domino and Olds, 1968; Jung and Boyd, 1966; Stark and Boyd,

TABLE 9.2. Effects of cholinergic drugs on ICSS

Physostigmine	Central anticholinesterase	ICSS inhibition
Physostigmine + atropine	Muscarinic blocker	Inhibition reversed
Neostigmine	Peripheral anticholinesterase	No change
Nicotine	Nicotinic agonist	Facilitation
Nicotine + mecamylamine	Nicotinic blocker	Facilitation reversed
Scopolamine	Muscarinic blocker	Facilitation
Pilocarpine	Muscarinic agonist	Inhibition

1963). However, nicotine, a nicotonic agonist, also facilitated ICSS, and this was blocked by mecamylamine, a nicotinic blocking agent (Newman, 1972; Pradhan and Bowling, 1971). Consistent with the concept of a muscarinic inhibition of ICSS and a nicotinic facilitation of ICSS are the results by Newman (1972) showing that scopolamine, a muscarinic blocker, facilitated ICSS. Pilocarpine, however, a muscarinic agonist, blocked ICSS (Newman, 1972). Whether these results will be as robust with threshold measures of reward remains to be determined. Also, the specific neural systems that may be involved remain unknown.

Animal Models of Depression and Reward

Chronic Amphetamine

Chronic administration of large doses of amphetamine is accompanied by a variety of psychopathological effects, not the least of which are a hypomania, a sense of power, hyperactivity, and in some cases, psychoses. Withdrawal from psychomotor stimulants is accompanied by depression that may be of clinical significance (Watson et al., 1972; Connell, 1958). This depression appears to be highly variable in its severity from individual to individual (Angrist, 1985). These drug-induced depressions are of interest not only because they may be of etiologic significance in stimulant abuse, but also because they may provide insight into the pathophysiology of nondrug-induced depression.

The mood-altering effects of psychostimulant drugs have long been associated with a facilitation of intracranial self-stimulation behavior (Stein, 1964) and a decrease in thresholds for ICSS reward (Stein and Ray, 1960). Thus, it follows that the mood-altering effects of withdrawal from indirect sympathomimetics might similarly reflect a decrease in ICSS behavior or an increase in reward thresholds. Studies using chronic administration of amphetamine to rats have supported this hypothesis.

Chronic administration of d-amphetamine over a four-day period (3 times per day, 1.0 to 12.0 mg/kg in increasing doses) produced a significant decrease in self-stimulation during a test session 48 hours after the last injection of amphetamine (Leith and Barrett, 1976). In this study, the animals were capable of responding during the two days of withdrawal since the high current levels generated high rates 'of responding within the rate-intensity functions. Similar results were observed by Simpson and Annau (1977) using a 24-hour continuous access to ICSS paradigm. In this study d-amphetamine was administered chronically 3 times per day at three doses (2.5, 5.0, and 10 mg/kg) for five days. Chronic d-amphetamine produced a dose-dependent decrease in responding for four days post-treatment.

Although cessation of chronic amphetamine administration has been reported to lead to a nonspecific reduction of behavior in rats (Herman et al., 1971), others have shown that measures of brain stimulation reward may be a more sensitive indication of the depression associated with amphetamine withdrawal. In a study with mice, withdrawal following chronic amphetamine (2 times daily, 10 mg/kg total per day, for 10 days) produced dramatic decreases in ICSS and increased immobility in a forced swim test (Kokkinidis, et al., 1986). However, this chronic treatment had little effect on shuttle box escape performance, the acoustic startle response, or locomotor activity. Similarly as noted above, marked increases in threshold for ICSS using a "rate-free" descending rate-intensity measure were observed 24 to 48 hours after the last injection of amphetamine (3 times per day for 4 days starting at 1 mg/kg and increasing 1 mg/kg with each injection, reaching 12 mg/kg on the last injection) (Leith and Barrett, 1980, 1981). In another elegant study using a two-lever, concurrent operant threshold procedure where the dependent variable was an increase in postreinforcement pauses, Cassens and colleagues showed that chronic administration of d-amphetamine (escalating doses of 1 mg/kg 3 times per day for 4 days up to 12 mg/kg) produced increases in reward threshold from 24 to 96 hours post-treatment (Cassens et al., 1981). These changes in ICSS behavior can be of long duration. d-Amphetamine administered for 14 days (5 mg/kg for 7 days and 10 mg/kg for 7 days) showed a significant elevation of reward threshold with no recovery for 18 subsequent days (Leith and Barrett, 1980).

Further support for this poststimulant reward change as an animal model of depression is the observation that chronic treatment with the tricyclic antidepressants imipramine and amitryptyline (both 10 mg/kg two days prior to each test session) attenuated the postamphetamine depression of ICSS rates (Kokkinidis et al., 1980). Note, however, that others have observed increases in ICSS in nonamphetamine-treated rats produced by chronic tricyclic antidepressant treatment (Fibiger and Phillips, 1981).

Cocaine and Reward

Cocaine has stimulant effects similar to those of amphetamine. Cocaine administration increases motor behavior in rats (Kelly and Iversen, 1976). Cocaine also facilitates ICSS (Crow, 1970) and reduces reward thresholds (Kornetsky and Esposito, 1981). Cocaine is readily self-administered by rats and man, and the postdrug withdrawal from cocaine also consists of symptoms similar to endogenous depression, including "anhedonia" (Blum, 1976; Wesson and Smith, 1977). However, to date little is known about the state of brain reward systems following a single bout of cocaine self-administration or what is the minimal amount of cocaine that will elicit a post-drug "anhedonic response."

To answer these questions, preliminary studies have been initiated in the author's laboratory using an animal model of drug self-administration. Rats were implanted both with intracranial self-stimulating electrodes (aimed at the posterior lateral hypothalamus) and intravenous jugular catheters. ICSS thresholds were determined using a discrete-trials threshold procedure, modified from that of Kornetsky and Esposito (1979).

Thresholds were determined in rats trained to turn a wheel manipulandum to obtain electrical stimulation (AC current). At the start of each trial, rats were presented with a noncontingent electrical stimulus of a 0.5-second duration. Subjects were given 7.5 seconds to turn the wheel manipulandum one-quarter of a rotation in order to obtain a contingent stimulus identical to the previously delivered noncontingent stimulus, after which the trial was terminated. If responding did not occur within the 7.5 seconds, the trial was also terminated. Trials were presented on the average of every 5 seconds with intertrial interval responding resulting in a 1-second delay before the start of the next trial. Ten trials were presented at each intensity level and stimulus intensities were delivered in a descending order with the intensity decreased by a step size of 5 μA. The descending series was initiated at a previously determined suprathreshold level. Five or more correct responses at each intensity level were scored as a plus, while fewer than five correct responses were scored as a minus. A descending series was continued until two successive minus scores were achieved. The threshold was defined as the midpoint in μA between the plus and the minus score.

After stable rates of cocaine self-administration (± 10 percent prior day's performance) were established, rats were tested for ICSS threshold before and after different length bouts of access to cocaine. Reward thresholds were determined at 1, 2, 3 6, and 18 hours postdrug.

Preliminary results for four rats are shown in Figure 9.5. Exposure to 12 and 18 hours of cocaine self-administration significantly increased reward threshold at 3 and 6 hours postdrug. These effects are similar to those observed by others using chronic administration of amphetamine; however, in the present study, the drug was self-administered intravenously by the rat (Table 9.3). After 12 hours and 18 hours the rats on average had intravenously self-administered 189 and 239 mg/kg of cocaine, respectively. Also of significance was the relatively short duration of exposure necessary to produce these changes in threshold.

It would be difficult to argue that the "anhedonia" observed in the present study had a major impact on the maintenance self-administration behavior of the rats, since the amount of drug taken was constant throughout the test session. However, it may be that the "crash" and subsequent "anhedonia"

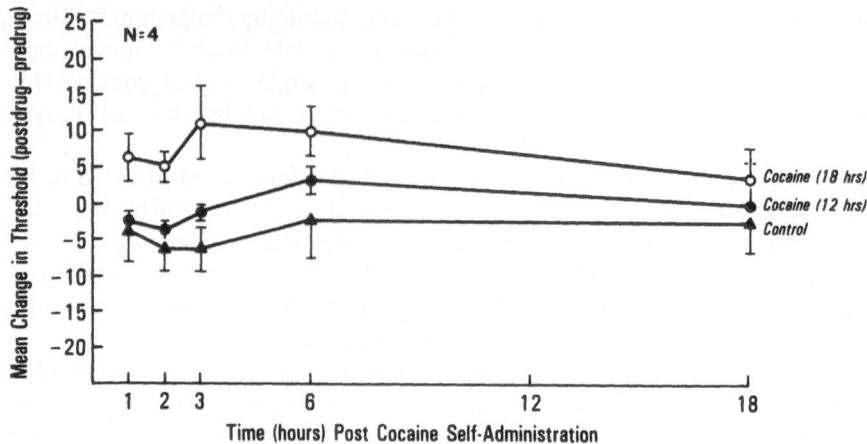

FIGURE 9.5. Changes in reward thresholds following a 12- or 18-hour bout of continuous cocaine self-administration. Rats were allowed unlimited access to cocaine (0.75 mg/kg/inj) intravenously with each injection. Reward thresholds were determined at different time points postself-administration using a modification of the Kornetsky and Esposito (1979) discrete-trials wheel manipulandum procedure.

may be important for the reinitiation of bouts with more long-term access. More important for the present discussion, these results suggest that relatively small amounts of self-administered stimulants can produce changes in reward thresholds postdrug. Thus, psychostimulants may act as significant "stressors" in individuals possibly already susceptible to changes in reward thresholds, or "anhedonia."

These results showing a change in reward thresholds following a prolonged bout of cocaine self-administration may also be attributed simply to the phenomenon of incentive contrast given that cocaine itself is a powerful reinforcer. It may be that the effects of ICSS are temporarily suppressed simply by the process of comparison of a small reinforcer to a previous large

TABLE 9.3. Total drug injections for four rats self-administering cocaine*

		3 hr	6 hr	9 hr	12 hr	15 hr	18 hr	Total
12 hr	X	56.30	42.53	39.51	47.97	—	—	189.00
Cocaine bout	SEM	13.93	5.61	6.53	15.36			20.05
18 hr	X	43.20	37.76	40.77	36.50	36.27	44.01	238.50
Cocaine bout	SEM	8.87	2.15	2.58	3.02	2.65	13.70	30.32

*Rats implanted with chronic indwelling jugular catheters and intracranial self-stimulating electrodes in the lateral hypothalamus were trained to lever press on an FR-1 schedule for intravenous cocaine (0.75 mg/kg/injection) in daily 3-hour sessions. After baseline stabilization (±10 percent variation in responding over a 3-day period) the rats were allowed more prolonged access to the drug (12 and 18 hours, respectively).

reinforcer, i.e. "elation/depression" effect (Crespi, 1942). This, indeed, may be a viable explanation particularly after short exposures to the drug and for short periods following drug cessation, since contrast effects are relatively short-lived. However, one may also speculate that the neural systems involved in contrast effects are the same systems involved in more prolonged changes in the brain reward systems associated with depression.

Stress and Reward

In some very interesting recent work, Zacharko, Anisman, and colleagues (Zacharko et al., 1983, 1984, 1987; also see chapter Zacharko of this volume) showed that inescapable shock applied in a yoked paradigm reduced responding for ICSS in mice from electrodes in the nucleus accumbens, medial forebrain bundle, and the ventral tegmental area, but not the substantia nigra (Zacharko et al., 1983). Repeated administration of desmethylimipramine over a 10-day period markedly reduced this stressor-induced reduction in ICSS (Zacharko et al., 1984). These authors have also shown some strain differences in the ICSS response to inescapable shock with DBA/2J mice showing a pronounced disturbance of responding for ICSS from the nucleus accumbens but no disturbance, even a facilitation, in mice from the BALB/c ByJ strain (Zacharko, et al., 1987; also see chapter Zacharko of this volume).

Problems and Promises

Validation to Date

Animal behavioral models of psychopathology have been characterized in various ways such as predictive, isomorphic, and homologous (see chapters McKinney and Kornetsky of this volume). In predictive models, the behavioral change may predict a treatment effect such as the well-known use of a conflict test to predict clinically useful "anti-anxiety" drugs. Isomorphic models may reproduce a behavioral change observed in a given psychopathology, but the etiology may be different. For example, pharmacological induction of immobility in animals mimics the immobility observed in schizophrenia, but the etiology is presumably different. In homologous models, both the etiology and the expression of the psychopathology are the same. For example, MPTP-induced Parkinson's disease in man and monkeys has an identical etiology and phenomenology.

How does the anhedonia model fit into this scheme? The use of chronic stimulant administration to produce an increase in reward thresholds post-drug has face validity and certainly is isomorphic with the human condition since a decrease in the ability to experience pleasure is possibly the single most important symptom of endogenous depression (American Psychiatric Association, 1980). In the case of intravenous self-administration of cocaine, there may be a clear parallel with the depression observed following cocaine abuse in humans (Blum, 1976; Wesson and Smith, 1977). This approaches a homologous model, at least for this specific case. Interestingly, for chronic stimulants and ICSS, the least explored construct is that of predictive validity. Only a few studies to date have shown that tricyclic antidepressants reverse the postamphetamine decrease in self-stimulation rates of responding (Kokkinidis et al., 1980).

Future Needs

Clearly more work is needed to validate this model both at the predictive and construct levels. Potential pharmacological treatments, such as other antidepressant drugs, that could reverse the increase in reward thresholds are still unexplored. Acute versus chronic effects of antidepressants should be examined since chronic treatment with antidepressants is usually required for significant therapeutic effects.

At the construct level, similar behavioral assays using natural rewards could be considered. Operant means to assess motivational value of stimuli with high incentive value and low drive should be developed and tested. In addition, other means of inducing depression-like states in animals should be combined with the anhedonia model. Only a few studies have been done on reward measures in animals subjected to chronic stress, learned helplessness, or separation (see chapter Zacharko and Anisman of this volume). Little work has explored possible genetic interactions that might exaggerate susceptibility to anhedonia or ever cause chronic long-term changes in reward threshold (again, see chapter Zacharko and Anisman of this volume). The availability of behavioral tests sensitive to change in hedonic value of "natural" rewards could allow a faster and systematic evaluation of these questions.

Acknowledgments. This chapter was prepared with support in part from NIDA grants DA04398 and DA04043 and NIAAA grant AA06420. The author gratefully thanks Ms. Athina Markou (Dept. of Psychology, University of California, San Diego), Dr. Trevor Robbins (Dept. of Experimental Psychology, University of Cambridge, Cambridge England), and Dr. Ben Williams (Dept. of Psychology, University of California, San Diego) for their helpful criticisms and suggestions on the manuscript. The author is particularly grateful to Dr. Carol Hubner who performed the preliminary experiments described in Figure 9.5.

References

Adams WJ, Lorens SA, Mitchell CL (1972): Morphine enhances lateral hypothalamic self-stimulation in the rat. *Proc Soc Exp Biol Med* 140:770-771
American Psychiatric Association (1980): *Diagnostic and Statistical Manual of Mental Disorders.* 3rd Ed., Washington DC: APA 278-279
Angrist B (1985): Personal Communication.

Baltzer JH, Levitt RA, Furby JE (1977): Etorphine and shuttle box self-stimulation in the rat. *Pharmacol Biochem Behav* 7:413-416

Belluzzi JD, Stein L (1977): Enkephalin may mediate euphoria and drive-reduction reward. *Nature* 226:556-558

Blum K (1976): Depressive states induced by drugs of abuse: clinical evidence, theoretical mechanisms and proposed treatment. *J Psychedelic Drugs* 8:235-262

Cassens GP, et al., (1981): Amphetamine withdrawal: Effects on threshold of intracranial reinforcement. *Psychopharmacology* 73:318-322

Cassens GP, Mills AW (1973): Lithium and amphetamine: Opposite effects on threshold of intracranial reinforcement. *Psychopharmacologia* 30:282-290

Connell PH (1958): Amphetamine psychosis. In: *Maudsley Monographs No. 5* London: Oxford University Press.

Cooper BR, Black WC, Paolino RM (1971): Decreased self-forebrain and lateral hypothalamic reward after alpha methyl-p-tyrosine. *Physiol Behav* 6:425-429

Crespi L (1942): Quantitative variation of incentive and performance in the white rat. *Am J Psychol* 15:467-517

Crow TJ (1970): Enhancement by cocaine of intracranial self-stimulation in the rat. *Life Sci* 9:375-381

Crow TJ, Spear PJ, Arbuthnott GW (1972): Intracranial self-stimulation with electrodes in the region of the locus coeruleus. *Brain Res* 36:275-287

De Villiers P (1977): Choice in concurrent schedules and a quantitative formulation of the law of effect. In: *Handbook of Operant Behavior,* Honig WK, Staddon JER, eds. Englewood Cliffs: Prentice-Hall 233-287

Domino EF, Olds ME (1968): Cholinergic inhibition of self-stimulation behavior. *J Pharmacol Exp Ther* 164:202-211

Dresse A (1966): Influence de 15 neuroleptiques (butyrophenones et phenothiazines) sur les variations de la teneur du cerveau en noradrenaline et activite du rat dans le test d'autostimulation. *Arch Int Pharmacodyn Ther* 159:353-365

Esposito R, Kornettsky C (1977): Morphine lowering of self-stimulation thresholds: Lack of tolerance with long-term administration. *Science* 95:189-191

Ettenberg A, Koob GF, Bloom FE (1981): Neuroleptic-induced anhedonia may be a motor artefact. *Science* 213:357-359

Fawcett J, et al. (1983): Assessing anhedonia in psychiatric patients: The pleasure scale. *Arch Gen Psychiatry* 40:79-84

Fibiger HF, Phillips AG (1981): Increased intracranial self-stimulation in rats after long-term administration of desipramine. *Science* 214:683-685

Gallistel CR (1983): Self-stimulation In: *The Physiological Basis of Memory,* Deutsch JA ed. New York, Academic Press

Gallistel CR, Karras D (1984): Pimozide and amphetamine have opposing effects on the reward summation function. *Pharmacol Biochem Behav* 20:73-77

German DC, Bowden DM (1974): Catecholamine systems as the neural substrate for intracranial self-stimulation: a hypothesis. *Brain Res* 73:381-420

Gibson S, McGeer EG, McGeer PL (1970): Effects of selective inhibitors of tyrosine and tryptophan hydroxylases on self-stimulation in the rat. *Exp Neurol* 27:283-290

Glick SD, Rapaport G (1974): Tolerance to the facilitatory effect of morphine on self-stimulation of the medial forebrain bundle in rats. *Res Commun Chem Pathol Pharmacol* 9:647-652

Goodall EB, Carey RJ (1975): Effects of d- versus 1-amphetamine, food deprivation, and current intensity on self-stimulation of the lateral hypothalamus, substantia nigra, and medial frontal cortex of the rat. *J Comp Physiol Psychol* 89:1029-1045

Hamilton AL, Stellar JR, Hart EB (1985): Reward, performance, and the response strength method in self-stimulating rats: Validation and neuroleptics. *Physiol Behav* 35:897-904

Hawkins JD, Pliskoff SS (1964): Brain-stimulation intensity, rate of self-stimulation and reinforcement strength: An analysis through chaining. *J Exp Anal Behav* 7:285-288

Hebb DO (1972): *Textbook of Psychology.* Philadelphia: WB Saunders

Herman ZS, et al. (1971): The influence of prolonged amphetamine treatment and amphetamine withdrawal on brain biogenic amine content and behavior in the rat. *Psychopharmacology* 21:74-81

Herrnstein RJ (1970): On the law of effect. *J Exp Anal Behav* 13:243-266

Herrnstein RJ (1971): Quantitative hedonism. *J Psychiatr Res* 8:399-412

Holtzman SG (1976): Comparison of the effects of morphine, pentazocine cyclazocine and amphetamine on intracranial self-stimulation in the rat. *Psychopharmacologia* 46:223-227

Ichitani Y, Iwasaki T, Satoh T (1985): Effects of naloxone and chlordiazepoxide on lateral hypothalamic self-stimulation in rats. *Physiol Behav* 34:779-782

Jung OH, Boyd ES (1966): Effects of cholinergic drugs on self-stimulation response rates in rats. *Am J Physiol* 210:432-434

Kelly PH, Iversen SD (1976): Selective 6-OHDA-induced destruction of mesolimbic dopamine neurons: Abolition of psychostimulant-induced locomotor activity in rats. *Eur J Pharmacol* 40:45-56

Klein DF (1974): Endogenomorphic depression: A conceptual and terminological revision. *Arch Gen Psychiatry* 31:447-454

Kling JW, Schrier Am (1971): Positive reinforcement. In: *Woodworth and Schlosberg's Experimental Psychology,* ed. 3, Kling JW, Riggs LA, eds. 3rd New York: Holt Rinehart and Winston 615-702

Kokkinidis L, Zacharko RM, Predy PA (1980): Post-amphetamine depression of self-stimulation responding from the substantia nigra: Reversal by tricyclic antidepressants. *Pharmacol Biochem Behav* 13:379-383

Kokkinidis L, Zacharko RM, Anisman H (1986): Amphetamine withdrawal: A behavioral evaluation. *Life Sci* 38:1617-1623

Koob GF (1977): Incentive shifts in intracranial self-stimulation produced by different series of stimulus intensity presentations. *Physiol Behav* 18:131-135

Koob GF, Spector NH, Meyerhoff JL (1975): Effects of heroin on lever pressing for intracranial self-stimulation, food and water in the rat. *Psychopharmacologia* 42:231-234

Koob GF, et al. (1977): Effects of d-amphetamine on concurrent self-stimulation of forebrain and brainstem loci. *Brain Res* 137:109-126

Kornetsky C, Esposito RU (1979): Euphorigenic drugs: Effects on reward pathways of the brain. *Fed Proc* 38:2473-2476

Kornetsky C, Esposito RU (1981): Reward and detection thresholds for brain stimulation: Dissociative effects of cocaine. *Brain Res* 209:496-500

Leith NJ, Barrett RJ (1976): Amphetamine and the reward system: Evidence for tolerance and post-drug depression. *Psychopharmacologia* 46:19-25

Leith NJ, Barrett RJ (1980): Effects of chronic amphetamine or reserpine on self-stimulation responding: Animal model of depression. *Psychopharmacology* 72:9-15

Leith NJ, Barrett RJ (1981): Self-stimulation and amphetamine: Tolerance to d and l isomers and cross tolerance to cocaine and methylphenidate. *Psychopharmacology* 74:23-28

Levitt RA, Stilwell DJ, Evers TM (1978): Morphine and shuttle box self-stimulation in the rat: Tolerance studies: *Pharmacol Biochem Behav* 9:567-569

Liebman JM, Butcher LL (1974): comparative involvement of dopamine and noradrenaline in rate-free self-stimulation in substantia nigra, lateral hypothalamus and mesencephalic central gray. *Naunyn-Schmidebergs Arch Pharmacol* 284:167-194

Lippa AS, et al. (1973): Neurochemical mediation of reward: A significant role for dopamine? *Pharmacol Biochem Behav* 1:23-28

Lorens SA, Mitchell CL (1973): Influence of morphine on lateral hypothalamic self-stimulation in the rat. *Psychopharmacologia* 32:271-277

Marcus R, Kornetsky C (1974): Negative and positive intracranial reinforcement thresholds: Effects of morphine. *Psychopharmacologia* 38:1-13

Maroli AN, Tsany W-K, Stutz RM (1978): Morphine and self-stimulation: Evidence for action on a common neural substrate. *Pharmacol Biochem Behav* 8:119-123

Miliaressis E, Bouchard A, Jacobowitz DM (1975): Strong positive reward in median raphe: Specific inhibition by parachlorophenylalanine. *Brain Res* 98:194-201

Newman LM (1972): Effects on cholinergic agonists and antagonists on self-stimulation behavior in the rat. *J Comp Physiol Psychol* 79:394-413

Olds J, Milner P (1954): Positive reinforcement produced by electrical stimulation of septal area and other regions of rat brain. *J Comp Physiol Psychol* 47:419-427

Olds J, Travis RP (1960): Effects of chlorpromazine, meprobamate pentobarbital and morphine on self-stimulation. *J Pharmacol Exp Ther* 128:394-404

Olds J, Killam KF, Bach-y-Rita P (1956): Self-stimulation of the brain used as a screening method for tranquilizing drugs. *Science* 124:265-266

Phillips AG, Fibiger HC (1973): Dopaminergic and noradrenergic substrates of positive reinforcement: Differential effects of d and 1-amphetamine. *Science* 179:575-577

Phillips AG, LePiane FG (1986): Effects of pimozide on positive and negative incentive contrast with rewarding brain stimulation. *Pharmacol Biochem Behav* 24:1577-1582

Phillips AG, Brook SM, Fibiger HC (1975): Effects of amphetamine isomers and neuroleptics on self-stimulation from the nucleus accumbens and dorsal noradrenergic bundle. *Brain Res* 85:13-22

Phillips AG, Carter DA, Fibiger HC (1976): Dopaminergic substrates of intracranial self-stimulation in the caudate-putamen. *Brain Res* 104-221-232

Pliskoff SS, Wright JE, Hawkins JD (1965): Brain stimulation as a reinforcer: Intermittent schedules. *J Exp Anal Behav* 8:75-88

Pradhan SN, Bowling C (1971): Effects of nicotine on self stimulation in rats. *J Pharmacol Exp Ther* 176:229-243

Robbins TW, Koob GF (1978): Pipradrol enhances the reinforcing properties of stimuli paired with brain stimulation. *Pharmacol Biochem Behav* 8:219-221

Roll SK (1970): Intracranial self-stimulation and wakefulness: Effect of manipulating ambient brain catecholamines. *Science* 168:1370-1372

Rolls ET, Kelly PH, Shaw SG (1974): Noradrenaline dopamine and brain stimulation reward. *Pharmacol Biochem Behav* 2:735-740

Sahakian BJ, Koob GF (1978): The relationships between pipradrol-induced responding for electrical brain stimulation, stereotyped behavior and locomotor activity. *Neuropharmacology* 17:363-366

Schaefer GJ, Holtzman SG (1979): Free-operant and auto-titration brain stimulation procedures in the rat: A comparison of drug effects. *Pharmacol Biochem Behav* 10:127-135

Schaefer GJ, Michael RP (1981): Threshold differences for naloxone and naltrexone in the hypothalamus and midbrain using fixed ratio brain self-stimulation in rats. *Psychopharmacology* 74:17-22

Sidman M, et al. (1955): Reward schedules and behavior maintained by intracranial self-stimulation. *Science* 122:830-831

Simon H, LeMoal M, Cardo B (1976): Intracranial self-stimulation from the dorsal raphe nucleus of the rat: Effects of the injection of parachlorophyenylalanine and of alpha-methyl-para-tyrosine. *Behav Biol* 16:353-364

Simpson DM, Annau Z (1977): Behavioral withdrawal following several psychoactive drugs. *Pharmacol Biochem Behav* 7:59-64

Stapleton JM, et al. (1979): Naloxone reduces pressing for intracranial stimulation of sites in periaqueductal gray area, nucleus accumbens substantial nigra, and lateral hypothalamus. *Physiol Psychol* 7:427-436

Stark P, Boyd ES (1963): Effects of cholinergic drugs on hypothalamic self-stimulation response rates of dogs. *Am J Physiol* 205:745-748

Stark P, et al. (1969): Sensitivity and specificity of positive reinforcing areas of neurosedatives, antidepressants and stimulants. *J Pharmacol Exp Ther* 166:163-169

Stark P, et al. (1970): Dissociation and the effects of p-chlorophenylalanine on self-stimulation and on brain serotonin. *Life Sci* 9:41-48

Stein L (1964): Self-stimulation of the brain and the central stimulant actions of amphetamine. *Fed Proc* 23:836-850

Stein L (1968): Chemistry of reward and punishment. In: *Psychopharmacology, A Review of Progress (1957-1967), Public Health Service Publication No. 1836,* Efron D, eds. Washington, DC, US Government Printing Office, 105-123

Stein L, Ray OS (1960): Brain stimulation reward "thresholds" self-determined in rat. *Psychopharmacologia* 1:251:256

Steiner SS, Stokely SN (1973): Methamphetamine lowers self-stimulation thresholds. *Physiol Psychol* 1:161-164

Stellar J, Kelley A, Corbett D (1983): Effects of peripheral and central dopamine blockade on lateral hypothalamic self-stimulation: Evidence for both reward and motor deficits. *Pharmacol Biochem Behav* 18:433-442

Stinus L, Thierry AM, Cardo B (1976): Effects of various inhibitors of tyrosine hydroxylase and dopamine-beta-hydroxylase on rat self-stimulation after reserpine treatment. *Psychopharmacologia* 45:287-294

Tyce FA (1968): Influence of methylphenidate hydroxychloride on self-stimulation of the brain by rats. *Psychol Rep* 23:379-385

Valenstein ES (1964): Problems of measurement and interpretation with reinforcing brain stimulation. *Psychol Rep* 71:415-437

Van der Kooy D, LePiane FG, Phillips AG (1977): Apparent independence of opiate reinforcement and electrical self-stimulation systems in rat brain. *Life Sci* 20-981-986

Van der Kooy D, Fibiger HC, Phillips AG (1978): An analysis of dorsal and median raphe self-stimulation: Effects of parachlorophenylalanine. *Pharmacol Biochem Behav* 8:441-445

Watson R, Hartmann F, Schildkraut JJ (1972): Amphetamine withdrawal: Affective state, sleep patterns, and MHPG excretion. *Am J Psychiatr* 129:263-269

Wauquier A, Niemegeers CJE (1972): Intracranial self-stimulation in rats as a function of various stimulus parameters. II. Influence of haloperidol, pimozide and pipamperone on medial forebrain bundle stimulation with monopolar electrodes. *Psychopharmacologia* 27:191-202

Wauquier A, Niemegeers CJE (1974): Intracranial self-stimulation in rats as a function of various stimulus parameters. IV. Influence of amphetamine on medial forebrain bundle stimulation with monopolar electrodes. *Psychopharmacologia* 34:265-274

Wauquier A, Niemegeers CJE, Lal H (1974): Differential antagonism by naloxone of inhibitory effects of haloperidol and morphine on brain self-stimulation. *Psychopharmacologia* 37:303-310

Wesson DR, Smith DE (1977): Cocaine: Its use for central nervous system stimulation including recreational and medical uses. In: *Cocaine: 1977 NIDA Research Monograph #13*, Peterson RC, Stillman RC, eds. Washington, DC, US Government Printing Office

White N (1976): Strength-duration analysis of the organization of reinforcement pathways in the medial forebrain bundle of rats. *Brain Res* 110:575-591

Williams B (1988): Reinforcement, choice and response strength. In: *Steven's Handbook of Experimental Psychology*, 2nd edition Atkinson RC, Herrnstein RJ, Lindzey G, Luce RD, eds., John Wiley & Sons, 107-244

Wise CD, Stein L. (1969): Facilitation of brain self-stimulation by central administration of NE. *Science* 163:299-301

Zacharko RM, et al. (1983): Region-specific reductions of intracranial self-stimulation after uncontrollable stress: Possible effects on reward processes. *Behav Brain Res* 9:129-141

Zacharko RM, et al. (1984): Prevention of stressor-induced disturbances of self-stimulation by desmethyl imipramine. *Brain Res* 321:175-179

Zacharko RM, et al. (1987): Strain-specific effects of inescapable shock or intracranial self-stimulation from the nucleus accumbens. *Brain Res* 426:164-168

Zarevics P, Setler PE (1979): Simultaneous rate-independent and rate-dependent assessment of intracranial self-stimulation: Evidence for the direct involvement of dopamine in brain reinforcement mechanisms. *Brain Res* 169:499-512

Section III   Pharmacologic Models

10

Models of Depression Used in the Pharmaceutical Industry

JAMES L. HOWARD, ROBERT M. FERRIS,
BARRETT R. COOPER, FRANCIS E. SOROKO,
CHING M. WANG, AND GERALD T. POLLARD

Introduction

An animal model is a representation of some aspects of a human disease. Modeling mental disease is problematic in that the symptoms may be poorly defined and the underlying pathophysiology poorly understood. This is especially true of depression, which is the name given to a heterogeneous group of disorders having in common the disturbance of mood (Cronholm, 1984; Klerman, 1984; Baldessarini, 1985). Models of depression are made primarily for two reasons: to allow experimental manipulation of behavioral and biochemical variables that might provide insight into the etiology and underlying pathophysiology of the disease and to allow prediction of how a variable such as an antidepressant drug might affect the disease (Everitt and Keverne, 1979).

Models have been evaluated according to degree of homology with the disease (Abramson and Seligman, 1977; McKinney and Bunney, 1969):

1. Similarity of inducing conditions
2. Similarity of behavioral states produced
3. Similarity of underlying neurobiological mechanisms
4. Reversal by clinically effective treatments

No model of depression adequately meets criteria 1 and 3, because little is known of the etiology and neurobiological mechanisms. Criterion 2 is not very useful, because behaviors that appear to be similar can have vastly different antecedents between and within species (Bond, 1984). Some models meet criterion 4 at least to a degree. Depression itself would not meet all of these criteria (McKinney, 1974): The causes are unknown (1 and 3) and the symptoms vary (2).

Models can be evaluated also according to the kinds of validity they are thought to have (Russell and Overstreet, 1984):

1. Content validity: how well the model covers a representative sample of the essential attributes of the disease. Because the nosology of depression

is ill-defined, no model can have high content validity. At best, aspects of depression may be abstracted and compared to aspects of a model (Bond, 1984).

2. Construct validity: how well the model represents a theoretical framework. Because all theories of depression are still uncertain, no model will gain much support on grounds of construct validity, although the correspondence between model and theory may be great.
3. Concurrent validity: how well effects in the model relate to established knowledge. One of the few widely accepted facts about depression is that certain drugs reverse it (Baldessarini, 1985); therefore, a model can be assessed according to whether proven antidepressants produce specific effects in it. This approach has been the most fruitful: Discovery of the commonalities of antidepressant drug action led to biochemical theories of depression, and the existence of effective drug therapy led to the development of models (Baldessarini, 1985; Cowen and Grahame-Smith, 1983; Weissman and Koe, 1987).
4. Predictive validity: how well the model predicts outcomes in the disease. No model has been very successful by this criterion. Given the state of knowledge of depression, only concurrent validity seems completely relevant.

The primary goal of research in the pharmaceutical industry is the discovery of drugs that are more effective or safer than existing therapies (Maxwell, 1984). This goal imposes pragmatic constraints on the design, evaluation, and use of models. Concurrent validity is emphasized: A model may be elegant, homologous, and intellectually gratifying; it may lead to explication of the biological bases of normal and abnormal mental function; but if it does not identify antidepressants and reject non-antidepressants better than other models, then it is not useful for finding new compounds, examining the structure-activity relationships of a series of compounds, and justifying clinical trials. Furthermore, it must be rapid and economical enough to screen a large number of compounds for activity, and it must be quantitatively reliable enough to allow comparison among compounds (Weatherall, 1985; Weissman and Koe, 1987). The result of these constraints is that the industrial researcher requires not a model of depression but a model of antidepressant drug action. There is a danger that uncritical application of this approach could hinder the discovery of novel compounds; therefore, there is a need to develop homologous models and screening methods in order to gain understanding of the disease and facilitate discovery of remedies.

This chapter describes how industrial researchers view and use models of depression. Neurochemical models receive special consideration because of their historical and current importance. The chapter is based on a rapid and selective review of the literature and is informed by conversations with selected colleagues. It is necessarily biased by what the authors do and whom they know.

Neurochemical Models

In the 1950s two effective antidepressants were discovered: Imipramine was the first of the tricyclic antidepressants (TCAs), and iproniazid was the first of the monoamine oxidase inhibitors (MAOIs). These discoveries were serendipitous: Imipramine was discovered during a search for a better antipsychotic agent (Kuhn, 1958); iproniazid was found to elevate mood and produce euphoria while being studied for use in tuberculosis. It was also known in the 1950s that reserpine, a drug used to treat hypertension, produced depression in about 15% of the patients receiving high doses (Bunney and Davis, 1965). Studies of the mechanisms of action of these and other agents revealed a commonality: Agents such as reserpine, which depleted serotonin (5-HT) and the catecholamines, gave rise to depression in both animal models and humans, while antidepressants such as imipramine and iproniazid, which increased the availability of these neurotransmitters in the synaptic cleft, reversed depression. These observations led to the hypothesis that depression was caused by a deficiency of monoamines, particularly norepinephrine (NE), at postsynaptic receptors.

The original monoamine hypothesis is based on the premise that the therapeutic action of antidepressants depends on their ability to increase the availability of biogenic amines in the snyaptic cleft—in the case of TCAs by inhibiting the high-affinity reuptake of NE and/or 5-HT, and in the case of MAOIs by inhibiting the oxidative deamination of NE and/or 5-HT by monoamine oxidase (Charney et al., 1981). These ideas influenced the development of both *in-vitro* and *in-vivo* model systems used to screen potential antidepressants from the late 1950s to the mid-1970s and continue to do so to varying degrees.

Neurochemical models (procedures, techniques, methods) used by industry in the search for antidepressants are based on increasing the levels or effectiveness of amines, especially NE and 5-HT, in the synaptic cleft and include inhibition of the breakdown of amines by MAO, inhibition of the high-affinity reuptake system, increase in the release of amines into the cleft, direct stimulation of postsynaptic receptors, direct antagonism of presynaptic receptors, or a combination of these actions.

Inhibition of MAO

Inhibition of MAO leads to increases of NE and 5-HT in brain, which presumably results in more of these neurotransmitters being present in the synaptic cleft. Johnston (1968) reported the presence of two forms of MAO, which he termed Type A and Type B. Type A uses NE, 5-HT, octopamine, tyramine, and dopamine (DA) as substrates; it is inhibited by the clinically effective antidepressant clorgyline. Type B uses benzylamine, phenylethylamine, tyramine, and DA as substrates; whether selective inhibition of Type B produces antidepressant action is unclear. Selective MAO inhibition remains a useful model for finding novel antidepressants. However, MAOIs have limited usefulness in the clinic, because most of them can produce significant side effects.

Inhibition of High-Affinity Reuptake

TCAs possess varying degrees of selectivity as high-affinity reuptake inhibitors of NE and 5-HT. It was generally held that tertiary amine TCAs such as imipramine selectively inhibit 5-HT reuptake, whereas secondary amine TCAs such as desmethylimipramine selectively inhibit NE reuptake (Maas, 1975; Maj et al., 1984). However, there are discrepancies between results *in vivo* and *in vitro* (Maj et al., 1984; Sulser, 1982; Richelson and Pfenning, 1984). Recently, selective inhibitors have been described: Fluoxetine, zimelidine, fluroxamine, femoxetine, paroxetine, and alaproclate are highly selective for 5-HT and weak for NE or DA (Fuller, 1986); maprotiline and amoxapine selectively inhibit uptake of NE but have little effect on 5-HT or DA. The fact that agents inhibit reuptake and are useful in the treatment of depression supports the contention that reuptake inhibition is a useful model for finding potential antidepressant drugs. On the other hand, some inhibitors of monoamine reuptake—e.g., cocaine and amphetamine—are not clinically effective antidepressants.

Release of Neurotransmitters

Theoretically agents that release monoamines should elevate the concentrations of these neurotransmitters in the synaptic cleft, which in turn should enhance activation of postsynaptic receptors. However, such agents have not proven to be clinically effective antidepressants, even if they also inhibit reuptake. Carlsson (1984) has offered a possible explanation for the lack of antidepressant activity by catecholamine-releasing agents such as cocaine and amphetamine: Release of DA leads to dopaminergic hyperactivity, which in turn may interfere with an antidepressant response. The authors' screening experience supports this contention. Fenfluramine, which releases 5-HT, is an anorectic and not an antidepressant. Screening for agents that release amines, particularly DA, is not, in the authors' opinion, a useful way to discover new antidepressants.

Direct Effects on Receptors

Over the past quarter century the original monoamine deficiency hypothesis has been challenged by the advent of second-generation antidepressants such as iprindole, mianserin, and bupropion, which do not block reuptake of monoamines or inhibit MAO. In addition, the temporal discrepancy between the onset of known biochemical actions, which occur within hours, and the onset of clinical action, which takes 2 to 3 weeks, suggested that the drugs' effects on amine reuptake and metabolism could not completely account for antidepressant efficacy.

Receptor-binding techniques that facilitated studies on drug-induced changes in receptor populations were developed at about the same time. As a result, by the mid-1970s emphasis had shifted from presynaptic phenomena to postsynaptic, receptor-mediated events (Vetulani et al., 1976; Frazer et al., 1974; Frazer et al., 1985). The concepts that emerged are reviewed by Sulser (1978) and briefly are as follows: Psychotropic drugs that precipitate depression (e.g., reserpine, 6-hydroxydopamine) increase the sensitivity of the limbic noradrenergic receptor-coupled adenylate cyclase system, while treatments that alleviate depression (e.g., TCAs, MAOIs, iprindole, and electroconvulsive shock [ECS]) produce the opposite effect on this system. In the revised catecholamine hypothesis, depression is seen as a state of pathologic hypersensitivity of catecholamine receptors (Mandell et al., 1975). Chronic (but not acute) antidepressant treatment desensitizes enhanced noradrenergic receptor function, thus reducing the amplification process that ultimately translates sensory input into physiological and behavioral events (Sulser et al., 1978; Sulser, 1982).

With this revised theory, research shifted from the study of acute presynaptic events to chronic postsynaptic receptor-mediated events. The use of techniques such as receptor-binding assays, single-unit electrophysiological recordings, brain adenylate cyclase measurements, and various behavioral tests brought about a new understanding of the effects of acute and chronic administration of antidepressants on pre- and postsynaptic receptors and their adaptive responses to such effects. An antidepressant could act directly as an agonist or antagonist of pre- or postsynaptic receptors, or it could act in previously hypothesized ways such as inhibiting reuptake; the result is a change in the receptor to compensate for the perturbation (Green, 1985).

In neurotransmitter binding assays, many antidepressants show high affinities for cholinergic muscarinic, histaminergic H_1 and H_2, noradrenergic α_1 and α_2, 5-HT_1 and 5-HT_2, and even dopaminergic D_2 receptors. Attempts to correlate these acute effects with antidepressant activity have not been successful. Instead, most of the acute receptor affinities shared by antidepressants are associated with clinical side effects. This information is nevertheless valuable to the industry. For example, blockade of muscarinic receptors is probably responsible for the marked anticholinergic side effects observed clinically with the TCAs, while blockade of H_2 receptors may be responsible for drowsiness and sedation. Blockade of peripheral α_1 receptors elicits postural hypotension and other cardiovascular effects, and blockade of central α_1 receptors may be responsible for sedation (Fuller and Wong, 1985). *In-vitro* studies are useful for profiling the side effects of potential antidepressants but have yielded little information directly relevant to antidepressant drug development. However, specific receptor binding assays could be useful for finding potential antidepressants that interact directly and selectively with noradrenergic or serotonergic receptors.

NE, 5-HT, and DA nerve terminals have presynaptic receptors. Blockade of these receptors is believed to increase neurotransmitter release, which leads to increased concentrations of neurotransmitter in the synaptic cleft. Agonists of presynaptic receptors should elicit the opposite effect, decreased concentrations. A highly selective NE or 5-HT presynaptic receptor blocker might be a useful antidepressant. Mianserin may offer an example of antidepressant action by this mechanism: It blocks α_2 receptors and in this manner is believed to bring about an increase in the release of NE into the cleft (Engberg and Svensson, 1980). Only further studies will show how useful presynaptic models are.

Direct agonists of postsynaptic NE or 5-HT receptors should theoretically lead to antidepressant activity by mimicking the effects elicited by an elevation of these neurotransmitters in the cleft. In fact, direct agonists of postsynaptic β receptors such as salbutamol (Lecrubier et al., 1980) have been reported to have antidepressant properties. The 5-HT agonist m-chlorophenylpiperazine, a metabolite of trazodone, may play a role in the antidepressant activity of trazodone (Sansone et al., 1983).

Adaptive Changes in Receptor Sensitivity after Chronic Treatment

In animal studies many types of receptors have been shown to change in sensitivity after chronic administration of antidepressants—e.g., α_1, α_2, β, 5-HT$_1$, 5-HT$_2$, DA, GABA-B, and benzodiazepine (for a comprehensive review see Charney et al., 1981). In most cases receptor number but not affinity is decreased; this phenomenon is termed down-regulation. However, the opposite effect has been reported, in some cases for the same drug tested in different laboratories. The most characteristic and reproducible change after chronic (14 to 21 days) but not acute (1 day) antidepressant treatment is down-regulation of β receptors (Vetulani et al., 1974, 1976; Frazer et al., 1974; Sulser et al., 1978). Decrease in 5-HT$_2$ receptor density has also been reported, but the effect is not as consistent as with β receptors.

Not only does chronic treatment with TCAs, MAOIs, most second-generation antidepressants, and ECS usually down-regulate β receptors, it also desensitizes NE-stimulated adenylate cyclase and thus decreases the production of the second messenger cyclic adenosine 3',5'-monophosphate (cyclic AMP) in limbic forebrain. The molecular basis for desensitization is presumed to be an increase in the concentration of the neurotransmitter at its receptor and in the length of time the agonist is in contact with the receptor (for a review see Richelson and El-Fakahany, 1982; Sulser, 1982). In fact a combination of down-regulation of β receptors and desensitization of adenylate cyclase has been proposed as a model of antidepressant action (Vetulani et al., 1976; Sulser et al., 1978). Support for this model comes also from the finding that a large number of drugs that act on the central nervous system—e.g., barbiturates, anticonvulsants, benzodiazepine anxiolytics, antipsychotics, and antihistamines—do not alter adenylate cyclase activity or down-regulate β receptors (Sellinger-Barnette et al., 1980). Mianserin and zimelidine can reduce NE-stimulated adenylate cyclase activity without decreasing β receptor number (Mishra et al., 1980). These data suggest that the receptor has been dissociated from the enzyme. Such phenomena are not uncommon in other cells (Sulser, 1982). However, measurement of cyclase activity seems to offer a more comprehensive model for finding potential antidepressants than measurement of only changes in β receptor number.

Serotonin seems to be involved in the antidepressant activity of at least TCAs, but the process is not yet fully understood. Janowsky et al. (1982), Brunello et al. (1982), and Racagni et al. (1983) have shown that a functional serotonergic system is necessary for chronically administered desmethylimipramine to down-regulate postsynaptic β receptors. The β agonists salbutamol and clenbuterol enhance the behavioral response of rats given 5-HT agonists (Cowen et al., 1982). Green (1985) showed that clenbuterol in mice enhanced behavioral responses elicited by 5-HT but did not produce adaptive changes in 5-HT receptors. Thus there appears to be a functional link between 5-HT and NE neuronal circuits in brain, and both could be involved in the expression of antidepressant activity. This link must be kept in mind when designing models for discovering potential antidepressants.

Some aspects of the revised monoamine hypothesis are being challenged. The necessity of 14- to 21-day administration of drug to elicit adaptive changes in β receptors is no longer tenable for at least desmethylimipramine (Frazer et al., 1985). Bupropion appears not to alter β receptors or to desensitize the cyclase after chronic administration in rats (Ferris et al., 1981; Ferris and Beaman, 1983), findings that suggest there may be yet unknown sites at which antidepressants have their pharmacological effects. The probability is high that research into the mechanisms regulating the amplification of the second messenger system (such as cyclic AMP-mediated phosphorylation), the relationship between 5-HT and NE neuronal function, steroid hormone influences on NE receptor-coupled adenylate cyclase activity, phospholipid methylation (for review see Sulser, 1982), and more recently GABA-B involvement (Lloyd and Pilc, 1984) will give rise to new theories for antidepressant activity and new biochemical models for the discovery of novel antidepressants.

Pharmacological Interactional Models

Models in which antidepressant activity is predicted by the interaction of a target compound with a challenge substance are among the oldest and most widely used in industry. Contemporary theoretical knowledge, such as the observation that reserpine produces symptoms of depression, determined the selection of challenge substances. The models have survived because of their concurrent validity—their capacity to identify new drugs with mechanisms of action different from those of the drugs used to develop the models.

The reserpine- and tetrabenazine-antagonism models (Rubin et al., 1957; Vernier et al., 1962) assess the capacity of a target compound to reverse ptosis, hypothermia, or sedation produced by reserpine or tetrabenazine. The yohimbine model (Gershon et al., 1962; Malick, 1983) assesses the target compound's capacity to potentiate yohimbine's autonomic and behavioral effects in dogs (Sanghvi et al., 1969) or its lethality in mice (Quinton, 1963). Other interactional models include potentiation of amphetamine, potentiation of L-DOPA (Everett, 1967), and reversal of the behavioral depression produced by 5-HTP (Aprison et al., 1978; Aprison and Hingtgen, 1981).

Although different investigators have used different procedures and emphasized different dependent measures, the authors have attempted to summarize the results of interactional models in Table 1. Yohimbine potenti-

TABLE 10.1. Pharmacological interaction models*

	Reserpine Antagonism	Tetra-Benazine Antagonism	Yohimbine Potentiation	Amphetamine Potentiation	5-HTP Reversal	L-DOPA Potentiation
TCAs	+	+	+	+−	+	+
MAOIs	+	+	+	+		
Bupropion	+	+	+	+		+
Fluoxetine	−	−	+		−	
Iprindole	−	+	+		+	−
Maprotiline	+		+			
Mianserin	−+	−+	+	−+	+	−
Nomifensine	+	+	+	+		
Salbutamol	+−		+			
Trazodone	+−		−	+−	+	−
Viloxazine	+	+	+	+	+?	
Zimelidine	+	+	+	−		
ECT		−	−			
Antipsychotics	+	−	−+	−		
Antihistamines	+	−	+	−+		+
Anticholinergics		−	+	−+		+
Anxiolytics		−	−	+−		
Stimulants	+	+	+	+		
Fenfluramine		−				
Mazindol						
Clonidine		−				
Methysergide		−	−	+	+	
Cyproheptidine		−	+	+		
GABAergics		−				
Quipazine		+	+			
Aspirin	+	−				
Propranolol		−				

*Data for this table were abstracted from Nagayama et al., 1981; Malick, 1983; Jesberger and Richardson, 1985; Leonard, 1984; Howard et al., 1981; de Graaf et al., 1985; Willner, 1984; and Delini-Stula, 1980.

+ = active; − = inactive.

ation is sensitive to TCAs, MAOIs, and atypical antidepressants with the exception of trazodone; ECS also is inactive; false positives include some antipsychotics, antihistamines, anticholinergics, stimulants, cyproheptadine, and quipazine. Tetrabenazine antagonism is sensitive to TCAs, MAOIs, and all atypical antidepressants tested except mianserin and fluoxetine; it yields false positives for only DA receptor agonists and quipazine; it is in general quite selective. The other interactional models are less sensitive and less selective. Yohimbine potentiation and tetrabenazine antagonism meet the criteria of simplicity and economy.

Interactional models have been criticized for having no face validity (Willner, 1984) and little utility in the search for new antidepressants. However, these tests have identified compounds with novel mechanisms of action—e.g., iprindole (Gluckman and Baum, 1969) and bupropion (Ferris and Beaman, 1983; Maxwell, 1983). These drugs have in turn helped to force revision of theories about the mechanisms of action of antidepressants. However, uncritical screening with interactional models could result in selective identification of compounds similar to existing drugs. Novel compounds can be identified by imposition of additional criteria such as structural, pharmacological, and biochemical uniqueness (Maxwell, 1984).

Observational Models

An observational model is defined here as one in which a behavior is observed following imposition of a non-pharmacological stimulus condition. Results obtained in these models have been reviewed extensively (Howard et al., 1981; de Graaf et al., 1985; Willner, 1984; Jesberger and Richardson, 1985).

TCAs, MAOIs, and atypical antidepressants (except salbutamol) inhibit muricidal behavior, as does ECS (Horovitz, 1966; Vogel, 1975). However, stimulants, anticholinergics, antihistamines, GABAergic compounds, and anorectics also inhibit muricide. Poor selectivity renders this model useful only as part of a battery.

TCAs, MAOIs, and some atypical antidepressants reverse immobility in the behavioral despair model (Porsolt et al., 1977; Browne, 1979). Chlorimipramine, trazodone, zimelidine, and salbutamol are inactive. Stimulants, anticholinergics, antihistamines, convulsants, L-aspartic acid, pentobarbital, and some opiates yield false positives. The test is easy to run but misses some active antidepressants and has poor selectivity. Therefore it is useful only as part of a battery. Muricide and behavioral despair have been widely used in industry.

The following four models are not widely used: In the olfactory bulbectomy model (Cairncross, 1984; Van Riezen et al., 1977), removal of the olfactory bulbs in the rat produces a deficit in passive avoidance and an increase in behavioral reactivity. In general, chronic but not acute administration of TCAs and atypical antidepressants reverses the deficits. Some atypical antidepressants are effective acutely also. MAOIs appear to be inactive. Baclofen and quipazine yield false positives. Antipsychotics and anxiolytics do not reverse the deficit in passive avoidance, but they do reverse the hyperreactivity. This remains an interesting model, because many antidepressants are active only following subchronic or chronic treatment. However, there is some question about sensitivity and selectivity, and the model would probably be too time-consuming and expensive as a primary method for identifying putative antidepressants among a large number of candidates.

In the learned helplessness model (Maier, 1984), subjecting an animal to inescapable shock produces a constellation of effects, including deficits in the ability to acquire a shuttlebox escape response. Chronic but not acute administration of TCAs, MAOIs, several atypical antidepressants, and ECS reverses helplessness, whereas chronic administration of anxiolytics, antipsychotics, stimulants, or depressants is ineffective. However, L-DOPA, apomorphine, clonidine, scopolamine, and intrahippocampal GABA yield false positives. A wider range of atypical antidepressants should be tested. The lack of complete selectivity and the time- and labor-intensiveness of this test suggest that it would not be useful as a primary screen. Also there are methodological differences between laboratories, and no one version of the paradigm shows all the features of interest (Weiss, this volume).

A recently developed model in which a 3-week period of chronic unpredictable stress produces behavioral deficits (Katz, 1981) has been shown to be sensitive to the chronic administration of some typical and atypical antidepressants and to ECS. The need for chronic administration has been established only for imipramine; few antidepressants have been tested; tranylcypromine was inactive. Results have been limited to a single laboratory, and the model appears to be time-consuming.

Behavioral change that follows separation of animals that have formed affectional bonds (McKinney, 1977; Crawley, 1984) has been proposed as a model possessing face validity. However, the change in question does not always occur (Porsolt, 1983). The paucity of data does not allow evaluation of the sensitivity to or selectivity for antidepressants. The expense and time involved preclude industrial use (Howard and Pollard, 1983).

Operant Behavioral Models

Antidepressants appear not to have any specific effects on schedule-controlled behavior (e.g., Rastogi and McMillan, 1985), nor are they distinguished as a class in the drug discrimination paradigm (Jones et al., 1980). Two tasks are worthy of comment, intracranial self-stimulation (ICSS) and differential reinforcement of low rate >72 (DRL>72).

Since the discovery of ICSS (Olds and Milner, 1954), many explanations of reward and pleasure have been offered. Because a defining condition of depression is a decrease in the capacity to experience pleasure, ICSS has been proposed as a tool in understanding the disease, and the effects of antidepressants have been investigated (e.g., Willner, 1984). Two factors argue against any simple use of ICSS to model depression. First, there is great controversy over methodology and interpretation of response rate data (Atrens, 1984; Liebman, 1983; Koob, this volume). Second, antidepressants do not cause any consistent effects in the various ICSS paradigms in which they have been tested. Attempts have been made to increase the usefulness of ICSS by testing interactions with amphetamine (e.g., Goldstein and Malick, 1983), producing deficits in response rate by withdrawal from chronic amphetamine (Leith and Barrettt, 1980), and producing deficits in response rate by lesioning the internal capsule (Cornfeldt et al., 1982; Koob, this volume). None of these procedures has been sufficiently validated with known antidepressants.

Over the past six years Seiden and coworkers (e.g., Seiden and O'Donnell, 1985) have attempted to show that the behavior of rats pressing a lever on a DRL>72 schedule of reinforcement is differentially sensitive to antidepressants. TCAs, MAOIs, and several atypical antidepressants increase the number of reinforcers earned during an operant session. Data have been offered to show that a limited number of non-antidepressants are negative. Sensitivity is imperfect: Bupropion and nomifensine are inactive. Selectivity is questionable: Other investigators (see Pollard and Howard, 1986) have found that antipsychotics and manipulation of motivation produce effects similar to those produced by antidepressants.

Electrophysiological Models

Electrophysiological models of depression have developed in parallel with the evolution of the monoamine hypothesis. The most common procedure is to use microelectrode recordings in single neurons known to function by a specific neurotransmitter—primarily NE in the locus coeruleus and 5-HT in the raphé nucleus—to study changes produced by an antidepressant on spontaneous or evoked activity. These models, which are relatively expensive to use in the industrial laboratory, provide quantitative analysis of compounds selected by the more economical behavioral and neurochemical screens. They offer the possibility of elucidating the nature of depression and the therapeutic actions of antidepressants.

The most common preparation is an anesthetized rat or cat placed in a stereotaxic apparatus and used once. Drug is administered systemically or iontophoretically onto the target neuron. Systemic TCAs block reuptake and, through a feedback mechanism, decrease spontaneous firing in locus coeruleus and raphé. Buspirone also decreases firing rate in locus coeruleus by an undetermined mechanism. The electrophysiological data confirm the neurochemical data: Tertiary TCAs that block 5-HT reuptake preferentially block spontaneous 5-HT firing; secondary TCAs that block NE reuptake preferentially block spontaneous NE firing (Sheard et al., 1972; Nybäck et al., 1975; Scuvée-Moreau and Dresse, 1979).

Changes in pre- and postsynaptic function induced by long-term treatment with antidepressants (1 to 4 weeks) have been investigated. Chronic TCAs either decreased or did not change responsiveness of inhibitory presynaptic α_2 receptors, enhanced excitatory postsynaptic α_1 receptors, and either decreased or did not change sensitivity of β receptors. Chronic TCAs enhanced sensitivity of postsynaptic 5-HT receptors in dorsal raphé to iontophoretically applied 5-HT but did not change sensitivity of 5-HT autoreceptors (Charney et al., 1981). These data suggest that TCAs may produce a slowly developing modulation of NE and 5-HT receptor activity that could be responsible for clinical antidepressant effect.

Anesthetization potentiates responsiveness of CNS neurons to many drugs (Trulson, 1984). The face validity of electrophysiological models would be greater in the unanesthetized cerveau isolé preparation, the peripherally paralyzed and ventilated preparation, or, preferably, the awake, freely moving animal. However, in these cases the greater presumed validity entails prohibitive difficulty and expense in the industrial laboratory except for a few targeted compounds.

The use of brain slices in place of whole-animal preparations has gained popularity (Dingledine et al., 1980; Dingledine, 1984; Kerkut and Wheal, 1981). This technique offers some advantages: Both electrophysiological and biochemical experiments can be done on the same tissue; concentrations of drugs in the tissue bath can be controlled precisely; neuroanatomical connections can be preserved; and intracellular recording is easier and more consistent. Loci examined in the whole-animal preparation have been examined in slices, and comparable results have been obtained (Aghajanian and Lakoski, 1984; Williams et al., 1985). Too little is known at present to judge the usefulness of slice preparations.

Summary

The authors have attempted to show that the goal of developing new drugs obliges the industrial researcher to evaluate models pragmatically. The most important criteria are that the model (test, screen, procedure) be sensitive to and selective for clinically active antidepressants and that it be simple and economical. No single model satisfies the first criterion; therefore, batteries of tests must be used (Cooper et al., 1983). Although this approach, applied uncritically, can lead to the production of "me too" drugs, it has identified new antidepressants with novel mechanisms of action, a result that in turn has forced revision of theories of depression and has influenced the development of more refined models. Progress depends upon enhanced cooperation among academic, industrial, and clinical researchers.

References

Abramson LY, Seligman MEP (1977): Modelling psychopathology in the laboratory: History and rationale. In: *Psychopathology: Experimental Models,* Maser JD, Seligman MEP, eds. San Francisco: Freeman

Aghajanian GK, Lakoski JM (1984): Hyperpolarization of serotonergic neurons by serotonin and LSD: Studies in brain slices showing increased K^+ conductance. *Brain Res* 305:181-185

Aprison MH, Hingtgen JN (1981): Hypersensitive serotonergic receptors: A new hypothesis for one subgroup of unipolar depression derived from an animal model. In: *Serotonin—Current Aspects of Neurochemistry and Function,* Haber B, Gabay S, Alivisatos S, Issidorides M, eds. New York: Plenum

Aprison MH, Takahashi R, Tachiki K (1978): Hypersensitive serotonergic receptors involved in clinical depression—A theory. In: *Neuropharmacology and Behavior,* Haber B, Aprison MH, eds. New York: Plenum

Atrens DM (1984): Self-stimulation and psychotropic drugs: A methodological and conceptual critique. In: *Animal Models in Psychopathology,* Bond NW, ed. New York: Academic Press

Baldessarini RJ (1985): Drugs and the treatment of psychiatric disorders. In: *Goodman and Gilman's The Pharmacological Basis of Therapeutics,* Gilman AG, Goodman LS, Rall TW, Murad F, eds. New York: Macmillan

Bond NW (1984): Animal models in psychopathology: An introduction. In: *Animal Models in Psychopathology,* Bond NW, ed. New York: Academic Press

Browne RG (1979): Effects of antidepressants and anticholinergics in a mouse "behavioral despair" test. *Eur J Pharmacol* 58:331-334

Brunello N, Barbaccia ML, Chuang DM, Costa E (1982): Down-regulation of β-adrenergic receptors following repeated injections of desmethylimipramine: Permissive role of serotonergic axons. *Neuropharmacology* 21:1145-1149

Bunney WE Jr, Davis JM (1965): Norepinephrine in depressive reactions: A review. *Arch Gen Psychiatry* 13:483-494

Cairncross KD (1984): Olfactory bulbectomy as a model of depression. In: *Animal Models in Psychopathology,* Bond NW, ed. New York: Academic Press

Carlsson A (1984): Current theories on the mode of action of antidepressant drugs. *Adv Biochem Psychopharmacol* 39:213-221

Charney DS, Menkes DB, Heninger GR (1981): Receptor sensitivity and the mechanism of action of antidepressant treatment. *Arch Gen Psychiatry* 38:1160-1180

Cooper BR, Howard JL, Soroko FE (1983): Animal models used in prediction of antidepressant effects in man. *J Clin Psychiatry* 44:63-66

Cornfeldt M, Fisher B, Fielding S (1982): Rat internal capsule lesion: A new test for detecting antidepressants. *Fed Proc* 41:1066

Cowen PJ, Grahame-Smith DG (1983): Introduction. In: *Preclinical Psychopharmacology,* Grahame-Smith DG, Cowen PJ, eds. Amsterdam: Excerpta Medica

Cowen PJ, Grahame-Smith DG, Green AR, Heal DJ (1982): β-Adrenoceptor agonists enhance 5-hydroxytryptamine-mediated behavioural responses. *Br J Pharmacol* 76:265-270

Crawley JN (1984): Preliminary report of a new rodent separation model of depression. *Prog Neuropsychopharmacol Biol Psychiatry* 8:447-457

Cronholm, B (1984): The concept of depression: Diagnosis and Classification. In: *Frontiers in Biochemical and Pharmacological Research in Depression,* Usdin E, Asberg M, Bertilsson L, Sjöqvist F, eds. New York: Raven Press

De Graaf JS, Van Riezen H, Berendsén HHG, Van Delft AML (1985): A set of behavioural tests predicting antidepressant activity. *Drug Dev Res* 5:291-301

Delini-Stula A (1980): Drug-induced alterations in animal behavior as a tool for the evaluation of antidepressants: Correlation with biochemical effects. In: *Psychotropic Agents, Part I: Antipsychotics and Antidepressants,* Hoffmeister F, Stille G, eds. Berlin: Springer-Verlag

Dingledine R, ed. (1984): *Brain Slices.* New York: Plenum

Dingledine R, Dodd J, Kelly JS (1980): The *in vitro* brain slice as a useful neurophysiological preparation for intracellular recording. *J Neurosci Methods* 2:323-362

Engberg G, Svensson TH (1980): Mianserin: Direct alteration of brain norepinephrine neurons by blocking alpha$_2$-receptors. *Commun Psychopharmacol* 4:233-239

Everett GM (1967): The dopa response potentiation test and its use in screening for antidepressant drugs. In: *Antidepressant Drugs,* Garattini S, Dukes M, eds. Amsterdam: Excerpta Medica

Everitt BJ, Keverne EB (1979): Models of depression based on behavioral observations of experimental animals. In: *Psychopharmacology of Affective Disorders,* Paykel ES, Coppen A, eds. New York: Oxford University Press

Ferris RM, Beaman OJ (1983): Bupropion: A new antidepressant drug, the mechanism of action of which is not associated with down-regulation of postsynaptic β-adrenergic, serotonergic (5-HT$_2$), α_2-adrenergic, imipramine and dopaminergic receptors in brain. *Neuropharmacology* 22:1257-1267

Ferris RM, White HL, Cooper BR, Maxwell RA, Tang FLM, Beaman OJ, Russell A (1981): Some neurochemical properties of a new antidepressant, bupropion hydrochloride (Wellbutrin®). *Drug Dev Res* 1:21-35

Frazer A, Lucki I, Sills M (1985): Alterations in monoamine-containing neuronal function due to administration of antidepressants repeatedly to rats. *Acta Pharmacol Toxicol* 56 (Suppl 1):21-34

Frazer A, Pandey G, Mendels J, Neeley S, Kane M, Hess ME (1974): The effects of tri-iodothyronine in combination with imipramine on [³H]-cyclic AMP production in slices of rat cerebral cortex. *Neuropharmacology* 13:1131-1140

Fuller RW (1986): Pharmacologic modification of serotonergic function: Drugs for the study and treatment of psychiatric and other disorders. *J Clin Psychiatry* 47 (Suppl 4):4-8

Fuller RW, Wong DT (1985): Effects of antidepressants on uptake and receptor systems in the brain. *Prog Neuropsychopharmacol Biol Psychiatry* 9:485-490

Gershon S, Holmberg G, Mattsson E, Mattsson N, Marshall A (1962): Imipramine hydrochloride: Its effects on clinical, autonomic and physiological functions. *Arch Gen Psychiatry* 6:96-101

Gluckman MI, Baum T (1969): The pharmacology of iprindole, a new antidepressant. *Psychopharmacologia* 15:169-185

Goldstein JM, Malick JB (1983): An automated descending rate-intensity self-stimulation paradigm: Usefulness for distinguishing antidepressants from neuroleptics. *Drug Dev Res* 3:29-35

Green AR (1985): Antidepressant treatments and serotonin receptor number and function. *Acta Pharmacol Toxicol* 56 (Suppl 1):128-137

Horovitz ZP (1966): The relationship of the amygdala to the mechanism of action of two types of antidepressants. In: *Recent Advances in Biological Psychiatry,* Wortis J, ed. New York: Plenum

Howard JL, Pollard GT (1983): Are primate models of neuropsychiatric disorders useful to the pharmaceutical industry? In: *Ethopharmacology: Primate Models of Neuropsychiatric Disorders.* Miczek KA, ed. New York: Alan R. Liss

Howard JL, Soroko FE, Cooper BR (1981): Empirical behavioral models of depression, with emphasis on tetrabenazine antagonism. In: *Antidepressants: Neurochemical, Behavioral, and Clinical Perspectives,* Enna SJ, Malick JB, Richelson E, eds. New York: Raven Press

Janowsky A, Okada F, Manier DH, Applegate CD, Sulser F, Steranka L (1982): Role of serotonergic input in the regulation of the β-adrenergic receptor-coupled adenylate cyclase system. *Science* 218:900-901

Jesberger JA, Richardson JS (1985): Animal models of depression: Parallels and correlates to severe depression in humans. *Biol Psychiatry* 20:764-784

Johnston JP (1968): Some observations upon a new inhibitor of monoamine oxidase in brain tissue. *Biochem Pharmacol* 17:1285-1297

Jones CN, Howard JL, McBennett ST (1980): Stimulus properties of antidepressants in the rat. *Psychopharmacology* 67:111-118

Katz RJ (1981): Animal models and human depressive disorders. *Neurosci Biobehav Rev* 5:231-246

Kerkut GA, Wheal HV, eds. (1981): *Electrophysiology of Isolated Mammalian CNS Preparations.* New York: Academic Press

Klerman GL (1984): History and development of modern concepts of affective illness. In: *Neurobiology of Mood Disorders,* Post RM, Ballenger JC, eds. Baltimore: Wllllamis and Wllklns

Kuhn R (1958): The treatment of depressive states with G 22355 (imipramine hydrochloride). *Am J Psychiatry* 115:459-464

Lecrubier Y, Puech AJ, Jouvent R, Simon P, Widloch D (1980): A beta adrenergic stimulant (salbutamol) versus clomipramine in depression: A controlled study. *Br J Psychiatry* 136:354-358

Leith NJ, Barrett RJ (1980): Effects of chronic amphetamine or reserpine on self-stimulation responding: Animal model of depression? *Psychopharmacology* 72:9-15

Leonard BE (1984): Pharmacology of new antidepressants. *Prog Neuropsychopharmacol Biol Psychiatry* 8:97-108

Liebman JM (1983): Discriminating between reward and performance: A critical review of intracranial self-stimulation methodology. *Neurosci Biobehav Rev* 7:45-72

Lloyd KG, Pilc A (1984): Chronic antidepressants and GABA synapses. *Neuropharmacology* 23:841-842

Maas JW (1975): Biogenic amines and depression: Biochemical and pharmacological separation of two types of depression. *Arch Gen Psychiatry* 32:1357-1361

Maier SF (1984): Learned helplessness and animal models of depression. *Prog Neuropsychopharmacol Biol Psychiatry* 8:435-446

Maj J, Przegalinski E, Mogilnicka E (1984): Hypotheses concerning the mechanism of action of antidepressant drugs. *Rev Physiol Biochem Pharmacol* 100:1-74

Malick JB (1983): Potentiation of yohimbine-induced lethality in mice: Predictor of antidepressant potential. *Drug Dev Res* 3:357-363

Mandell AJ, Segal DS, Kuczenski R (1975): Metabolic adaptation to antidepressant drugs: Implications for pathophysiology and treatment in psychiatry. In: *Catecholamines and Behavior. 2. Neuropsychopharmacology,* Friedhoff AJ, ed. New York: Plenum

Maxwell RA (1983): Second generation antidepressants: The pharmacological and clinical significance of selected examples. *Drug Dev Res* 3:203-211

Maxwell RA (1984): The state of the art of the science of drug discovery—an opinion. *Drug Dev Res* 4:375-389

McKinney WT Jr, Bunney WE Jr (1969): Animal model of depression. I. Review of evidence: implications for research. *Arch Gen Psychiatry* 21:240-248

McKinney WT Jr (1974): Animal models in psychiatry. *Perspect Biol Med* 17:529-541

McKinney WT Jr (1977): Biobehavioral models of depression in monkeys. In: *Animal Models in Psychiatry and Neurology,* Hanin I, Usdin E, eds. Oxford: Pergamon Press

Mishra R, Janowsky A, Sulser F (1980): Action of mianserin and zimelidine on the norepinephrine receptor coupled adenylate cyclase system in brain: Subsensitivity without reduction in β-adrenergic receptor binding. *Neuropharmacology* 19:983-988

Nagayama N, Hingtgen JN, Aprison MH (1981): Postsynaptic action by four antidepressive drugs in an animal model of depression. *Pharmacol Biochem Behav* 15:125-130

Nybäck HV, Walters JR, Aghajanian GK, Roth RH (1975): Tricyclic antidepressants: Effects on the firing rate of brain noradrenergic neurons. *Eur J Pharmacol* 32:302-312

Olds J, Milner P (1954): Positive reinforcement produced by electrical stimulation of septal area and other regions of rat brain. *J Comp Physiol Psychol* 47:419-427

Pollard GT, Howard JL (1986): Similar effects of antidepressant and non-antidepressant drugs on behavior under an interresponse-time >72-s schedule. *Psychopharmacology* 89:253-258

Porsolt RD, Le Pichon M, Jalfre M (1977): Depression: A new animal model sensitive to antidepressant treatments. *Nature* 266:730-732

Porsolt RD (1983): Failure of repeated peer separations to induce depression in infant rhesus monkeys. *Drug Dev Res* 3:567-572

Quinton RM (1963): The increase in the toxicity of yohimbine induced by imipramine and other drugs in mice. *Br J Pharmacol* 21:51-66

Racagni G, Mocchetti I, Calderini G, Battistella A, Brunello N (1983): Temporal sequence of changes in central noradrenergic system of rat after prolonged antidepressant treatment: Receptor desensitization and neurotransmitter interactions. *Neuropharmacology* 22:415-424

Rastogi SK, McMillan DE (1985): Effects of some typical and atypical antidepressants on schedule-controlled responding in rats. *Drug Dev Res* 5:243-250

Richelson E, El-Fakahany E (1982): Changes in the sensitivity of receptors for neurotransmitters and the actions of some psychotherapeutic drugs. *Mayo Clin Proc* 57:576-582

Richelson E, Pfenning M (1984): Blockade by antidepressants and related compounds of biogenic amine uptake into rat brain synaptosomes: Most antidepressants selectively block norepinephrine uptake. *Eur J Pharmacol* 104:277-286

Rubin B, Malone MH, Waugh MH, Burke JC (1957): Bioassay of Rauwolfia roots and alkaloids. *J Pharmacol Exp Ther* 120:125-136

Russell RW, Overstreet DH (1984). Animal models in neurobehavioral toxicology. In:*Animal Models in Psychopathology,* Bond NW, ed. New York: Academic Press

Sanghvi I, Bindler E, Gershon S (1969): The evaluation of a new animal method for the prediction of clinical anti-depressant activity. *Life Sci* 8:99-106

Sansone M, Melzacka M, Hano J, Vetulani J (1983): Reversal of depressant action of trazodone on avoidance behaviour by its metabolite m-chlorophenylpiperazine. *J Pharm Pharmacol* 35:189-190

Scuvée-Moreau JJ, Dresse AE (1979): Effect of various antidepressant drugs on the spontaneous firing rate of locus coeruleus and dorsal raphe neurons of the rat. *Eur J Pharmacol* 57:219-225

Seiden LS, O'Donnell JM (1985): Effects of antidepressant drugs on DRL behavior. In: *Behavioral Pharmacology: The Current Status*, Seiden LS, Balster RL, eds. New York: Alan R. Liss

Sellinger-Barnette MM, Mendels J, Frazer A (1980): The effect of psychoactive drugs on beta-adrenergic receptor binding sites in rat brain. *Neuropharmacology* 19:447-454

Sheard MH, Zolovick A, Aghajanian GK (1972). Raphe neurons: Effect of tricyclic antidepressant drugs. *Brain Res* 43:690-694

Sulser F (1978): Functional aspects of the norepinephrine receptor coupled adenylate cyclase system in the limbic forebrain and its modification by drugs which precipitate or alleviate depression: Molecular approaches to an understanding of affective disorders. *Pharmakopsychiatry Neuropsychopharmakol* 11:43-52

Sulser F (1982): Regulation and adaptation of central norepinephrine receptor systems: Modification by antidepressant treatments. *Psychiatr J Univ Ottawa* 7:196-203

Sulser F, Vetulani J, Mobley PL (1978): Commentary: Mode of action of antidepressant drugs. *Biochem Pharmacol* 27:257-261

Trulson ME (1984): Pharmacological investigation of CNS unit responses in awake, freely moving animals. *Trends Pharmacol Sci* 5:287-289

Van Riezen H, Schnieden H, Wren AF (1977): Olfactory bulb ablation in the rat: Behavioural changes and their reversal by antidepressive drugs. *Br J Pharmacol* 60:521-528

Vernier VG, Hanson HM, Stone CA (1962): The pharmacodynamics of amitriptyline. In: *Psychosomatic Medicine*, Nodine JH, Moyer JH, eds. Philadelphia: Lea and Febiger

Vetulani J, Dingell JV, Sulser F (1974): Effect of chronic treatment with desipramine (DMI) and iprindole (IP) on the norepinephrine (NE) sensitive adenylate cyclase system in slices of the rat limbic forebrain (LFS). *The Pharmacologist* 16:287

Vetulani J, Stawarz RJ, Dingell JV, Sulser F (1976): A possible common mechanism of action of antidepressant treatments: Reduction in the sensitivity of the noradrenergic cyclic AMP generating system in the rat limbic forebrain. *Naunyn Schmiedebergs Arch Pharmacol* 293:109-114

Vogel JR (1975): Antidepressants and mouse-killing (muricide) behavior In: *Antidepressants*, Fielding S, Lal H, eds. New York: Futura

Weatherall M (1985): How are drugs discovered? In: *Pharmaceutical Medicine*, Burley DM, Binns TB, eds. London: Edward Arnold

Weissman A, Koe BK (1987): Contributions of industrial research to basic neuropharmacology: Pre-clinical screening and discovery. In: *Psychopharmacology: The Third Generation of Progress*, Meltzer HY et al., eds. New York: Raven Press

William JT, Henderson G, North RA (1985). Characterization of α_2-adrenoceptors which increase potassium conductance in rat locus coeruleus neurones. *Neuroscience* 14:95-101

Willner P (1984): The validity of animal models of depression. *Psychopharmacology* 83:1-16

11

Pharmacological, Biochemical, and Behavioral Analyses of Depression: Animal Models

ROBERT M. ZACHARKO AND HYMIE ANISMAN

It has been posited that stressful events may precipitate depression in humans (Abramson et al, 1978; Akiskal and McKinney, 1973; Anisman and Zacharko, 1982a). Alternatively, it is not unlikely that such events may simply exacerbate symptoms in already depressed individuals (Slater and Roth, 1969), or that the response to stressors may be symptomatic of an already existant depression (Hudgens et al. 1967; Morrison et al., 1968). Although it is clear that stressful events may profoundly influence the behavior of animals in various testing paradigms, the mechanisms subserving these behavioral alterations remain to be fully elucidated. Furthermore, there is still some question as to whether these behavioral changes can legitimately be considered as valid models of human depression (Willner, 1985). Among other things, several forms of depression exist, and even in one type of depression the symptoms presented may vary considerably across individuals. Indeed, the view has been expressed that depression may be a biochemically heterogeneous illness, wherein the symptoms may be a consequence of serotonin (5-HT) or norepinephrine (NE) neuronal dysfunction and possibly dopamine (DA) variations as well (Jimerson and Post, 1984; Schildkraut, 1978; van Praag, 1978, 1984). In light of these considerations, it is probably inappropriate (and perhaps even counterproductive) to develop unitary animal models that focus primarily on one neurotransmitter at the expense of others. In the sections that follow, an overview will be presented concerning the effects of stressors on behavioral processes in animals, as well as an analysis of pharamacological treatments on these behavioral disturbances.

Neurochemical Concomitants of Uncontrollable Stressors

Several investigators (Anisman, 1984; Anisman and Zacharko, 1982a; Weiss et al., 1976; Weiss and Goodman, 1985) have suggested that the behavioral disturbances associated with stressors may result from neurochemical alterations. It is assumed that when an organism is exposed to an environmental stressor, it will attempt behavioral methods of contending with the aversive stimuli. Concurrently, several neurochemical changes may occur, presumably as an adaptive response to meet environmental demands. For instance, these neurochemical changes may

1. enable the organism to both initiate and maintain appropriate defensive strategies,
2. increase vigilance or attention to biologically significant environmental stimuli,
3. blunt the psychological or physical impact of the stressor, and
4. provoke hormonal and immunological alterations that protect the organism from potentially harmful antigens.

If these adaptive changes do not occur, then the organism may be more vulnerable to behavioral pathology.

Norepinephrine

Figure 11.1 depicts a schematic representation of the norepinephrine (NE) alterations that occur under various stressor conditions. It is assumed that the impact of stressors will vary across strains, and even within a strain the effects of the stressors will be dependent upon a host of experiential,

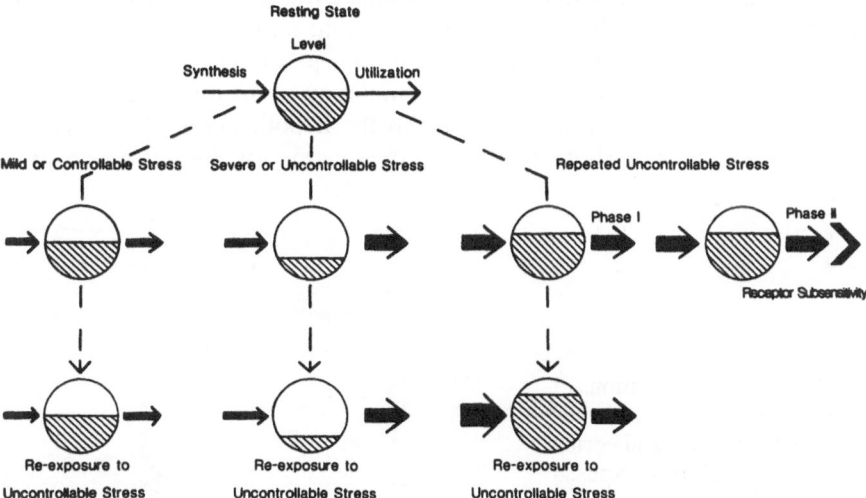

FIGURE 11.1. Schematic representation of the changes of NE synthesis, utilization, and levels under various conditions of aversive stimulation (acute controllable or uncontrollable stress, as well as chronic stress) and upon reexposure to a stressor at some time following the initial insult. Reprinted with permission of the New York Academy of Sciences from Anisman and Zacharko, *Ann New York Acad Sci* 476:213, 1986.

organismic, and environmental factors. Thus, in considering the potential consequences of environmental insults, it is essential that the effects of the stressor be considered against the backdrop upon which they are superimposed. When an organism is exposed to an environmental challenge that can be dealt with through behavioral means, the utilization of NE is increased, as is its synthesis, and amine concentrations remain fairly stable. When behavioral methods of contending with the stressor are unavailable (or are perceived as being unavailable), NE utilization increases, ultimately exceeding the rate at which the transmitter is synthesized, and consequently amine concentrations are reduced. It is thought that the reduced levels of the transmitter may render the organism more vulnerable to illness (Anisman, 1984).

In evaluating the contribution of coping factors to NE variations, several factors must be considered. As shown in Figure 11.2, exposure to escapable and inescapable shock have been shown to influence NE concentrations differentially, with only the latter treatment reducing the amine concentrations (see also Weiss et al., 1979). Thus, it might be assumed that the NE reductions are a consequence of animals "learning" that they cannot control stressor offset (i.e., animals learn that they are helpless; Maier and Seligman, 1976). Alternatively, the view might be entertained that any stressor, if sufficiently severe or protracted, will ultimately result in reductions of amine concentrations, but as animals gain mastery over the response contingencies, the rate of amine utilization will be reduced, and the decline of amine concentrations will be limited. In fact, using a free operant-avoidance procedure, Tsuda and Tanaka (1985) reported that both escapable and yoked inescapable shock applied over a 3- or 6-hour period increased the utilization of NE (as determined from the accumulation of the metabolite, MHPG) and reduced the concentrations of NE in the hypothalamus. Indeed, it appeared that escapable shock was more effective in provoking these changes.

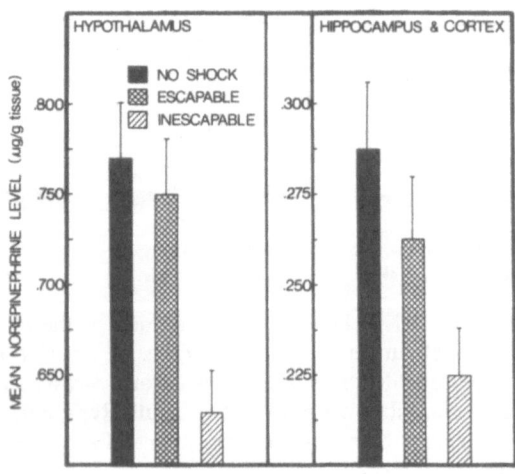

FIGURE 11.2. Concentrations of NE (± SEM) in hypothalamus and in hippocampus and cortex of mice that had been exposed to either 60 escapable shocks, yoked inescapable shock, or no shock. Reprinted with permission of Elsevier Science Publishers B.V. Amsterdam from Anisman et al., *Brain Research* 191:585, 1980.

Following 21 hours of training, so that the response was well established in the escapable group, the utilization of NE was less marked than it was in the yoked group. Likewise, in animals that had previously been trained to emit the avoidance response, subsequent exposure to either escapable or inescapable shock did not lead to reduced NE concentrations, although utilization of the amine was more pronounced in rats of the yoked inescapable group than in animals reexposed to escapable shock.

A second factor that warrants consideration concerns the potential interaction between coping ability and the specific response required of an animal. Although it is typically accepted that lack of control invariably lends itself to amine alterations whereas controllable stressors do not, there are occasions when the controllable stressor may actually lead to more pronounced amine variations than the uncontrollable stressor. As already indicated in the Tsuda and Tanaka (1985) study the NE reduction associated with three hours of escapable shock was greater than an equivalent amount of inescapable shock. It seems that engaging in the escape response may initially result in greater transmitter utilization than does exposure to shock per se. Only when mastery of the response is established does amine utilization drop off. Provisional data collected in this laboratory have also revealed that when animals are required to emit contraprepared responses in an aversive situation, the bio chemical changes associated with the stressor may be equal to or greater than those associated with an inescapable stressor (Prince et al., 1986a).

The effectiveness of stressors in provoking amine variations also appears to be dependent on the severity of the stressor employed, as well as the brain region examined. For instance, increased MHPG accumulation is readily apparent in hypothalamus and the amygdala, and reductions of the amine are evident even after moderate stressor application. In regions such as the hippocampus, cortex, and thalamus the altered NE utilization does not appear unless the stressor severity is increased, and only after fairly protracted sessions are amine concentrations reduced (Tanaka et al., 1982). Likewise, it has been reported that following a stressor of moderate severity the NE variations persist for a longer period of time in hypothalamus (3 hours) than in the hippocampus (.25 hour) (Anisman et al., 1987). Moreover it was reported that following a fairly pronounced stressor NE reductions in the locus coeruleus (LC) are apparent after a 72-hour interval, while in hypothalamus they persist only over a 48-hour period (Weiss et al., 1981). These findings raise several important considerations with respect to the behavioral sequelae of uncontrollable stressors. In attempting to assess the neurochemical and behavioral correlates of uncontrollable aversive events, it may not be sufficient to examine simply the magnitude of a particular biochemical change. Rather, it may be more appropriate to evaluate the time required for the neurochemical system to return to control values. A marked amine reduction in an intact animal may persist for only a brief period and behavioral disturbances may not be evident. In an animal that has been compromised in some way (e.g., owing to previous stressor experiences, pharmacological

FIGURE 11.3. Concentrations of
NE (± SEM) in hypothalamus
and hippocampus as a function
of shock treatment. Mice received either no shock on each
of 2 days (N/N), 60 shocks of
150 uA on 1 day followed by no
shock on the second day (60/N),
no treatment on the first day followed by 10 shocks on the
second day (N/10), or 60 shocks
on the first day followed by
10 shocks on the second day
(60/10). Reprinted with permission of Academic Press, London
from Anisman et al., *Theory in
Psychopharmacology*, Vol 1,
p 75, from Anisman et al., 1981.

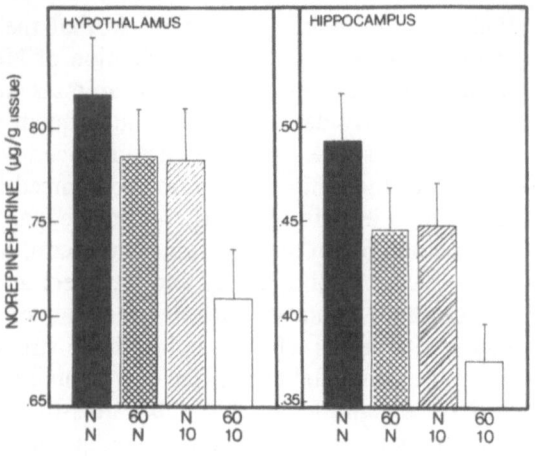

treatments, or age) amine depletions may be equally large, but may persist for lengthy periods, with behavioral disturbances more pronounced. In
effect, the period of amine (or hormone) recovery may reflect the adaptability or preparedness of the system, and hence may be a more appropriate
indicant of vulnerability to pathology.

Although the NE variations associated with an acute stressor are relatively
transient, it appears that exposure to an uncontrollable stressor may result
in the sensitization or conditioning of the mechanisms subserving the NE
variations. In particular, it has been demonstrated that in mice that had
previously been exposed to inescapable shock, subsequent exposure to even
a limited amount of shock increased the utilization of NE, and resulted
in a reduction of the amine concentrations (see Figure 11.3). Likewise,
Cassens and co-investigators (1980) reported that upon exposure to cues
that had previously been associated with footshock the accumulation of
MHPG increased, indicating that the enhanced NE utilization induced by
footshock is subject to conditioning effects. These data raise the possibility
that stressors may have behavioral or affective repercussions long after
the stressor was initially encountered. A traumatic event that results in
the sensitization of a neurochemical mechanism may render the organism
more vulnerable to behavioral disturbances, and even seemingly innocuous
environmental events may ultimately influence behavioral output. It was
recently demonstrated as well that exposure to a stressor may enhance the
response to subsequent drug treatments such as amphetamine or cocaine
(Anisman et al., 1985; Antelman and Chiodo, 1983; MacLennan and Maier,
1983). Accordingly, the possibility exists that aversive events may also
contribute to the enhancement of drug-induced psychotic behaviors.

In contrast to the NE reduction associated with an acute stressor, NE concentrations comparable to those of nonstressed animals are evident following repeated exposure to an aversive stimulus. For instance, the NE depletion associated with 20 minutes of restraint stress is absent following 180 minutes of restraint applied within a single session (Kvetnansky et al., 1976), and the NE reduction evident after one session of inescapable shock is not apparent in animals that received 15 sessions of the stressor on successive days (Anisman et al., 1987; Irwin, et al., 1986; Weiss et al., 1976). It seems that following repeated stressor application the utilization of NE is increased just as it is after an acute stressor, but a compensatory increase of synthesis is provoked, thereby preventing the reduction of amine concentrations (Kvetnansky et al., 1976; Weiss et al., 1976). Additionally, Weiss et al. (1976) reported that in animals exposed to repeated footshock NE uptake was reduced, thereby increasing the efficiency of transmitter that had been released.

The increased NE utilization associated with repeated footshock declines rapidly following stressor termination, and within 24 hours MHPG accumulation is comparable to that of nonstressed animals (e.g., Irwin et al., 1986). However, it appears that 24 hours following a chronic stress regimen the increased synthesis of NE persists. Indeed, at this time synthesis may actually exceed that seen soon after stressor application (e.g., Roth et al., 1982; Thierry et al., 1968), and NE concentrations may be increased above those evident in nonstressed animals (Anisman et al., 1987; Irwin et al., 1986; Roth et al., 1982). The reduction of utilization, coupled with the sustained increase of NE synthesis, presumably is responsible for the augmented amine levels and might represent an adaptive change to assure that transmitter levels are sufficient not only to deal with current stressors, but also with impending stressors.

As in the case of acute stressors, it appears that chronic stressors may have long-term repercussions. As shown in Figure 11.4, in animals that had been exposed to an acute stressor, reexposure to a limited amount of footshock two weeks later resulted in a reduction of NE concentrations. In animals exposed to a chronic stressor regimen (i.e., two-week period), subsequent reexposure to the stressor increased NE release and also resulted in increased concentrations of the amine. These effects were assumed to reflect the conditioning or sensitization of NE synthesis. Accordingly, it was proposed that having undergone neurochemical adaptation in response to a particular stressor, the potential impact of a subsequent stressor experience was minimized (Irwin et al., 1986).

It has been the author's contention that acute stressor application may increase vulnerability to affective changes, but that these would be relatively short-lived. Of course, those variables that maximize the degree or duration of the amine alterations (e.g., genetic or ontogenetic factors, as well as the organism's previous stressor history) would similarly influence vulnerability to affective disturbances. Following repeated stressor application affective

FIGURE 11.4. Hypothalamic NE and MHPG concentrations as a percent (± SEM) of control values in mice exposed to no shock, acute inescapable shock, or repeated shock over 15 days followed 2 weeks later by either limited shock exposure or placement in the apparatus without being shocked. Reprinted with permission of Elsevier Science Publishers B.V. Amsterdam from Irwin et al., *Brain Research* 379:101, 1986.

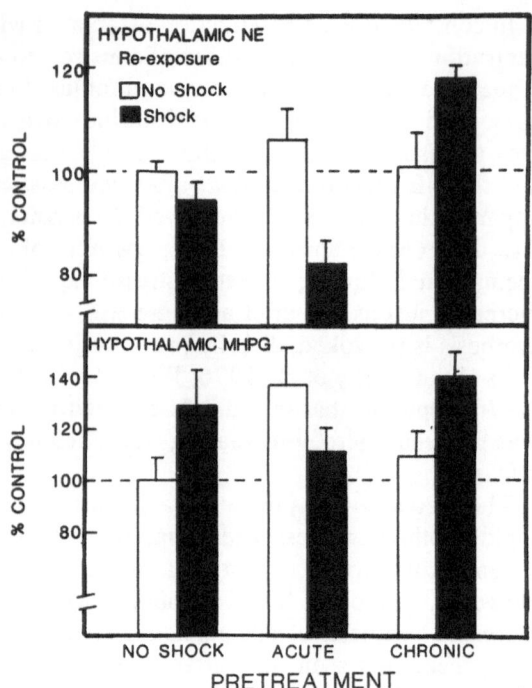

disturbances might not be expected, provided the adaptive changes actually occurred. It is conceivable, however, that the adaptive changes would not progress comparably across individuals, and those factors that hinder adaptive neurochemical variations would favor the development of the behavioral or affective disturbances. In such individuals, pharmacological intervention (e.g., antidepressants) would be necessary to promote the adaptive variations.

In examining the effects of chronic stressors, some investigators have reported that repeated exposure to environmental insults may result in several changes beyond the alterations of NE turnover. In particular, it was reported such stressor regimens were associated with β-NE receptor subsensitivity and a diminution in the activity of NE-sensitive cAMP (Nomura et al., 1981; Stone, 1983; Stone et al., 1984; U'Prichard and Kvetnansky, 1980). It seems as well that the receptor and cAMP variations may be independent of one another, since the receptor alterations typically are evident soon after the last stressor session, whereas the cAMP variations are apparent 24 hours after stressor termination (U'Prichard and Kvetnansky, 1980; Stone, 1979). In light of the proposition that the clinical effectiveness of antidepressants may likewise stem from the down regulation of β-NE coupled adenylate cyclase system, rather than alterations of NE concentrations (Sulser, 1978, 1982), the view was offered that the alterations of NE receptor activity might be fundamental to the behavioral alterations associated with uncontrollable aversive events. In particular, it was suggested that the development of the receptor down regulation was essential in order for behavioral adaptation to occur, and such receptor changes might actually be indicative of increased efficiency of other aspects of neurotransmission (Stone, 1983).

In accordance with the aforementioned view, it was demonstrated that the diminution of some of the behavioral or physiological concomitants of chronic stressors (e.g., anorexia and gastric lesions) was paralleled by β-NE receptor subsensitivity (Stone and Platt, 1982). However, it should be underscored that this relationship is not necessarily a casual one. The β-NE subsensitivity may be indicative of other adaptive changes that may have occurred following a chronic stressor regimen, and is not directly responsible for the behavioral adaptation that occurs with repeated stressor application. In fact, it has been shown that β-NE receptor blockers (e.g., propranolol) do not eliminate behavioral deficits induced by stressors in animals (Anisman et al., 1981c) are ineffective as antidepressants in humans, and may in fact, induce depressive symptoms (Noll et al., 1985).

In their analysis of the behavioral and neurochemical consequences of stressors, Weiss and Goodman (1985) have emphasized the importance of NE variations in the dorsal bundle in the promotion of behavioral disturbances. It was suggested that following chronic stressor application concentrations of NE decline within the LC, and consequently diminished concen trations of the amine will be available to stimulate autoreceptors present on the NE cell bodies. This, in turn, will result in increased synthesis and release of NE at the terminal regions (e.g., hippocampus) and might be associated with the development of β-NE receptor subsensitivity. In effect, a suggestion is offered to accommodate both the original hypotheses concerning the effects of stressors on NE concentrations and the subsequent suppositions concerning the receptor down regulation that occurs following repeated stressor administration. As will be recalled, some investigators have maintained that it was premature to attribute beneficial effects to the β-NE subsensitivity, since the stressor-provoked down regulation, as well as that associated with antidepressant treatment, might simply be an index of adaptive changes associated with the treatment (Anisman and Zacharko, 1982a; Stone et al., 1984). Furthermore, it has been reported that chronic antidepressant treatment is not only associated with β receptor down regulation, but also results in alpha NE receptor up regulation (Heninger and Charney, 1987; Vetulani, 1983, 1984; Vetulani et al., 1984). Thus, the clinical effectiveness of the drug treatment may be due to the alpha-NE receptor variations rather than to the more commonly assumed β-NE down regulation.

Dopamine

The initial reports concerning the DA variations stemming from stressor application assessed relatively large brain regions, in which concentrations of DA were found to increase or remained unchanged (see Anisman et al.,

1981b; Dunn and Kramarcy, 1984). Subsequent studies revealed, however, that stressors increased the turnover of DA and reduced concentrations of the amine, but unlike the NE alterations, these changes occurred in only selected brain regions.

Although stressors generally do not reduce hypothalamic DA concentrations, discrete analysis of DA in individual hypothalamic nuclei revealed that stressors would, in fact, engender reductions of the amine within the arcuate nucleus (Kobayashi et al., 1976). It was subsequently demonstrated that several stressors would provoke mesolimbic and mesocortical DA variations, but would not produce nigrostriatal DA alterations (Blanc et al., 1980; Herve et al., 1979; Thierry et al., 1976; Tissari et al., 1979). For instance, as seen in Figure 11.5, footshock (30 or 360 shocks of 150 uA) increased the accumulation of the DA metabolite DOPAC, and reduced absolute concentrations of DA within the mesolimbic frontal cortex, and to a lesser extent DOPAC accumulation was increased in the nucleus accumbens. In substantia nigra, however, variations of DA activity were absent (Anisman and Zacharko, 1986). In general, it seems that the frontal cortex is particularly sensitive to the effects of stressors, while the nucleus accumbens is less so. However, it may be the case that some stressors (e.g., restraint) may be particularly effecting in altering DA activity within the nucleus accumbens (Watanabe, 1984). In the substantia nigra and striatum DA variations are typically ab-

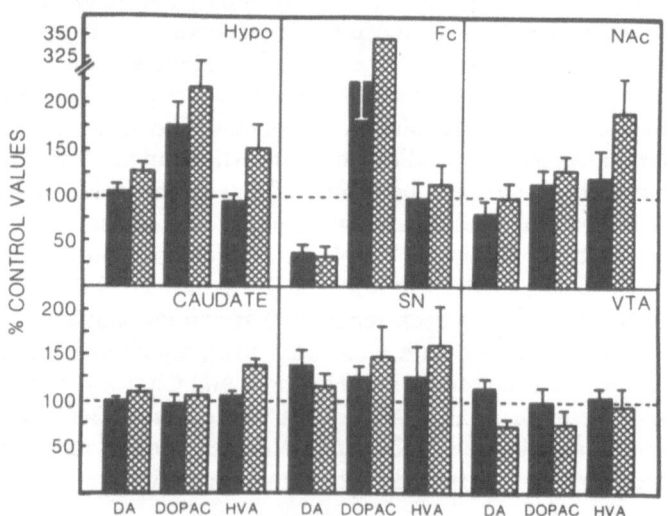

FIGURE 11.5. Dopamine (DA), DOPAC, and HVA as a percent of control values (± SEM) in hypothalamus (hypo), frontal cortex (Fc), nucleus accumbens (NAc), caudate, substantia nigra (SN), and ventral tegmentum (VTA) of mice that received either 30 or 360 shocks of 2-second duration (150 uA). Reprinted with permission of the New York Academy of Sciences from Anisman and Zacharko, *Ann New York Acad Sci* 476:206, 1986.

sent, although several contradictory reports are available (e.g., Beley et al., 1976; Dunn and File, 1983). In these studies, however, the stressor was one which involved cold exposure (e.g., exposure to 4° C, cold water swim), and the DA alterations might reflect a thermoregulatory response, as opposed to a stressor effect per se.

The mesocortial DA alterations associated with stressors appear to be subject to conditioning like effects. In particular, Herman and co-investigators (1982) reported that exposure to footshock increased DA turnover in several regions, including the mesocortex, nucleus accumbens, amygdaloid complex and olfactory tubercle. Upon subsequent exposure to cues that had been paired with footshock the increased DOPAC accumulation was again evident in the mesolimbic frontal cortex. The fact that this effect was not apparent in other regions might suggest that frontal cortex was uniquely sensitive to conditioning processes or that this region is more sensitive to stressor effects in general. More recently, Deutch and colleagues (1985) reported that conditioned alterations of DOPAC accumulation were not only evident in the prefrontal cortex, but also in the ventral tegmentum, with the magnitude of the DOPAC accumulation in the VTA being approximately half that of the cortex. These investigators suggested that the pronounced increase of DA activity in the prefrontal cortex may be due to the lack of autoreceptors, and hence these neurons remain active for longer periods following stressors than do those of other DA neurons originating in the VTA. Regardless of the source for these alterations, these data raise the possibility that the behavioral effects of stressors that involve DA alterations might be apparent even at lengthy intervals following initial exposure to the environmental insult.

The increased accumulation of DOPAC in the frontal cortex and in the nucleus accumbens was shown to be antagonized by pretreatment with the anxiolytic diazepam (Fadda et al., 1978; Fekete et al., 1981). Inasmuch as the β-NE inhibitor chlorpropanol also antagonized the stressor-provoked DOPAC accumulation, the suggestion was offered that the variations of DA activity may involve an NE influence (Fekete et al., 1981). Of course, it is conceivable that the alterations of DA associated with stressors may be secondary to peripheral (or central) concomitants associated with anxiety. It will be recalled that variations of NE in the locus coeruleus have been implicated as fundamental to the induction of anxiety (e.g., Weiss and Goodman, 1985), and it will be seen later that the development of anxiety may also be critical for the development of behavioral disturbances that have been attributed to alterations of NE activity.

In contrast to the reduction of DA concentrations in frontal cortex and the nucleus accumbens evident after an acute stressor, it has been reported that DA levels exceed control values following repeated stressor administration (Richardson, 1984). Interestingly, Herman and colleagues (1984) reported that the increased DOPAC accumulation evident after a single shock session was very much reduced following 10 shock sessions, an effect that has been observed in our laboratory as well (Ahluwalia et al., 1985). Thus, while the adaptation evident in NE neurons following chronic stressor application is a result of a compensatory increase in synthesis, in DA neurons the adaptation arises because of a decrease in the rate of utilization.

Behavioral Consequences of Uncontrollable Stressors

Of the various stress paradigms that have been considered, one that has received greatest attention concerns the interference of escape performance that occurs in animals that had previously been exposed to inescapable shock. In particular, it was demonstrated that, although escapable shock did not appreciably influence later escape performance, exposure to an identical amount of inescapable shock applied in a yoked paradigm severely retarded later performance. Indeed, animals that had received the uncontrollable treatment appeared to accept the shock passively without apparent attempts to escape or avoid. Inasmuch as animals initially exposed to controllable and to uncontrollable shock had received an identical amount of aversive stimulation, it appears that the behavioral disturbances are attributable to lack of behavioral coping methods, rather than to the shock treatment per se.

Maier and Seligman (1976) offered a cognitive hypothesis to account for the performance deficits induced by uncontrollable stressors. Specifically, it was suggested that during the inescapable shock session animals learn that responses and outcomes are independent of one another (learned helplessness). Accordingly, upon subsequent encounters with aversive stimuli these animals do not make attempts to either escape or to avoid. Additionally, it was suggested that stressors induced high levels of arousal and had the effect of disrupting the animal's ability to form response-outcome associations. Thus, even if an animal was successful in escaping from the stressor on a given trial, it did not form the association between the response and stressor offset, and on ensuing trials the behavioral disturbance persisted. This view was subsequently extended to account for human depression, and a reformulated view of the helplessness hypothesis was offered (Abramson et al., 1978). According to this formulation humans make attributions for failure experiences, which in turn lead to expectancies of future performance. Some of these negative expectancies and attributions favor the development of depression.

The helplessness hypothesis is an intuitively appealing one and perhaps, not surprisingly, this position has received considerable attention. Support for this proposition, however, has not been unanimous, and several alternative hypotheses have been offered. These hypotheses are not only able to account for the data derived from animal experimentation, but also lend themselves to analyses of human depression wherein stressors are given a central role. In particular, it was suggested that uncontrollable stressors resulted in disturbances in the animal's ability to initiate active responses in the face of weak stimuli (e.g. a conditioned stimulus) and in maintaining active responses in the face of strong stimuli (i.e., during the primary aversive stimulus). These motor deficits were hypothesized to develop as either learned competing response tendencies (e.g., through adventitious reinforcement of an immobility response, as in the case of shock offset occurring when an animal was inactive) or as an unconditioned response possibly reflecting reduced concentrations of NE, DA, and/or 5-HT (Anisman et al., 1978; Anisman et al., 1979b; Weiss and Glazer, 1975; Weiss et al., 1976; Anisman et al., 1980b).

In accordance with the view that the behavioral deficits were associated with difficulties in maintaining active responses, it was demonstrated in rats and mice that the interference effect produced by inescapable shock was apparent only in motorically demanding tasks. For instance, Glazer and Weiss (1976a, b) reported that escape deficits were more pronounced in a shuttle task involving a high hurdle than when a low hurdle was used. Moreover, the interference was absent in a simple task that necessitated a nose-poke response. Along the same line, Anisman et al. (1978) reported that the interference was more pronounced in a task that required animals to maintain active responses for several seconds prior to escape being possible (i.e., in this instance by delaying opening of the gate separating the two compartments of the shuttle box) than in one where escape was immediately possible upon shock onset. Moreover, it was shown that manipulations that activated the animal (e.g., sounding a buzzer or interrupting the shock train) minimized the escape interference (Anisman and Zacharko, 1982b). In accordance with these findings, Irwin and colleagues (1980) reported that the interference produced by inescapable shock in a swim escape task was greater in animals tested in a paradigm that necessitated prolonged swimming than in a task where the response could be completed quickly. Likewise, the interference became more pronounced as the water temperature was reduced, which presumably also increased the difficulty in maintaining a sustained active response.

The view that these deficits were subserved by stressor-provoked neurochemical changes was supported by several lines of evidence. As indicated earlier, the escape interference, like the depletion of NE, was evident in animals that had been exposed to uncontrollable footshock and not in animals that received controllable shock (see Anisman and Zacharko, 1986). It has been argued (Weiss et al., 1981) that the NE depletion induced by stressors in the locus coeruleus was fundamental to the performance deficits induced by inescapable shock (in a swim task), since the time course for the deficits was paralleled by NE variations within the locus coeruleus but not in other regions (e.g., hypothalamus). Particularly impressive evidence implicating neurochemical alterations in the interference effect is derived from the finding that the increased NE concentrations associated with chronic inescapable shock is accompanied by elimination of the shuttle escape deficits that are ordinarily evident after an acute uncontrollable stressor (Weiss et al., 1975).

If the escape deficit resulted from the development of helplessness, then the magnitude of the interference should have been more pronounced after a chronic stressor regimen than after a single session of inescapable shock. The fact that this was not the case is inconsistent with a cognitive hypothesis, although these findings do not necessarily imply that the altered NE turnover or levels were responsible for the amelioration of the escape deficits. In fact, Platt and Stone (1982) demonstrated that diminution of motor deficits in a swim task following repeated stressor administration coincided with the development of β-NE receptor subsensitivity. Thus these investigators argued that these receptor variations were actually responsible for the performance change.

Further evidence in favor of a neurochemical accounting for the escape interference is derived from pharmacological studies. It was reasoned that if the neurochemical changes induced by inescapable shock were, in fact, responsible for the interference effect, then drug treatments that reduced NE, DA, or both, or blocked DA receptors should mimic the effects of inescapable shock. That is, these treatments should disrupt performance in paradigms where inescapable shock has such an effect, and conversely the drug treatments should be without effect in those paradigms in which stressors do not adversely affect performance. In accordance with this position, it was shown that drugs that inhibited dopamine-β-hydroxylase activity (FLA-63), tyrosine hydroxylase activity (α-MpT) or blocked DA receptors (e.g., pimozide) did not affect shuttle performance in a task where escape was possible immediately upon shock onset. However, as seen in Figure 11.6, paralleling the inescapable shock effects, when the shuttle escape response was briefly prevented (i.e., by delaying access to the safe compartment) thus necessitating sustained active responding, profound performance disturbances were engendered (Anisman et al., 1979b). Treatment with the cholinesterase inhibitor physostigmine was also found to mimic the effects of inescapable shock, thus implicating a role for acetylcholine in the escape interference. This effect was not apparent following treatment with the peripheral cholinesterase inhibitor neostigmine and could not be antagonized by the peripherally acting anticholinergic methylscopolamine, thus indicating that the effects of physostigmine were not due to peripheral cholinergic changes (Anisman et al., 1981a; Anisman et al., 1979b). However, as will be seen shortly, the contribution of ACh and catecholamines could be distinguished from one another.

Just as catecholamine-depleting agents or antagonists elicited an interference effect, treatment with compounds which stimulate catecholamine activity effectively eliminated the escape interference. As depicted in Figure 11.7, administration of L-dopa (in conjunction with an extracerebral decarboxylase inhibitor) prior to escape testing eliminated the effects of previously administered inescapable shock (Anisman et al., 1979b; Anisman et al., 1979a). Likewise, treatment with L-dopa prior to the inescapable shock session prevented the subsequent appearance of the escape interference (Anisman et

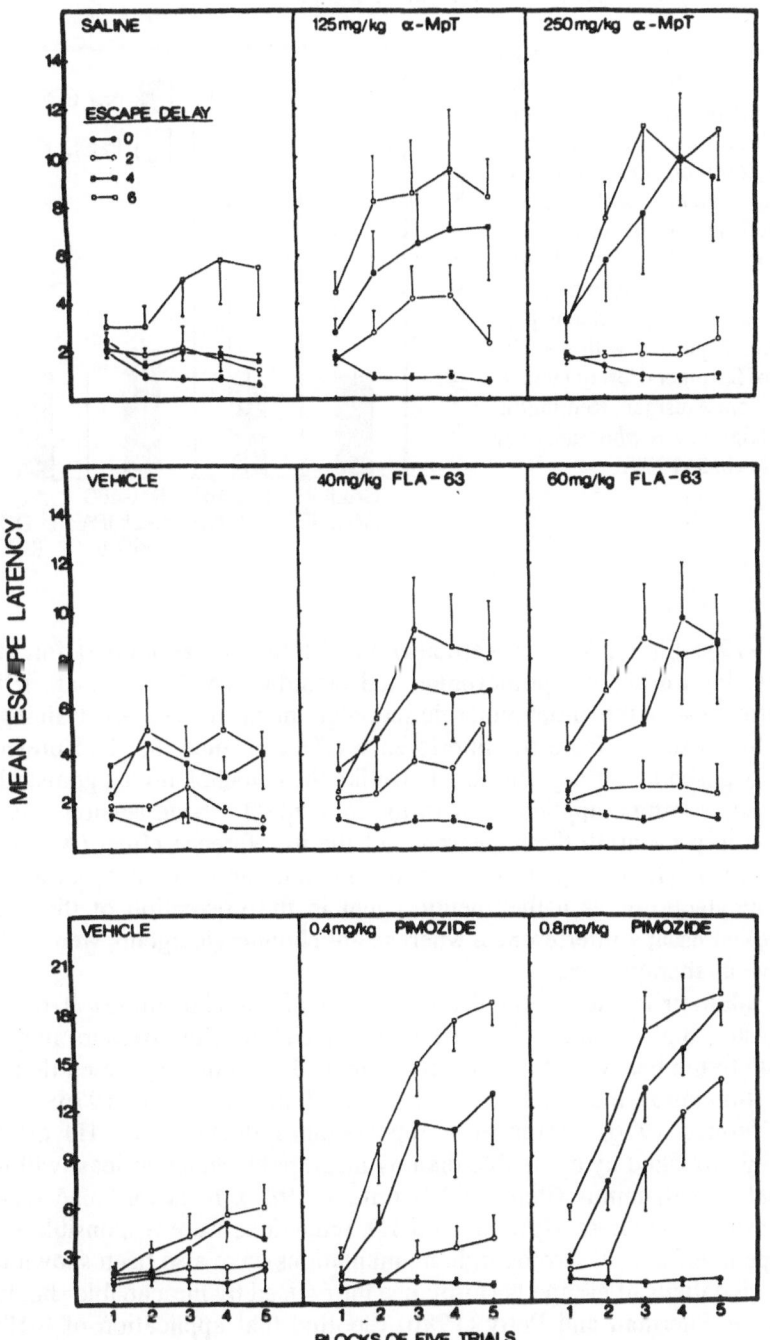

FIGURE 11.6. Mean escape latencies (± SEM) in mice tested in a shuttle task where escape was delayed for either 0, 2, 4, or 6 seconds following shock onset. Mice had been treated with either α-MpT (0, 125 or 250 mg/kg), FLA-63 (0, 40, or 60 mg/kg) or pimozide (0, 0.4, or 0.8 mg/kg). Reprinted with permission of Springer-Verlag Heidelberg from Anisman, Remington and Sklar, *Psychopharmacology* 61:111–115, 1979.

FIGURE 11.7. Mean escape latencies (± SEM) over 25 trials as a function of the shock condition and treatment L-dopa (400 or 600 mg/kg) or saline together with an extracerebral decarboxylase inhibitor (MK-486), or MK-486 alone. Testing was conducted using a six-second delay procedure. Reprinted with permission of Springer-Verlag Heidelberg from Anisman, Remington and Sklar, *Psychopharmacology* 61:111-115, 1979.

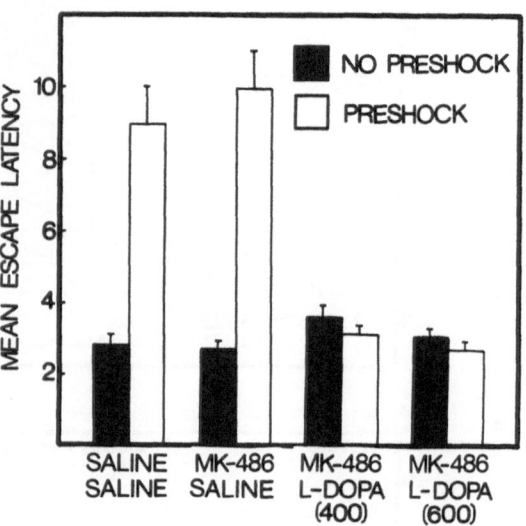

al., 1979a). Similar effects were also noted following administration of relatively high doses of apomorphine and clonidine (Anisman et al., 1980a). Treatment with the anticholinergic scopolamine prior to escape training was also found to eliminate the interference effect in mice that had previously been exposed to inescapable shock. Unlike the catecholamine agonists, however, scopolamine applied prior to the inescapable shock session was ineffective in preventing the appearance of the interference effect (Anisman et al., 1981a). Thus, it appears that catecholamine agonists may act as either a prophylactic or as a therapeutic agent in the prevention of the stressor provoked escape interference, whereas anticholinergic agents were only effective as therapeutics.

In addition to the potential involvement of catecholamines, it has been suggested that serotonergic variations associated with stressors may contribute to the behavioral deficits. It has indeed been demonstrated that aversive stimulation increases 5-HT turnover (Palkovits et al., 1976; Telegdy and Vermes, 1976). Moreover, it appears that reductions of 5-HT are more readily provoked by escapable than by inescapable shock at least within the lateral hypothalamus (Petty and Sherman, 1982). In accordance with the suggestion that stressor-provoked 5-HT reductions were responsible for the escape interference, serotonergic manipulations have also been shown to affect the deficits of escape performance introduced by inescapable shock. For instance, Sherman and Petty (1980) reported that application of 5-HT directly to the septum or frontal cortex effectively eliminated the interference in previously stressed animals, whereas the same treatment applied prior to the uncontrollable shock session was without effect. Contrary to this report, Brown and co-investigators (1982) observed that systemic treatment with the 5-HT depleting agent, parachlorophenylalanine (PCPA) induced an escape interference, whereas preloading with 1-tryptophan enhanced escape performance. Work conducted in this laboratory revealed that the 5-HT releasing agent p-chloroamphetamine eliminated the escape interference induced by inescapable shock, whereas the 5-HT antagonist, methysergide, successfully induced an escape interference (Hamilton et al., 1986). Although several attempts were made to alter the effects of inescapable shock with PCPA and with 5-hydroxytryptophan, we were uniformly unsuccessful in modifying escape performance (Anisman et al., 1979a).

The suggestion that the stressor-induced escape interference may serve as a model of human depression has received support from the finding that drug treatments that are effective therapeutics in humans are also effective in eliminating the escape interference in animals tested in a shuttle task. For instance, it was found that repeated treatment with agents such as imipramine, desmethylimipramine, nortriptyline and amitryptyline effectively reduced the interference ordinarily engendered by inescapable shock (Sherman et al., 1979; Sherman et al., 1982; Telner and Singhal, 1981). Yet, there is reason for discomfort with this suggestion. From the purely intuitive vantage, it is exceedingly difficult to accept that performance deficits in animals required to emit an operant to terminate shock are in some way similar to the cognitive, affective, and vegetative changes that typify human clinical depression. Additionally, even if one were to accept that aversive events may precipitate depression, the stressful events encountered by humans are most often chronic in nature, particularly if one considers the prolonged rumination that often follows a stressful event. Furthermore, although tricyclic antidepressants have been shown to alleviate the stressor provoked behavioral deficits in animals, it is also the case that compounds that are notoriously ineffective as antidepressants (e.g. L-dopa, scopolamine) were also successful in eliminating or preventing the escape interference associated with inescapable shock. Thus, despite the initial enthusiasm for this animal model of depression, at least in so far as the escape paradigm is concerned, this paradigm has some problems of specificity in pharmacological validation.

In light of the perceived shortcomings of the stress model of depression described this far, attempts have been made to provide improvements to the model, without necessarily rejecting the shuttle data. Among other things, increasingly greater attention has been devoted to evaluation of a wider range of behaviors, and particularly those which are, at least on the surface, reminiscent of the symptoms that characterize human depression. Moreover, greater effort has been expended in evaluating the impact of repeated stressors, as opposed to assessing the effects of a single session of inescapable shock. Although there is little doubt that it will always be difficult to know with certainty whether animals are subject to affective or cognitive changes that are apparent in humans, the more recent approaches to assessing the effects of stressors in animals do appear to have a greater degree of validity (see review in Willner, 1985).

Anxiety

It will be recalled that Weiss and Goodman (1985) offered the suggestion that stressor-induced NE variations within the locus coeruleus were particularly important in the development of the performance deficits. Moreover, since activation of NE neurons within the locus coeruleus might have the effect of promoting vigilance (or anxiety; Aston-Jones and Bloom, 1981; Aston-Jones et al., 1984), the stress paradigm might be a model for depression in which anxiety is a primary characteristic of the syndrome. There is indeed reason to believe that some behavioral effects of uncontrollable stressors may be modified by the anxiety engendered by the aversive treatment.

As indicated earlier, the time course for the behavioral disturbances associated with inescapable shock is dependent upon a number of organismic and treatment variables. In dogs, the interference of escape performance was shown to be relatively transient, disappearing within 72 hours of stressor termination (see Maier and Seligman, 1976). In contrast, in rodents the interference was shown to be either short-lived or very long lasting, depending on the stressor parameter employed. For instance, after intense shock of short duration the interference was apparent only at short intervals, whereas shock of relatively low intensity and long duration resulted in the interference effect being apparent only after relatively long intervals (Glazer and Weiss, 1976a, b). The authors likewise observed that the interference was least evident in mice soon after inescapable shock and became more pronounced within 24 hours after stressor termination; however, in mice this was the case irrespective of the shock train duration (Anisman et al., 1978).

Like Glazer and Weiss (1976a, b) the authors maintained that inescapable footshock favored the development of behavioral disturbances in a shock-escape task, but that some other consequence of the stressor, possibly a transient anxiety elicited by the aversive stimulus, prevented the expression of the interference effect at a brief test interval. Since exposure to inescapable shock may result in the development of learned competing responses that are manifested in a subsequent shock-escape task, the authors opted to examine the effects of inescapable shock on later performance in a swim paradigm where it was less likely that previously established competing response styles would appear (Prince and Anisman, 1984). Figure 11.8 depicts the amount of time mice engaged in active swimming immediately or 24 hours after exposure to either escapable shock, yoked inescapable shock or no treatment. As in the case of the shuttle-escape deficits, the amount of time engaged in active swimming 24 hours after inescapable shock was reduced relative to that of nonshocked mice or mice that had been exposed to escapable shock. In contrast, immediately after the initial stressor session the amount of time engaged in active swimming was increased, irrespective of whether mice had initially received escapable or inescapable shock. Thus, it was suggested that application of a stressor, regardless of the animal's ability to determine its offset, provoked anxiety that induced invigorated responding. Over time, the

FIGURE 11.8. Mean (± SEM) immobility time over three-minute periods in mice exposed to escapable (ES), yoked inescapable (YIS), or no shock (NS). Swim testing was conducted either immediately or 24 hours after shock (150 uA) application. Reprinted with permission of Academic Press, Orlando from Prince and Anisman, *Behav Neur Biol* 42:108, 1984.

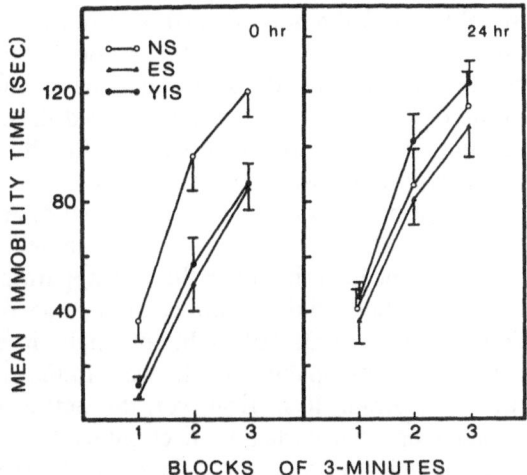

BLOCKS   OF   3-MINUTES

anxiety diminished, permitting expression of the response depression, which is typically evident in inescapably shocked animals. Commensurate with the proposition that the initial motor activation was related to anxiety, it was demonstrated (Prince et al., 1986b) that diazepam effectively reduced the excitation in mice tested immediately after exposure to inescapable shock.

It was recently reported that administration of diazepam prior to inescapable shock effectively limited the interference effect in rats tested 24 hours later. Interestingly, in rats that received the drug prior to testing rather than prior to inescapable shock the performance deficit was not modified (Drugan et al., 1984). Thus, it was suggested that the development of anxiety may be essential in the provocation of the neurochemical changes that subserve the interference effect. By preventing the anxiety reaction during inescapable shock, these secondary neurochemical variations may have been prevented as well. However, once these neurochemical alterations were provoked, the anxiolytic would be ineffective in modifying the performance deficits.

Associative Disturbances and Perseveration

It was hypothesized that, in addition to cognitive disturbances, exposure to uncontrollable stressors resulted in associative deficits wherein animals were deficient in forming response-outcome associations (Maier and Seligman, 1976). In accordance with this view, it was demonstrated that in a response choice Y-maze discrimination task (i.e., always turn right in order to escape) rats that had been exposed to inescapable shock acquired the discrimination more slowly than did previously nonshocked animals (Jackson et al., 1980). Contrary to these findings there are several reports indicating that uncontrollable shock did not affect stressor motivated discrimination performance. For instance, Irwin et al. (1980) reported that inescapable shock disrupted escape performance in a swim task, the extent of which was directly related to the motoric difficulty of the task and the energy expenditure required of the animal. Irrespective of the task difficulty, however, position discrimination performance was unaffected by the stressor.

Using a paradigm very much like that of Jackson et al. (1980), but where trial onset and termination were computer controlled (rather than by an experimenter hovering over the test apparatus) investigators found that inescapable shock did not affect either cue or response-choice discrimination (Anisman et al., 1984). Interestingly, in an ambiguous response-choice paradigm, inescapable shock was found to disrupt performance in a reversal-learning task. However, the performance deficit could not be ascribed to associative disturbances, given that inescapable shock did not affect either acquisition or reversal learning in other paradigms. It was proposed that when the task was a relatively ambiguous one (i.e., where external cues did not differentiate correct from incorrect choices), animals adopted a perseverative tendency wherein they persisted in adopting a previously acquired response style (or one towards which they were predisposed).

In addition to this response perseveration, it was subsequently demonstrated that under some conditions animals exposed to an uncontrollable stressor would also exhibit a stimulus perseveration. In particular, it was demonstrated that when animals were placed in an arena or Y-maze, one portion of which was illuminated, mice would spend a greater proportion of their time in the lit area. In animals that had been exposed to inescapable footshock the tendency to approach the illuminated region increased further, and animals also encountered a transient deficit in acquiring an escape response in which they were required to swim to the dark arm (Szostak and Anisman, 1985). Like the response invigoration associated with inescapable shock, the stimulus perseveration could be antagonized by an anxiolytic (Prince et al., 1986b).

In discussing the factors that contribute to avoidance learning, Bolles (1970) had suggested that stressors result in a narrowing of an animal's defensive style to species-specific defensive behaviors. In accounting for the behavioral effects of inescapable shock, it was likewise suggested that the animal's problem-solving strategies would be restricted to those responses for which the animals were highly prepared, particularly when the environmental cues were ambiguous or when the animals had an innate preference for particular stimuli (Anisman et al., 1984). In effect, it was argued that the restriction of the animal's coping strategies might serve the animal well since the highly prepared response styles were readily adopted, and in many instances might result in stressor termination. However, when the task required contraprepared responses the perseveration might come to hinder acquisition of the escape response.

Recently, Minor et al., (1984) offered an alternative accounting for the acquisition deficits associated with inescapable shock. Following several unsuccessful attempts to replicate the findings of Jackson et al. (1980), it was noted that when irrelevant cues were presented to the animal in the escape task, and when reinforcement for appropriate responses was delayed, acquisition deficits were induced in animals that had been exposed to inescapable shock. These investigators thus suggested that uncontrollable stressors disrupted attentional processes, possibly owing to dorsal bundle NE reductions, such that animals did not filter relevant from irrelevant information. In a sense, this position is not far removed from the proposition that stressors provoke a perseverative tendency, particularly when the stressful situation is ambiguous. After all, it might reasonably be expected that as the saliency of relevant cues increased, the impact of attentional deficits would be diminished. When the task was an ambiguous one, animals might begin to respond to irrelevant cues to a greater extent, and might thus persist in responding in a set fashion.

Reinforcement and Reward Mechanisms

One of the fundamental symptoms of depression is that of anhedonia (see chapter 10 by Koob this volume). Depressed individuals report that they no longer enjoy activities that they previously found rewarding, ranging from time spent with their families, shopping, social events, sexual activity, and so forth. Surprisingly, there is only a limited amount of data that evaluated the effects of controllable and uncontrollable stressors on behaviors involving positive reinforcement and on social behaviors. As indicated earlier, it is difficult to accept that performance in a shuttle escape task or in a swim paradigm is actually a valid model of depression. Accordingly, several investigators (Katz, 1981 a, b, 1982; Weiss et al., 1981; Willner, 1985; Zacharko et al., 1983) have attempted to evaluate the effects of stressors in a broader series of behavioral situations, including those that do not involve aversive stimulation.

The authors' initial attempts to evaluate the effects of stressors on subsequent performance in nonaversive situations involved assessment of spontaneous alternation in an eight-arm radial maze (Bruto and Anisman, 1983). Typically, when animals are placed in a novel maze, they tend to explore in a systematic fashion wherein they enter those arms that were least recently visited (spontaneous alternation). In mice tested in an eight-arm radial maze, the authors noted that animals successively entered adjacent arms (adjacent alternation). Neither escapable nor inescapable shock reliably affected spontaneous alternation performance; however, the inescapable shock treatment was found to disrupt the adjacent alternation tendency. This effect was evident immediately after the shock treatment, but did not persist over a 24-hour period. If animals were reexposed to a limited amount of shock (which in itself was without effect) 24 hours after the initial shock session, then the disruption of adjacent alternation was again induced. It will be recalled that the neurochemical changes induced by stressors of moderate severity are likewise transient, but can be reinduced upon reexposure to the stressor. It was proposed that the inescapable shock treatment modified performance by disrupting the systematic style that animals adopt in exploring their environment.

As in the case of the exploratory tendencies, it appears that stressors may influence performance in food-motivated learning paradigms. For instance, it was demonstrated that acquisition of a lever-pressing response was retarded (as determined from increased inter-response times) in animals that had received inescapable shock (Rosellini, 1978). When a one second delay of reinforcement was introduced following a lever-press response, the effect of inescapable shock was somewhat more pronounced, being apparent over 15 as opposed to 5 trials. It will be recognized that these findings are consistent with those seen in a shuttle test in that the magnitude of the interference was enhanced as task difficulty increased. At the same time, however, it is clear that the effects seen in an appetitive situation are distinguishable from those seen in tasks involving aversive motivation, where the performance deficits may be much longer lasting. In a subsequent study (Rosellini et al., 1982) inescapable shock was shown to retard acquisition of a response-choice discrimination, as well as a response-choice reversal. Moreover, this effect persisted over several sessions. Later studies from this same laboratory revealed that in animals trained to respond for food on a random interval schedule, exposure to inescapable shock had a greater disruptive effect on subsequent performance than did escapable shock or no shock treatment (Rosellini et al., 1984). On the basis of these findings it was argued that inescapable shock results in increased sensitivity to expectancies of response-reinforcer independence. In effect, the organism exposed to a stressor is presumably more sensitive to situations where behavioral control is not possible.

The studies by Rosellini and associates suggest that the stressor did not affect either motivational processes or the rewarding value of the primary reinforcer. Recent studies from this laboratory (e.g., Zacharko et al., 1983; Zacharko et al., 1984b) and from that of Katz (1981 a, b) and Willner (1984) suggest that this is not the case. The studies from the latter two laboratories revealed that the preference for saccharin over tap water ordinarily demonstrated by rats was reduced in animals that had been exposed to a regimen of different stressors applied on successive days. Accordingly, the view was entertained that relatively mild unpredictable stressors, when chronically applied, would reduce the reinforcement value associated with rewarding stimuli. Moreover, it was shown (Katz, 1982; Katz and Sibel, 1982) that the reduction in the saccharin preference was eliminated with chronic antidepressant drug treatments, indicating that this paradigm might be useful as an animal model of depression. It might be noted, as well, that a preliminary series of studies conducted in our laboratory revealed that in mice a single session of inescapable shock or a chronic schedule involving different types of stressors resulted in a diminution in the preference for saccharin ordinarily exhibited. The magnitude of the effect, however, was considerably less marked than that reported in rats. This was the case irrespective of whether or not mice had had previous experience with saccharin, the concentration of the saccharin employed (1 to 4 percent), or the interval between the stressor experience and exposure to the saccharin solution. In contrast to the effect of acute shock or an intermittent stressor, the authors observed that in mice exposed to a chronic, predictable footshock regimen (i.e., where the stressor was applied at the same time on successive days) the initial reduction in the saccharin preference diminished, and in fact, stressed animals exhibited an increased preference for this solution relative to non-stressed or acutely stressed animals. Thus, it appears that the effects of acute and chronic stressors in this paradigm parallel those seen in the shuttle test and in the neurochemical variations previously described.

The approach used most frequently to assess the motivational alterations associated with stressors is that of evaluating responding for electrical brain stimulation from various catecholamine-containing brain regions. In light of the finding that stressors such as footshock alter DA activity in some brain regions (e.g., mesolimbic areas) but not others (e.g., nigrostriatal regions), it was of interest to determine whether stressors would likewise differentially affect self-stimulation from these brain areas. As seen in Figure 11.9, escapable shock did not affect responding of CD-1 mice for brain stimulation

FIGURE 11.9. Mean (± SEM) rate of responding for intracranial self-stimulation (over a 10-minute period) from the medial forebrain bundle *(top)*, nucleus accumbens *(middle)*, and substantia nigra *(bottom)* in mice that had been exposed to escapable shock, yoked inescapable shock, or no shock. Baseline rates represent responding on the day prior to shock treatment. Mice were tested immediately (0), and again 24 and 168 hours after the shock session. Reprinted with permission of Elsevier Science Publishers B.V. Amsterdam from Zacharko et al., *Behav Brain Res* 9:135, 1983.

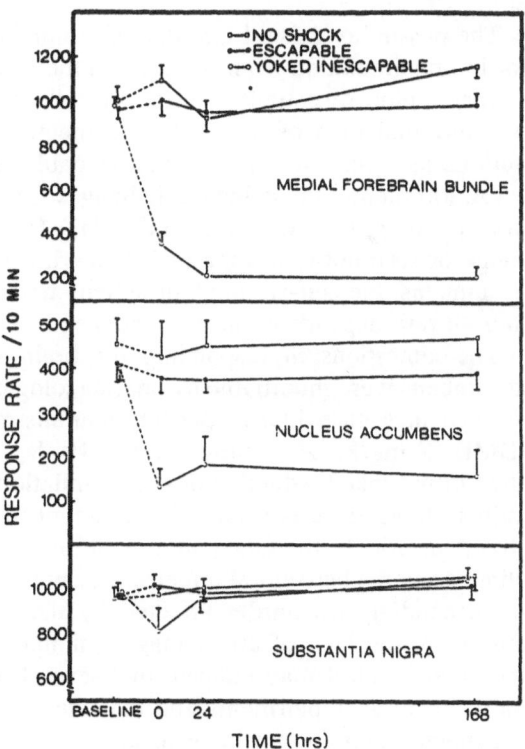

from the nucleus accumbens (Nas), substantia nigra pars compacta (SN), or the medial forebrain bundle (MFB) through which DA fibers of passage traverse. In contrast, an equivalent amount of inescapable footshock applied in a yoked paradigm reduced responding from the Nas and MFB by about 80 percent in mice tested immediately after the stressor treatment. This effect was a long lasting one, being pronounced 168 hours after the inescapable shock session provided that animals had initially been tested immediately after stressor application. Contrary to the changes in the Nas and MFB, inescapable shock did not affect responding from the substantia nigra (Zacharko et al., 1983).

The stressor-induced alterations of responding for brain stimulation from the mesolimbic system cannot be attributed to motoric changes, since comparable changes in locomotor activity were evident irrespective of the region examined. Moreover, if motoric changes contributed to reduced responding, altered self-stimulation rates would have been expected from the substantia nigra as well as the accumbens. In addition, it does not appear that the reduced responding from the accumbens was specific to the current intensity used for self-stimulation. Using a descending current paradigm, investigators observed that response rates from the accumbens declined across a wide range of intensities among shocked mice. Conversely, responding from the substantia nigra was not similarly affected, indicating that the lack of effect from this region was not specific to a particular current intensity (Bowers, et al., 1987).

The possibility existed that the differential effects of stress on responding for brain stimulation from the SN and the Nas were due to the fact that the former region essentially consists of cell bodies whereas the latter region is a terminal field of the ventral tegmental area (VTA). Accordingly, the authors assessed the effects of inescapable shock on responding from the VTA. Paralleling the behavioral changes from the Nas, a marked reduction in responding was evident from the VTA following exposure to inescapable shock or restraint (Anisman and Zacharko, 1986). As noted in the nucleus accumbens, the self-stimulation deficits noted from the VTA were evident in both rate-dependent and rate-independent paradigms.

The alterations in responding for brain stimulation from the nucleus accumbens were modifiable by pharmacological manipulation. In particular, as seen in Figure 11.10, repeated administration of desmethylimipramine (DMI) (5 mg/kg $\times$ 2 daily) over a 10-day period markedly reduced the stressor provoked reduction of self-stimulation (Zacharko et al., 1984b). The authors have since observed that although this effect is a robust one, it is strain specific. In the CD-1 mouse chronic DMI generally has the effect of eliminating the behavioral impairments, but in the DBA/2J mouse reductions of responding are hardly affected by the tricylic antidepressant. On the one hand, the lack of effect may be unique to the DBA/2J strain, but on the other hand, it may indicate that several different mechanisms subserve the behavioral impairments and that the effectiveness of tricyclics may be dependent upon the mechanisms underlying these impairments.

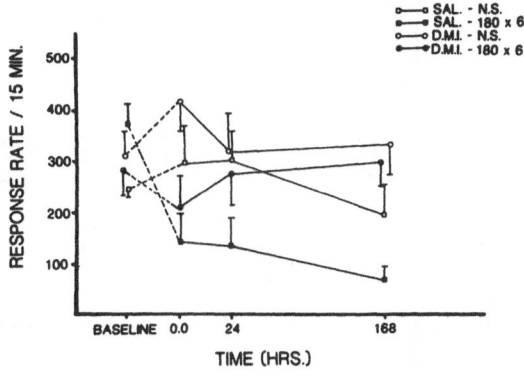

FIGURE 11.10. Mean (± SEM) number of responses for electrical brain stimulation from the nucleus accumbens over a 15-minute period in mice exposed to 180 shocks (180 × 6) or no shock (NS). Mice had received 15 days of treatment with either DMI (5.0 mg/kg × 2) or saline (SAL) prior to the stress treatment. Mice were tested for intracranial self-stimulation on three occasions following the stressor treatment (immediately, 24, or 168 hours). Reprinted with permission of Elsevier Science Publishers B.V. Amsterdam from Zacharko et al., *Brain Research* 321:177, 1984.

Like other stressor-induced behavioral changes, it appeared that the response reduction ordinarily evident after an acute session of inescapable shock was absent among mice exposed to a chronic shock regimen (Zacharko et al., 1984a). It will be noted, however, that considerable interindividual variability exists with respect to the time course of the adaptation (Bowers et al., 1985). Whereas a rapid and marked adaptation was evident in some animals, little or no adaptation was apparent in other mice that had been exposed to the chronic stressor. Interestingly, in those animals in which the adaptation was not evident, DMI effectively enhanced performance while disrupting responding among those animals that had adapted to the chronic stressor regimen. Accordingly the possibility was entertained that chronic stressor application ordinarily results in adaptive neurochemical changes that preclude the development of behavioral disturbances. However, in some animals these adaptive changes do not occur, rendering them most vulnerable to the behavioral (and affective) disturbances. In these animals tricyclic agents may alleviate the symptoms of depression by inducing neurochemical alterations similar to those ordinarily associated with effective adaptation.

Interindividual and Genetic Differences in Response to Stressors

In humans, the symptoms of depression may vary considerably across individuals, as does the therapeutic efficacy of various antidepressant agents. As indicated earlier, it seems that depression may be a biochemically hetero-

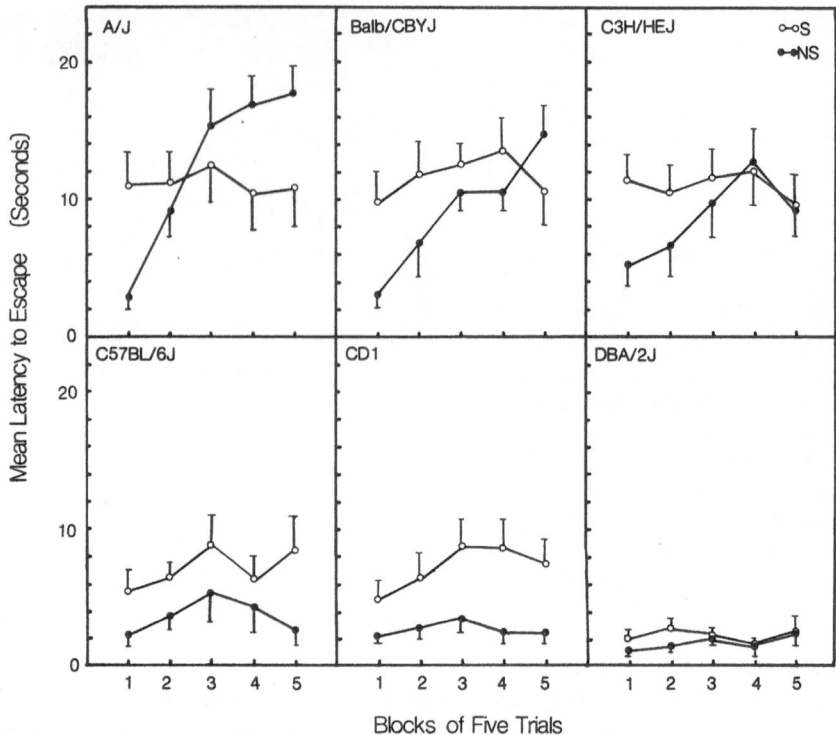

FIGURE 11.11. Mean (± SEM) latencies to escape from shock over blocks of five trials as a function of shock treatment in six strains of mice. Mice had been exposed to either 360 shocks, 150 uA or two-second duration (S), or to no shock treatment (NS) and tested 24 hours afterward. Reprinted with permission from Shanks and Anisman, *Behavioral Neuroscience,* Feb. 9 (in press) © 1989 American Psychological Association.

geneous illness, hence accounting for the differential effectiveness of drug treatments and possibly the diverse symptom profiles that exist. Likewise, in animals the behavioral effects of stressors vary appreciably. Whereas some animals exhibit marked behavior disturbances in a given task, other animals are seemingly unaffected by the aversive treatment (Anisman et al., 1979b; Shanks and Anisman, 1988). Recent data collected in this laboratory revealed marked interstrain behavioral differences following exposure to uncontrollable stressors. As seen in Figure 11.11, in some strains of mice (e.g., Balb/cByJ and C3H/HeJ) uncontrollable footshock engenders marked disturbances of subsequent shuttle escape performance, while in other strains (e.g., DBA/2J) the stressor hardly influences escape latencies. Thus, while the escape interference appears to be a fairly ubiquitous phenomenon appearing across a wide range of species (Maier and Seligman, 1976), within a species profound interstrain (and interindividual) differences apparently may occur in a behavioral task following uncontrollable stressor application.

Just as in the human condition where considerable diversity exists with respect to the symptoms characterizing depression, the authors have also observed that the strain differences associated with stressors also interact with the nature of the behavioral task in which animals were tested. For instance, strains of mice that exhibited marked deficits of escape performance following inescapable shock did not demonstrate disturbances of swim performance or exploratory style. Conversely, alterations of swim performance in a particular strain were not accompanied by disturbances of escape responding after exposure to inescapable shock (Shanks and Anisman, 1988). Indeed, as seen in Figure 11.12, DBA/2J mice that displayed particularly proficient escape responding after inescapable shock exhibited a pronounced disturbance of responding for brain stimulation from the nucleus accumbens. In contrast, in BALB/cByJ mice where escape performance was severely hampered by the inescapable shock treatment, responding for electrical brain stimulation was transiently enhanced (Zacharko et al., 1987).

As yet unpublished preliminary data collected in this laboratory have revealed that the strain-specific effects of stressors have not only been observed in behavioral tasks, but also with respect to neurotransmitter alterations. As in the case of the behavioral changes, it seems that a given strain may be particularly vulnerable to a specific neurochemical change in one brain region (e.g., DA alterations in the accumbens of the DBA/2J mouse), while a

FIGURE 11.12. Mean (± SEM) rate of responding for electrical brain stimulation from the nucleus accumbens in three strains of mice. Mice received a 15-minute test session 24 hours prior to treatment (baseline) and again immediately, 24 hours, and 168 hours following exposure to either inescapable shock (360 shocks, 150 uA of 2-second duration) or no shock. Reprinted with permission of Elsevier Science Publishers B.V. Amsterdam from Zacharko et al., *Brain Research* 426:167, 1987.

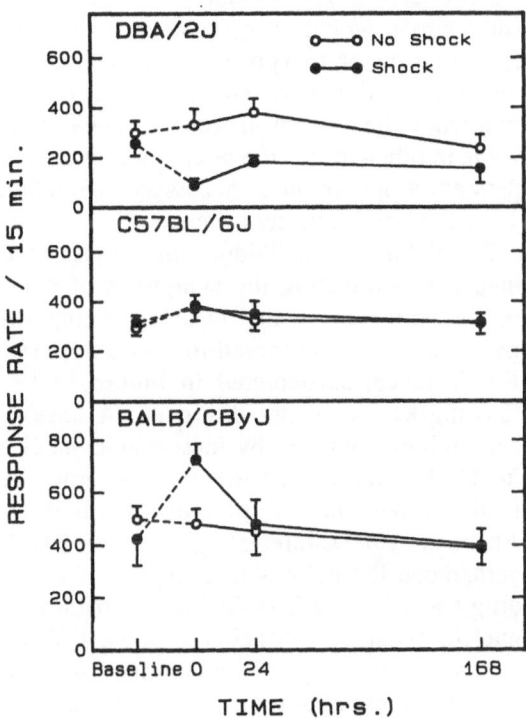

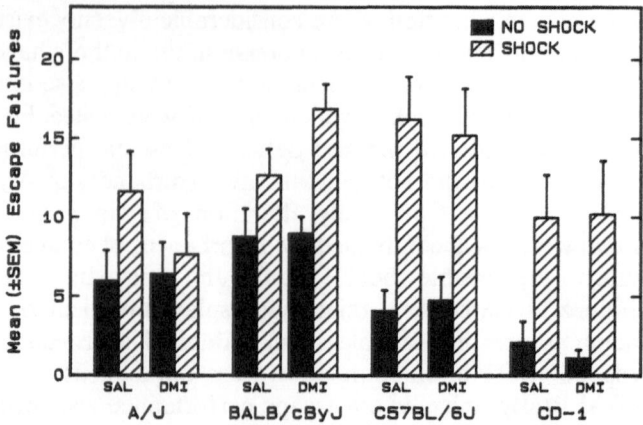

FIGURE 11.13. Mean frequency of escape failures ($\pm$ SEM) in four strains of mice that had been exposed to inescapable shock (360 shocks, 150 uA of 2-second duration) or no shock and tested 15 days afterward. Following the initial stress treatment mice received either saline or DMI treatment (5.0 mg/kg, twice per day) over 14 days.

second strain may be more vulnerable to transmitter change in other brain regions. Furthermore, while data are not yet available, it is conceivable that the course of the neurochemical adaptation associated with a chronic stressor regimen would progress at different rates across strains or that the effects of stressor reexposure on catecholamine alterations would vary with the strain variable. As such, it might be expected that behavioral adaptation would be evident in some strains following a chronic stressor regimen, while in other strains the adaptation would not be evident, and hence antidepressant drugs would be necessary to ameliorate the behavioral impairments associated with aversive events.

Paralleling the individual differences in the effectiveness of drug treatments in eliminating the symptoms of depression in humans, the efficacy of antidepressant agents in antagonizing stressor-provoked behavioral impairments is also observed to vary across strains of mice and test situations. For instance, as depicted in Figure 11.13, chronic treatment with DMI (5.0 mg/kg, twice per day over 14 days) successfully eliminated the escape failures induced by inescapable shock in A/J mice, but had no effect in BALB/cByJ, C57BL/6J or CD-1 mice. Amitriptyline, however, was effective in reducing escape failures in both A/J and CD-1 mice (Shanks and Anisman, 1987). Interestingly, in the CD-1 strain in which DMI did not antagonize the deficits of escape performance, it will be recalled that the drug was effective in reducing the disturbances of self stimulation from the nucleus accumbens (Zacharko et al., 1984).

Summarizing, the aforementioned data suggest that the use of any one behavioral paradigm or any single species or strain of animal may be inappropriate in the development of an animal model of depression. The symptom profile of the disorder, its underlying mechanisms, and the efficacy of drug treatments in ameliorating symptoms are too varied to be adequately assessed in a simple animal model. At the very least, it is necessary to assess a wide range of behaviors reflective of the symptoms of the disorder and to evaluate these behaviors across different strains of animals that may be differentially sensitive (behaviorally and neurochemically) to the effects of stressors. Moreover, the effectiveness of pharmacological intervention should be considered with respect to specific symptoms, rather than on the entire syndrome.

Overview

The fundamental view entertained here is that stressors induce several neurochemical changes which may be of adaptive significance. In response to a stressor of moderate severity the synthesis of central amines keeps pace with utilization, and levels of the transmitters remain stable. Under some conditions, however, utilization may exceed amine synthesis, resulting in a net decline of the amine levels. It is thought that this reduction may render the organism less well prepared to deal with further insults, and may increase vulnerability to behavioral disturbances. It follows that those variables that predispose the organism to the stressor-provoked amine reductions or increase the persistence of the reductions increase vulnerability to behavioral pathology. In addition to acute stressors, it is believed that chronic stressors may influence vulnerability to illness. Specifically, the amine reductions associated with acute stressors are absent following repeated stressor regimens owing to a compensatory increase in synthesis. However, if such adaptive changes do not occur, then vulnerability to illness is increased. Accordingly, analysis of the stress-depression topography requires, at the very least, identification of those variables that lend themselves to the amine alterations.

While some of the variables that influence the amine alterations have been identified (e.g., stressor controllability and chronicity, stress history, genetic factors, age, background conditions upon which stressors are superimposed), the animal models of depression that have most often been employed appear to be far too simplistic. Among other things, considerable interindividual (and interstrain) variability exists with respect to the amine alterations induced by stressors. Although a stressor may have profound effects on NE alterations in one organism, the same stressor may have more pronounced effects on DA or 5-HT activity in a second strain or individual. Similarly, interindividual variability may exist with respect to the brain region in which the stressor-provoked amine alterations are maximal. Accordingly, the symptom profile associated with the stressor may be expected to vary considerably across animals. Thus, in developing animal models of depression, it may be counterproductive to emphasize the effects of stressors on one neurotransmitter or the variations of one behavior at the expense of others.

In light of these considerations, the approach the authors adopted to model depression in animals has been to assess the effects of a range of stressors on several behaviors that are reminiscent of the symptoms associated with clinical depression. Although not dismissing the import of behaviors such as shuttle escape performance, the emphasis has been that alterations of reward processes, anxiety or vigilance, attentional mechanisms, as well as variations in vegetative functioning all need to be considered in order to appreciate more fully the impact of stressors and their relationship to the depressive syndrome.

References

Abramson LY, Seligman MEP, Teasdale JD (1978): Learned helplessness in humans: Critique and reformulation. *J Abnorm Psychol* 87:49-74

Ahluwalia P, Zacharko RM, Anisman H (1985): Dopamine variations associated with acute and chronic stressors. *Soc Neurosci Abst* 11:49

Akiskal HS, McKinney WT (1973): Depressive disorders: Toward a unified hypothesis. *Science* 182:20-29

Anisman H (1984): Vulnerability to depression: Contribution of stress. In: *Neurobiology of Mood Disorders,* Post RM, Ballenger JC, eds. Baltimore: Williams & Wilkins

Anisman H, Zacharko RM (1982a): Depression: The predisposing influence of stress. *Behav Brain Sci* 5:89-137

Anisman H, Zacharko RM (1982b); Stimulus change influences escape performance: Deficits induced by uncontrollable stress and by haloperidol. *Pharmacol Biochem Behav* 17:263-269

Anisman H, Zacharko RM (1986): Behavioral and neurochemical consequences associated with stressors. In: *Stress-induced Analgesia,* Kelley DD, ed. *Ann New York Acad Sci* 467:205-225

Anisman H, deCantanzaro D, Remington G (1978): Escape performance following exposure to inescapable shock: Deficits in motor response maintenance. *J Exp Psychol [Anim Behav]* 4:197-218

Anisman H, Glazier SJ, Sklar LS (1981a): Cholinergic influences on escape deficits produced by uncontrollable stress. *Psychopharmacology* 74:81-87

Anisman H, et al., (1985): Stressor invoked exacerbation of amphetamine-elicited perseveration. *Pharmacol Biochem Behav* 23:173-183

Anisman H, et al., (1987): Variations of norepinephrine concentrations following chronic stressor application. *Pharmacol Biochem Behav* 26:653-659

Anisman H, Hamilton M, Zacharko RM (1984): Cue and response choice acquisition and reversal after exposure to uncontrollable shock: Induction of response perseveration. *J Exper Psychol* 10:229-243

Anisman H, Irwin J, Sklar LS (1979a): Deficits of escape performance following catecholamine depletion: Implications for behavioral deficits induced by uncontrollable stress. *Psychopharmacology* 64:163-170

Anisman H, Kokkinidis L, Sklar LS (1981b): Contribution of neurochemical change to stress-induced behavioral deficits. In: *Theory in Psychopharmacology* Cooper SJ, ed. London: Academic Press

Anisman H, Pizzino A, Sklar LS (1980b): Coping with stress, norepinephrine depletion and escape performance. *Brain Research* 191:583-588

Anisman H, Remington G, Sklar LS (1979b): Effects of inescapable shock on subsequent escape performance: Catecholaminergic and cholinergic mediation of response initiation and maintenance. *Psychopharmacology* 61:107-124

Anisman H, Ritch M, Sklar LS (1981c): Noradrenergic and dopaminergic interactions in escape behavior: Analysis of uncontrollable stress effects. *Psychopharmacology* 74:263-268

Anisman H, Suissa A, Sklar LS (1980a): Escape deficits induced by uncontrollable stress; Antagonism by dopamine and norepinephrine agonists. *Behav Neur Biol* 28:34-47

Antelman SM, Chiodo LA (1983): Amphetamine as a stressor. In: *Stimulants: Neurochemical Behavioral and Clinical Perspectives*, Creese I, ed. New York: Raven Press.

Aston-Jones G, Bloom FE (1981): Norepinephrine-containing locus coeruleus neurons in behaving rats exhibit pronounced responses to non-noxious environmental stimuli. *J Neurosci* 1:887-900

Aston-Jones G, Foote SL, Bloom FE (1984): Anatomy and physiology of locus coeruleus neurons: Functional implications. In: *Norepinephrine: Clinical Aspects*, Ziegler MG, Lake CR, eds. Baltimore: Williams & Wilkins

Beley A, et al., (1976): Time dependent changes in the rate of noradrenaline synthesis in various rat brain areas during cold exposure. *Pflugers Arch* 368:225-229

Blanc G, et al., (1980): Response to stress of mesocortical-frontal dopaminergic neurons in rats after long term isolation. *Nature* 284:265-276

Bolles RC (1970): Species-specific defense reactions and avoidance learning. *Psychol Rev* 77:32-48

Bowers WJ, Zacharko RM, Anisman H (1985): Repeated stressor or desmethylimipramine effects on footshock induced depression of self-stimulation. *Soc Neurosci Abst* 11:51

Bowers WJ, Zacharko RM, Anisman H (1987): Evaluation of stressor effects on intracranial self-stimulation from the nucleus accumbens and the substantia nigra in a current intensity paradigm. *Behav Brain Res* 23:85-93

Brown L, Rosellini RA, Samuels OB, Riley EP (1982): Evidence for a serotonergic mechanism of the learned helplessness phenomenon. *Pharmacol Biochem Behav* 17:877-883

Bruto V, Anisman H (1983): Alterations of exploratory patterns induced by uncontrollable shock. *Behav Neural Biol* 37:302-316

Cassens G, et al., (1980): Alterations in brain norepinephrine metabolism induced by environmental stimuli previously paired with inescapable shock. *Science* 209:1138-1140

Deutch AY, Tam S-Y, Roth RH (1985): Footshock and conditioned stress increase 3,4-dihydroxyphenylacetic acid (DOPAC) in the ventral tegmental area but not substantia nigra. *Brain Research* 333:143-146

Drugan RC, et al., (1984): Librium prevents the analgesia and shuttlebox escape deficit typically observed following inescapable shock. *Pharmacol Biochem Behav* 21:749-754

Dunn AJ, File SA (1983): Cold restraint alters dopamine metabolism in frontal cortex, nucleus accumbens and neostriatum. *Physiol Behav* 31:511-513

Dunn AJ, Kramarcy NR (1984): Neurochemical responses in stress: Relationships between the hypothalamic-pituitary adrenal and catecholamine systems. In: *Handbook of Psychopharmacology,* Iversen LL, Iversen SD, Snyder SH, eds. New York: Plenum Press

Fadda F, et al., (1978): Stress induced increase in 3,4-dihydroxyphenylacetic acid (DOPAC) levels in the cerebral cortex and in nucleus accumbens: Reversal by diazepam. *Life Sci* 23:2219-2224

Fekete MIK, et al., (1981): Effects of anxiolytic drugs on the catecholamine and DOPAC (3,4-dihydroxyphenylacetic acid) levels in brain cortical areas and on corticosterone and prolactin secretion in rats subjected to stress. *Psychoneuroendo* 6:113-120

Glazer HI, Weiss JM (1976a): Long-term and transitory interference effects. *J Exper Psychol: Anim Behav Proc* 2:191-201

Glazer HI, Weiss JM (1976b): Long term interference effect: An alternative to "Learned Helplessness". *J Exp Psychol [Anim Behav]* 2:202-213

Hamilton ME, Zacharko RM, Anisman H (1986): Influence of p-chloroamphetamine and methysergide on the escape deficits provoked by inescapable shock *Psychopharmacology* 90:203-206

Heninger GR, Charney DS (1987): Mechanism of action of antidepressant treatments: Implications for the etiology and treatment of depressive disorders. In: *Psychopharmacology: The Third Generation of Progress,* Meltzer HY, ed. New York: Raven Press

Herman JP, et al., (1982): Differential effects of inescapable footshock and stimuli previously paired with inescapable footshocks on dopamine turnover in cortical and limbic areas of the rat. *Life Sci* 30:2207-2214

Herman JP, Stinus L, Le Moal M (1984): Repeated stress increases locomotor response to amphetamine. *Psychopharmacology* 84:431-435

Herve D, et al., (1979): Differences in the reactivity of the mesocortical dopaminergic neurons to stress in the Balb/C and C57/BL6 mice. *Life Sci* 25:1659-1664

Hudgens R, Morrison J, Barchha R (1967): Life events and onset of primary affective disorders. *Arch Gen Psychiatry* 16:134-145

Irwin J, et al., (1986): Central norepinephrine and plasma corticosterone following acute and chronic stressors: Influence of social isolation and handling. *Pharmacol Biochem Behav* 24:1151-1154

Irwin J, Ahluwalia P, Anisman H (1986): Sensitization of norepinephrine activity following acute and chronic footshock. *Brain Res* 376:98-103

Irwin J, Suissa A, Anisman H (1980): Differential effects of inescapable shock on escape performance and discrimination learning in a water escape task. *J Exp Psychol [Anim Behav]* 6:21-40

Jackson RL, Alexander RH & Maier SF (1980): Learned helplessness, inactivity and associative deficits: Effects of inescapable shock on response choice escape learning. *J Exp Psychol [Anim Behav]* 6:1-20

Jimerson DC, Post RM (1984): Psychomotor stimulants and dopamine agonists in depression. In *Neurobiology of Mood Disorders,* Post RM, Ballenger JC, eds. Baltimore: Williams & Wilkins

Katz RJ (1982): Animal model of depression: Pharmacological sensitivity of a hedonic deficit. *Pharmacol Biochem Behav* 16:965-968

Katz RJ (1981a): Animal models and human depressive disorders. *Neurosci Biobehav Rev,* 5:231-246

Katz RJ (1981b): Animal models of depression: Effects of electroconvulsive shock therapy. *Neurosci Biobehav Rev* 5:273-277

Katz RJ, Sibel M (1982): Further analysis of the specificity of a novel animal model of depression: Effects of an antihistaminic, antipsychotic and anxiolytic compound. *Pharmacol Biochem Behav* 16:979-982

Kobayashi RM, et al., (1976): Selective alterations of catecholamines and tyrosine hydroxylase activity in the hypothalamus following acute and chronic stress. In: *Catecholamines and Stress.* Usdin E, Kvetnansky R, Kopin IJ, eds. Oxford: Pergamon Press

Kvetnansky R, et al., (1976): Catecholamines in individual hypothalamic nuclei in stressed rats. In: *Catecholamines and Stress* Usdin E, Kvetnansky R, Kopin IJ, eds. Oxford: Pergamon Press

MacLennan AJ, Maier SF (1983): Coping and stress-induced potentiation of stimulant stereotypy in the rat. *Science* 219:1091-1093

Maier SF, Seligman MEP (1976): Learned helplessness: Theory and evidence. *J Exp Psychol [Gen]* 105:3-46

Minor TR, Jackson RL, Maier SF (1984): Effects of task-irrelevant cues and reinforcement delay on choice escape learning following inescapable shock: Evidence for a deficit in selective attention. *J Exp Psychol [Anim Behav]* 10:543-556

Morrison J, Hudgens R, Barchha R (1968): Life events and psychiatric illness. *B J Psychiatry* 114:423-432

Noll KM, Davis JM, DeLeon-Jones, F (1985): Medication and somatic therapies in the treatment of depression. In: *Handbook of Depression,* Beckham EE, Leber WR, eds. New York: Dorsey Press

Nomura S, et al., (1981): Stress and β-adrenergic receptor binding in the rat's brain. *Brain Res* 224:199-203

Palkovits M, et al., (1976): Effects of stress on serotonin and tryptophan hydroxylase activity of brain nucleis. In: *Catecholamines and Stress,* Usdin E, Kvetnansky R, Kopin IJ, eds. Oxford: Pergamon Press

Petty F, Sherman AD (1982): A neurochemical differentiation between exposure to stress and the development of learned helplessness. *Drug Devel Res* 2:43-45

Platt JE, Stone EA (1982): Chronic restraint stress elicits a positive antidepressant response on the forced swim test. *Eur J Pharmacol* 82:179-181

Prince CR, Ahluwalia P, Anisman H (1986a): Catecholamine and corticoid variations associated with prepared and contraprepared defensive responses. *Soc Neurosci Abs,* 12:1060

Prince CR, Anisman H (1984): Acute and chronic stress effects on performance in a forced-swim task. *Behav Neur Biol* 84:99-119

Prince CR, Collins C, Anisman H (1986b): Stressor-provoked response patterns in a swim task: Modification by diazepam. *Pharmacol Biochem Behav* 24:323-328

Richardson JS (1984): Brain part monoamines and the neuroendocrine mechanisms activated by immobilization stress in the rat. *Int J Neurosci* 23:57-68

Rosellini RA (1978): Inescapable shock interferes with the acquisition of a free appetitive operant. *Anim Learn Behav* 6:155-159

Rosellini RA, et al., (1984): Uncontrollable shock proactively increases sensitivity to response-reinforcer independence in rats. *J Exp Psychol [Anim Behav]* 10:346-359

Rosellini RA, DeCola JP, Shapiro NR (1982): Cross-motivational effects of inescapable shock are associative in nature. *J Exp Psychol [Anim Behav]* 8:376-388

Roth KA, Mefford IM, Barchas JD (1982): Epinephrine, norepinephrine, dopamine and serotonin: Differential effects of acute and chronic stress on regional brain amines. *Brain Res* 239:417-424

Schildkraut JJ (1978): Current status of the catecholamine hypothesis of affective disorders. In: *Psychopharmacology: A Generation of Progress,* Lipton MA, DiMascio A, Killam KF, eds. New York: Raven Press

Shanks N, Anisman H (1987): Strain specific behavioral effects of inescapable shock and desmethylimipramine. *Soc Neurosci Abstr* 13:660

Shanks N, Anisman H (1988): Stressor provoked disturbances in six strains of mice. *Behav Neurosci,* in press

Sherman AD, et al., (1979): A neuropharmacologically relevant animal model of depression. *Neuropharmacology* 18:891-893

Sherman AD, Petty F (1980): Neurochemical basis of the action of anti-depressants on learned helplessness. *Behav Neur Biol* 30:119-134

Sherman AD, Sacquitne JL, Petty F (1982): Pharmacologic specificity of the learned helplessness model of depression. *Pharmacol Biochem Behav* 16:449-454

Slater S, Roth M (1969): *Mayer Gross Clinical Psychiatry.* Baltimore: Williams & Wilkins

Stone EA (1979): Subsensitivity to norepinephrine as a link between adaptation to stress and antidepressant therapy: An hypothesis. *Research Comm Psychol Psychiat Behav* 4:241-255

Stone EA (1983): Problems with current catecholamine hypotheses of antidepressant agents. *Behav Brain Sci* 6:535-577

Stone EA, et al., (1984): Reduction of the cAMP response to norepinephrine in rat cerebral cortex following repeated restraint stress. *Psychopharmacology* 82:403-405

Stone EA, Platt JE (1982): Brain noradrenergic receptors and resistance to stress. *Brain Res* 237:405-414

Sulser F (1978): Functional aspects of the norepinephrine receptor coupled adenylate cyclase system in the limbic forebrain and its modification by drugs which precipitate or alleviate depression: Molecular approaches to an understanding of affective disorders. *Pharmako Neuropsycho* 11:43-52

Sulser F (1982): Antidepressant drug research: Its impact on neurobiology and psychobiology. In: *Typical and Atypical Antidepressants: Molecular Mechanisms,* Costa E, Racagni, G, eds. New York: Raven Press

Szostak C, Anisman H (1985): Stimulus perseveration in a water maze following exposure to uncontrollable shock. *Behav Neur Biol* 43:178-198

Tanaka M, et al., (1982): Time-related differences in noradrenaline turnover in rat brain regions by stress. *Pharmacol Biochem Behav* 16:315-319

Telegdy G, Vermes M (1976): Changes induced by stress in the activity of the serotonergic system in limbic brain structures. In: *Catecholamines and Stress.* Usdin E, Kvetnansky R, Kopin IJ, eds. Oxford: Pergamon Press

Telner J, Singhal RL (1981): Effects of nortriptyline treatment on learned helplessness in the rat. *Pharmacol Biochem Behav* 14:823-826

Thierry AM, et al., (1968): Effects of stress on the metabolism of norepinephrine, dopamine and serotonin in the central nervous system of the rat. *J Pharmacol Exper Therap* 163:163-171

Thierry AM, et al., (1976): Selective activation of the mesocortical DA system by stress. *Nature* 263:242-244

Tissari AH, et al., (1979): Footshock stress accelerates non-striatal dopamine synthesis without activating tyrosine hydroxylase. *Arch Pharmacol* 308:155-158

Tsuda A, Tanaka M (1985): Differential changes in noradrenaline turnover in specific region of rat brain produced by controllable and uncontrollable shocks. *Behav Neurosci* 99:802-817

U'Pritchard DC, Kvetnansky R (1980): Central and peripheral adrenergic receptors in acute and repeated immobilization stress. In: *Second International Symposium in Catecholamines and Stress*, Usdin E, Kvetnansky R, Kopin IJ, eds. New York: Elsevier-Dutton

van Praag HM (1978): Amine hypotheses of affective disorders. In: *Handbook of Psychopharmacology*, Iversen LL, Iversen SD, Snyder SH, eds. New York: Plenum Press

van Praag HM (1984): Depression, suicide, and serotonin metabolism in the brain. In: *Neurobiology of Mood Disorders*, Post RM, Ballenger JM, eds. Baltimore: Williams & Wilkins

Vetulani J (1983): Alpha-1 up, beta down: A counterproposal to Stone. *Behav Brain Sci* 4:560-561

Vetulani J (1984): Studies on the neurochemical basis of action of antidepressant drugs and electroconvulsive treatment. *Pol J Pharmacol Pharm* 36:101-116

Vetulani J, et al., (1984): Alpha-up beta-down adrenergic regulation—a possible mechanism of action of antidepressant treatments. *Pol J Pharmacol Pharm* 36:321-328

Watanabe H (1984): Activation of dopamine synthesis in mesolimbic dopamine neurons by immobilization stress in the rat. *Neuropharmacology* 23:1335-1338

Weiss JM, et al., (1975): Effects of chronic exposure to stressors on avoidance-escape behavior and on brain norepinephrine. *Psychosom Med* 37:522-534

Weiss JM, et al., (1979): Coping behavior and stress-induced behavioral depression: Studies of the role of brain catecholamines. In: *The Psychobiology of Depressive Disorders*, Depue RA, ed. New York: Academic Press

Weiss JM, Glazer HI (1975): Effects of acute exposure to stressors on subsequent avoidance-escape behavior. *Psychol Med* 37:499-521

Weiss JM, et al., (1981): Behavioral depression produced by an uncontrollable stressor: Relationship to norepinephrine, dopamine and serotonin levels in various regions of rat brain. *Brain Res Rev* 3:167-205

Weiss JM, Goodman PA (1985): Neurochemical mechanisms underlying stress-induced depression. In: *Stress and Coping*, Field T, McCabe P, Schneiderman N, eds. New Jersey: Lawrence Erlbaum

Weiss JM, Glazer HI & Pohorecky LA (1976): Coping behavior and neurochemical changes: An alternative explanation for the original "learned helplessness" experiments. In: *Animal Models in Human Psychobiology*, Serban G, Kling A, eds. New York: Plenum Press

Willner P (1984): The validity of animal models of depression. *Psychopharmacology* 83:1-16

Willner P (1985): *Depression: A Psychobiological Synthesis*. New York: John Wiley & Sons

Zacharko RM, et al., (1983): Region-specific reductions of intracranial self-stimulation after uncontrollable stress: Possible effects on reward processes. *Behav Brain Res* 9:129-141

Zacharko RM, et al., (1984b): Prevention of stressor-induced disturbances of self-stimulation by desmethylimipramine. *Brain Res* 321:175-179

Zacharko RM, et al., (1987): Strain-specific effects of inescapable shock on intracranial self-stimulation from the nucleus accumbens. *Brain Res* 426:164-168

Zacharko RM, Bowers WJ, Anisman H (1984a): Responding for brain stimulation: Stress and desmethylimipramine. *Prog Neurol Psychopharmacol Biol Psychiatry* 8:601-606

Acknowledgments. Supported by grants MT-6486 and MA 8130 from the Medical Research Council of Canada.

12

Pharmacological Probes in Primate Social Behavior

R. Francis Schlemmer, Jr., Jennifer E. Young, and John M. Davis

Introduction

The appropriateness of interaction with other individuals constitutes a significant factor in the determination of the diagnosis and treatment response in major psychiatric categories. Therefore, animal models of these disorders which include evaluation of social behavior add an important dimension to the model. Non-human primates have been used as subjects in most social studies because of their well documented social structure and noted behavioral similarities to humans.

A growing volume of literature detailing the use of primate social colonies as animal models has emerged over the past two decades. In most cases, the model has been induced by treatment of colony members with drugs which in humans produce a state that in some way resembles endogenous depression, mania, or psychosis. Unfortunately, the drug-induced state rarely induces a syndrome which is identical to the endogenous condition, but instead, induces important characteristics of the disorder. Therefore, similar drug treatment of monkeys or any other animal should not be expected to model all symptoms of a psychiatric disorder. Instead, attention should be focused on modeling individual characteristics of the disorder. Primate social colony models offer a larger number of strategic behaviors for study.

The purpose of this chapter is to review drug-induced models of affective disorders and psychosis using the primate social colony paradigm. In addition, available evidence regarding the mediation of behavioral changes which appear to be particularly relevant to the study of these disorders will be examined. This work has encompassed a number of primate species (Table 12.1).

Non-human Primate Social Colony Models of Mental Disorders

Models of Depression

Depression can be precipitated by a number of identifiable factors. Treatment with certain drugs can result in depressive episodes in nondepressed individuals. Several cases of severe depression have been associated with

TABLE 12.1 Primate species used in social studies using pharmacologic probes

Common name	Genus species
New world monkeys	
Common marmoset	*Callithrix jacchus*
Squirrel monkey	*Saimiri sciureus*
Old world monkeys	
Bonnet Macaque	*Macaca radiata*
Crab-eating or long-tailed Macaque	*Macaca fascicularis*
Japanese Macaque	*Macaca fuscata*
Pigtail Macaque	*Macaca nemestrina*
Rhesus monkey	*Macaca mulatta*
Stumptail macaque	*Macaca arctoides*
Talapoin monkey	*Miopithecus talapoin*
Vervet monkey	*Cercopithecus aethiops sabaeus*

the use of reserpine in hypertensive patients (Freis, 1954). McKinney et al. (1971) administered reserpine to three selected members of a social colony of juvenile rhesus monkeys for 81 days, which resulted in a syndrome bearing similar features to that seen in humans. Locomotion and visual exploration were decreased whereas self-huddling was increased in treated monkeys. Treated animals only responded to high levels of social stimuli. Ptosis, posturing, and tremor were also noted. Since the pharmacological action of reserpine is to deplete neuronal stores of norepinephrine, dopamine, and serotonin, the behavioral changes were attributed to disruption of monoamine transmission.

Redmond et al. (1971b) attempted to differentiate the contributions of catecholamine and serotonin systems in primate behavior. Four adult stumptail macaques from two social colonies received the catecholamine synthesis inhibitor alpha-methyl-para-tyrosine (AMPT), and four stumptail macaques received para-chlorophenylalanine (PCPA), an inhibitor of serotonin synthesis. AMPT-treated monkeys had a significant reduction in initiated social activity whereas PCPA-treated monkeys did not. Animals assumed an energy conserving slumped-over posture during AMPT treatment, and half of the treated monkeys lost rank within the colony during this period but remained healthy. Conversely, monkeys receiving PCPA maintained interest in colony activity despite deteriorating health. In both cases, non-treated monkeys interacted with and groomed treated monkeys suggesting that they tolerated the drug-induced condition. These results suggest that catecholamines play a more important role than serotonin in the behavioral changes in monkeys which resemble human depressive symptoms.

This hypothesis was further supported by a study conducted on two free-ranging adult rhesus macaque colonies in a natural setting using the neurotoxin 6-hydroxydopamine (6-OHDA). The monkeys were captured, surgically implanted with cannulae into the cerebral ventricles, received either drug treatment with 6-OHDA or artificial cerebral fluid, and released to rejoin their respective troops (Redmond et al., 1973). Two of the four

6-OHDA-treated monkeys failed to initially return to their social groups upon release, although they did not appear disoriented. Treated monkeys had significantly fewer initiated social interactions and reduced social and self-grooming when compared with sham or field controls within ten days after treatment and release. These animals assumed a slumped-over posture, had an expressionless face, and walked with a stiff gait. These results are in close agreement with those for AMPT and reserpine, indicating that disruption of catecholamine transmission produces behavioral changes in monkeys suggestive of symptoms of human depression, including social withdrawal, an apparent decrease in concern of group activity, decreased social and self-grooming, decreased locomotion, and energy conserving posturing.

There have been no reports of efforts to reverse or attenuate the behavioral changes induced in these primate models of depression with known antidepressant agents. Redmond et al. (1971a) attempted to reverse the behavioral changes induced by AMPT in one monkey with L-DOPA, but this attempt was unsuccessful.

Models of Psychosis

More laboratories have used primate social colonies to study models of psychosis than models of depression. This is probably due to the similarities found between the psychotic behavior induced in humans by psychomotor stimulants such as amphetamine, which is seen in paranoid schizophrenia (Connell, 1958; Snyder, 1972), and the well-documented behavioral syndrome induced by these drugs in primates (Miczek, 1983). Psychosis is also commmon during the manic phase of bipolar illness. Therefore, paranoid psychosis and mania may share common etiological mechanisms.

The authors have recently reviewed models of psychosis in primate social colonies induced by three classes of psychotomimetics—CNS stimulants, phencyclidine, and hallucinogens—and concluded that the behavioral syndrome induced by amphetamine appears to bear great similarity to endogenous human psychosis (Schlemmer and Davis, 1983) and, thus, possibly mania. There has been remarkable consistency in reports of major behavioral changes induced by amphetamine and methamphetamine from several laboratories despite differences in the primate species studied, acute versus chronic administration, and the route of administration. The most frequently reported behavioral changes induced by the amphetamines include: social withdrawal, increased submissive behavior, increased checking or visual scanning, bouts of intense scratching, and the induction of various forms of stereotyped behavior (Machiyama et al., 1970; Kjellberg and Randrup, 1972; Crowley et al., 1974; Garver et al., 1975; Miller and Geiger, 1976; Schiorring, 1977; Poignant and Avril, 1978; Haber et al., 1979; Bellarosa et al., 1980; Miczek and Yoshimura, 1982; Nielsen et al., 1983; Ridley and Baker, 1983; Schlemmer and Davis, 1983; Smith and Byrd, 1983).

In the amphetamine studies in the authors' laboratory, d-amphetamine was administered in time-release form for 12 consecutive days to selected members of adult stumptail macaque social colonies. This treatment regimen was selected because constant exposure to amphetamine was maintained over a long time period thereby mimicking the "amphetamine binge" from which most reports of amphetamine psychosis were generated. All experiments began with observation of undrugged behavior (baseline) followed by treatment of two to three female members of the colony (four to six monkeys) with d-amphetamine (1.6 mg base/kg) in time-release form, nasogastrically, every 12 hours for 12 days. Daily one-hour observations were conducted by trained observers who were unaware of the experimental conditions. They recorded the behavior of each monkey in the colony with a checklist of over 40 social and solitary behaviors for the species using the focal-sampling technique in which the behavior of individual monkeys within the colony is recorded for 30 seconds every five mintues in rotation. Recently, the observations have been entered directly into a computer; but the behaviors on the checklist and other scoring procedures have remained the same. (For more details about the drug administration and observation techniques, see Schlemmer and Davis, 1983). A total of 26 monkeys from 9 colonies have received amphetamine treatment in the study.

The most apparent behavioral changes induced by d-amphetamine were stereotyped behavior and increased checking (changes in visual field determined by head and eye movement). Over 20 distinct forms of stereotyped behavior were noted throughout the experiments. Self-grooming was the most common form of stereotypy. The stereotyped behavior was usually intense and persisted throughout amphetamine treatment. On the other hand, increased checking was usually most intense early in treatment but then diminished in intensity during the later stages. Although it has been suggested that the increased checking constitutes a form of hypervigilance (Garver et al., 1975), Ridley and Baker, (1983) have presented evidence to the contrary.

Despite the predominance of stereotypy and checking, other behavioral changes induced by amphetamine may have greater relevance to psychosis. For instance, social withdrawal, a primary symptom of schizophrenia, was noted in almost all amphetamine-treated monkeys. Most treated monkeys became spatially isolated from other members in the colony. In addition, initiated social grooming was significantly reduced or eliminated whereas self-grooming was typically increased. Therefore, there was a shift in the same motor activity (grooming) from a social to a solitary context. This shift is in contrast to models of depression where both forms of grooming were reduced. Another interesting behavioral change was the significant increase in submissive gestures given to other animals by treated monkeys, while the level of aggression directed toward treated animals remained unchanged or was decreased. These monkeys often overreacted (submissively) to approaching cage-mates. This response has been suggested as a model of human paranoia since the monkey apparently reacted to a non-threatening situation as if it was threatening (Schlemmer and Davis, 1981a). Interestingly, higher ranking females were more likely to respond in this manner than lower ranking females. This finding has been replicated by Wilson et al. (1983) in the same species, but it is the opposite of what Haber et al. (1979) reported in rhesus monkeys. In addition, methamphetamine decreased the dominance-to-submission ratio in pigtail macaques (Crowley et al., 1974).

Amphetamine also induced intense scratching in many treated monkeys similar to that reported in patients with amphetamine psychosis (Snyder, 1972). In these patients, the scratching response was invariably associated with tactile hallucinations. Therefore, the scratching elicited in monkeys by amphetamine may reflect a quantitative measure of hallucination (Schlemmer and Davis, 1983). Other bizarre behaviors were noted less frequently during amphetamine treatment in some monkeys, but may have application to the study of psychosis. These changes included tracking imaginary objects and picking into the air (similar to the changes reported in vervet monkeys by Nielsen et al. [1983]), continuous peering under the cage without apparent purpose, following or shadowing other colony members around the cage, and repeated threatening without subsequent chase or attack. Throughout the study, treated monkeys appeared alert and aware of their environment.

Antipsychotic drugs are effective in combatting many symptoms of both endogenous and drug-induced psychoses, including amphetamine psychosis (Angrist et al., 1974). Similarly, antipsychotic agents at least partially reversed key behavioral changes induced by chronic amphetamine treatment in monkeys. In the authors' laboratory, concomitant administration of haloperidol, pimozide, chlorpromazine, and clozapine antagonized or partially reversed amphetamine-induced social withdrawal, increased submissive behavior, induced intense scratching, and increased checking (Schlemmer and Davis, 1983). Each drug antagonized amphetamine stereotypy with the exception of clozapine, which also failed to do so in rodents (Burki et al., 1975). Conversely, the non-antipsychotic agents promethazine and diazepam were ineffective in reversing this syndrome. Stereotypy and increased checking were antagonized in all treated monkeys by the antipsychotics (except clozapine), whereas social withdrawal proved most difficult to reverse with approximately half of the treated monkeys responding favorably. Interestingly, negative symptoms of schizophrenia such as social withdrawal do not respond to antipsychotic treatment as consistently as positive symptoms, suggesting another similarity of the primate model to the human disorder. Comparable antagonistic action of antipsychotic agents on amphetamine-induced behavior in primate social groups has been reported in stumptail macaques with haloperidol (Miller and Geiger, 1976), in marmosets with haloperidol (Scraggs and Ridley, 1979), in squirrel monkeys with haloperidol and chlorpromazine (Miczek and Yoshimura, 1982), and in vervets with flupentixol (Nielsen et al., 1983); however, social interactions disrupted by amphetamine were not restored in any of these studies. In addition, amphetamine-induced behavioral changes in marmosets were not antagonized by diazepam, the

244 R. Francis Schlemmer, Jr., Jennifer E. Young, and John M. Davis

alpha-adrenergic antagonist aceperone, or the beta-blocker propranolol (Scraggs and Ridley, 1979).

Cocaine, another CNS stimulant which precipitates psychosis in humans, has also been tested in members of primate social colonies. Miczek and Yoshimura (1982) compared the effects of cocaine and d-amphetamine in squirrel monkeys. Like amphetamine, cocaine increased the frequency of locomotion and checking, and decreased social interaction.

Drugs from two other classes of psychotomimetics—phencyclidine (PCP), an arylcyclohexylamine, and the hallucinogens—have been tested in group-housed monkeys. PCP significantly reduced or eliminated social interaction initiated by treated rhesus (Miller et al., 1973) and stumptail macaques (Schlemmer et al., 1978). Treated monkeys appeared uninterested in cage activity and resting scores significantly increased (Schlemmer and Davis, 1983). Chronic administration of PCP induced stereotyped behavior which was reversed by the dopamine antagonist pimozide. Nine hallucinogens, including d-lysergic acid diethylamide (LSD) and mescaline, have been tested in stumptail macaques (Schlemmer and Davis, 1986). Although some differences between individual compounds were noted, all hallucinogens significantly disrupted social behavior. Interestingly, several behavioral changes were qualitatively similar to those seen with the psychomotor stimulants, including increases in distancing, submissive gestures, and checking, and reduction or elimination of initiated social grooming although the hallucinogen-induced changes were usually not as intense as with amphetamine or apomorphine. Hallucinogens also induced two emergent behaviors—limb jerks (myoclonic spasms of the extremities) and body shakes (wet dog shakes)—which appear to be specifically induced in this species by hallucinogenic drugs.

Separation Models

One of the most vigorous areas of primate behavioral research involves the separation syndrome. Separation has frequently been linked to serious affective disorders, yet some portions of the syndrome resemble other forms of psychopathology, including autism and schizophrenia (McKinney, 1974). Therefore, it is best not to view separation as a model of a specific disorder, but rather as a model of individual characteristics of psychopathology.

The separation syndrome in primates and its usefulness as an animal model of psychopathology is discussed in more detail elsewhere in this volume (see chapters by McKinney and Ehler et al.). Previous separation history can have a profound influence on drug response, which, in turn, is relevant to understanding the mediation of behavioral abnormalities in primates. These studies suggest that catecholamine systems are integral in the mediation of the separation response. Treatment with the catecholamine synthesis inhibitor AMPT potentiated the despair response to separation in peer-grouped monkeys, but mother-reared peer-grouped monkeys required a much larger dose of AMPT to produce comparable results (Kraemer and McKinney, 1979). AMPT was more potent in monkeys with more previous separation experience than in those with fewer previous separations. Conversely, inhibition of serotonin synthesis with PCPA did not appear to potentiate the despair response to a significant degree. Interestingly, however, the dopamine beta-hydroxylase inhibitor fusaric acid increased locomotion and reduced huddling during separation, which are opposite results to those found with AMPT. The results with AMPT and PCPA are in agreement with those of Redmond et al. (1971b), suggesting that catecholamine systems play a more important role in mediating both the separation and non-separation induced behavioral changes than do serotonin systems. However, the results with fusaric acid are somewhat surprising since they suggest that a reduction in norepinephrine levels without dopamine depletion at least partially reverses the separation response.

In addition to the studies with monoamine synthesis inhibitors, it has been demonstrated that previously separated rhesus monkeys are hypersensitive to the behavioral effects of the catecholamine agonist d-amphetamine (Kraemer et al., 1983). This finding further implicates catecholamines in the mediation of separation.

Summary

The primary aim of most of the preceding studies was to model a specific psychiatric disorder or a specific characteristic of psychopathology. Whenever primate social colonies are studied in this regard, the intent is usually to produce a syndrome in monkeys which bears a close resemblance to the human condition—i.e., homologous model (Kornetsky, 1977; also this volume). Clearly, none of the models discussed is homologous to either depression, mania, or psychosis. However, certain individual behavioral changes have been interpreted as having direct similarity to a symptom or symptoms of one or both of the disorders. The authors have attempted to identify these behavioral changes as "target behaviors" for each model (Table 12.2). Target behaviors are behavioral changes that (1) are induced in most of the primate social studies attempting to model depression or psychosis, (2) can be induced clinically by similar treatment regimens, and (3) bear direct resemblance or relevance to the human disorder.

Grooming is an important behavior in primates occupying a significant amount of the day's activity in most species. Therefore, it is not surprising that grooming is altered in models of both disorders. Initiated social grooming is typically decreased or eliminated in models of depression and psychosis; however, self-grooming is decreased in depression models, but increased in the amphetamine psychosis models. These grooming behaviors appear to parallel the hypoactivity and disinterest of depression and the "focusing on self" of schizophrenia. In addition, the decreased social grooming and spatial isolation reflect the social withdrawal. Another behavior common

TABLE 12.2. Target behaviors

Depression	Psychosis
Social groom ↓	Social groom ↓
Self-groom ↓	Self-groom ↑
Withdrawal ↑	Withdrawal ↑
Visual scanning ↓	Visual scanning ↑
Locomotion ↓	Submissive gestures ↑
Posture ↓	Scratching ↑
Self-cling ↑	Bizarre behavior ↑

to models of both disorders is visual scanning or checking. In the depression models, visual scanning is decreased, whereas it is increased greatly in the models of psychosis.

Behaviors which appear to be more relevant to depression include locomotion, self-cling, and postural changes. As expected, locomotion is decreased in each of the depression models. There is an increase in self-cling or self-huddle. Also, the slumped-over, energy conserving posture similar to that of depressed patients is apparent in some monkeys, as is the expressionless face.

Two additional behavioral changes that may be related to psychotic behavior are increases in scratching and submissive gestures, which have been suggested as models of tactile hallucination and paranoia, respectively. In addition, there have been numerous reports of bizarre behavior induced in primates by psychotomimetic drugs which may parallel similar behavior in schizophrenic patients. Some behaviors, such as stereotyped behavior, are common; but others, such as following or shadowing behavior, may only appear as isolated incidents. Nevertheless, these behaviors should be taken into consideration as yet another similarity to human psychotic behavior.

Selective Agonist and Antagonist Studies

A second area of interest using pharmacological probes in primates has been directed at determining neurotransmitter mediation of primate social behavior. In most cases, a relatively selective agonist or antagonist of a specific neurotransmitter system is administered to determine behavioral changes induced by stimulation or antagonism of specific systems. Some of these studies have gone hand in hand with attempts to model psychiatric disorders, while others have primarily focused on elucidating behavior mediated by a particular neurotransmitter. In either case, these results are relevant to the study of mechanisms underlying primate behavior which may be related to human psychopathology. These studies will be reviewed briefly, placing particular emphasis on the effect of these agents on the "target behaviors" for major psychiatric disorders.

Dopamine

Two selective dopamine (DA) agonists, apomorphine and piribedil have been tested in members of primate social colonies. Acute administration of apomorphine to stumptail macaques induced dose-dependent increases in submissive gestures, checking, stereotyped behavior, scratching, and vocalizations, as well as, reductions in initiated social grooming and food foraging (Schlemmer et al., 1980). The significant increase in submissive gestures was particularly noteworthy in that (as with amphetamine) the increase occurred in the absence of an increase in aggression directed toward the treated monkeys. This effect was seen regardless of social rank. Even the dominant male had a large increase in submissive gestures when treated with apomorphine (Schlemmer and Davis, 1981a). In another study, the effect of chronic treatment with apomorphine was compared with d-amphetamine in a cross-over design (Schlemmer and Davis, 1981b). Following baseline observation, d-amphetamine was administered nasogastrically for 12 consecutive days as described previously. One year later, the same monkeys received apomorphine (0.5 mg/kg) intramuscularly twice a day for 12 consecutive days. Strikingly similar behavioral responses were noted for apomorphine and amphetamine for several behaviors, including an increase in the target behaviors (submissive gestures, distancing, and checking) and a decrease in initiated social grooming. Differences in responses were noted for two behaviors—locomotion and self-grooming. Apomorphine significantly increased locomotion, but amphetamine did not. Conversely, only d-amphetamine increased self grooming. In marmosets, acute administration of apomorphine also decreased social interactions and increased checking (Scraggs et al., 1979). Piribedil, which has been the only other preferential dopamine agonist tested, disrupted social interactions in two squirrel monkeys in a complex, rank-dependent manner (Poignant and Avril, 1978). Scratching and locomotion were also significantly increased above baseline levels by piribedil.

Dubach and Bowden (1983) injected dopamine bilaterally into the ventral striatum of four crab-eating macaques. The most significant behavioral changes were an increase in lookout (checking) and examine, and the virtual elimination of both social grooming and self-grooming. There were no significant changes in presents, mounts, and locomotion. The behavioral changes reported in this study are in basic agreement with those reported for dopamine agonists and extend these findings to suggest that lookout (checking) and grooming are, at least in part, mediated through a striatal dopamine mechanism.

Several dopamine antagonists have been tested in primate social colony studies, primarily to evaluate the antagonists' antipsychotic action. Although these drugs share dopamine receptor blocking properties, they vary in their selectivity for the dopamine receptor (Peroutka and Snyder, 1980). Two of the more selective dopamine antagonists studied in primates have been haloperidol and trifluoperazine which have been administered to stumptail macaques. Both compounds potently reduced checking, locomotion, and both forms of grooming (Table 12.3). In addition, most treated monkeys showed evidence of bradykinesia, tremor, and rigidity, and would momentarily stop (posturing or freezing) in the midst of walking around the cage. The latter behavioral changes are consistent with movement abnormalities associated with the dopamine deficiencies in Parkinson's disease.

It is evident from these studies that dopamine systems play an important role in the mediation of both motor and social behaviors in monkeys. Dopamine agonists and antagonists disrupted both social and self-grooming. Also, distancing from other colony members was significantly increased by apomorphine, and to a lesser extent by dopamine antagonists. Dopamine agonists and antagonists appeared to have opposite effects on checking, locomotion, and submissive gestures. Stimulation of dopamine receptors by apomorphine resulted in large increases in all three behaviors, whereas dopamine blockade reduced each. To date, no studies have attempted to differentiate behaviors mediated by dopamine receptor subtypes in primate social colonies with the exception of the experiment with piribedil, a relatively selective D2 agonist. However, the behavioral effects of D1 and D2 agonists have been studied in individually caged primates (Woolverton et al., 1984).

Norepinephrine

Fewer studies have examined the role of noradrenergic (NE) systems in mediation of primate social behavior, probably due, in part, to the fact that there are fewer selective noradrenergic agonists and antagonists which enter the CNS. The studies which have been conducted have focused on the alpha$_2$-NE system where centrally active agonists and antagonists are available.

Two laboratories have studied the effects of the alpha$_2$-NE agonist clonidine in adult stumptail macaque social groups (Schlemmer et al., 1981; Young et al., 1986). The predominant effect of clonidine was a profound increase in feeding, which led to substantial weight gain when treatment was continued (Schlemmer et al., 1979). Social interaction and grooming were decreased (Table 12.3), which may have been secondary to the feeding response. Ptosis was present suggesting sedation although locomotion was unaffected. Vocalizations, predominantly food cries, were also increased by clonidine.

The authors have studied eight acute doses of the alpha$_2$-NE antagonist yohimbine in four members of a stumptail macaque social colony. The doses ranged from 0.03 to 5.0 mg/kg and were administered intramuscularly 15 minutes prior to behavioral observation. No consistent behavioral changes were noted in doses up to 1 mg/kg. One monkey appeared "restless" and had increases in locomotion in a non-dose-dependent manner. Higher doses of yohimbine (3 to 5 mg/kg) decreased social grooming and self-grooming and disrupted normal colony activity (Table 12.3). Although the selectivity

TABLE 12.3a. Effect of preferential agonists and antagonists on target behaviors in stumptail macaques

		Social Groom	Self-Groom	Distance >3'
DA				
Apomorphine	Base	3.64 ± 0.51	3.53 ± 0.75	4.25 ± 1.59
	Tx	0.56 ± 0.21	0.81 ± 0.41	9.50 ± 0.51**
Haloperidol	Base	5.84 ± 0.98	2.72 ± 0.72	4.04 ± 0.48
	Tx	0.08 ± 0.08**	0.08 ± 0.08*	7.24 ± 0.72
α_2-*NE*				
Clonidine	Base	2.25 ± 0.71	4.93 ± 1.32	7.25 ± 0.53
	Tx	0.00 ± 0.00	0.25 ± 0.25	9.25 ± 0.85
Yohimbine	Base	2.52 ± 1.15	4.58 ± 0.97	7.79 ± 1.03
	Tx	0.00 ± 0.00	0.25 ± 0.25	9.75 ± 0.25
5-HT				
Quipazine	Base	3.25 ± 0.51	5.15 ± 0.38	6.28 ± 0.54
	Tx	0.00 ± 0.00*	0.00 ± 0.00**	9.63 ± 0.75
Cinanserin	Base	5.84 ± 0.98	2.72 ± 0.72	4.04 ± 0.48
	Tx	4.76 ± 1.06	2.04 ± 0.85	4.92 ± 0.67
ACh				
Physostigmine + MS	Base	6.88 ± 1.83	3.36 ± 0.91	3.64 ± 0.94
	Tx	0.10 ± 0.10**	0.00 ± 0.00*	7.60 ± 1.14
Atropine	Base	6.88 ± 1.83	3.36 ± 0.91	3.64 ± 0.94
	Tx	5.10 ± 1.70	4.40 ± 1.23	3.70 ± 0.89

Each value represents the mean + SEM frequency for each behavior for 4 to 5 monkeys. Doses for each drug are: apomorphine, 0.5 mg/kg; haloperidol, 0.1 mg/kg; clonidine 0.1 mg/kg; yohimbine, 5 mg/kg; quipazine, 1.6 mg/kg; cinanserin, 5 mg/kg; physostigmine, 0.06 mg/kg, + methscopolamine (MS), 0.01 mg/kg; and atropine, 0.2 mg/kg. All drugs were administered intramuscularly.

Statistical significance is shown as:
*$p < 0.05$ when treatment (Tx) is compared with the respective baseline (Base).
**$p < 0.01$ when treatment (Tx) is compared with the respective baseline (Base).

of yohimbine for the alpha$_2$-NE receptor at these doses might be questioned, 5 mg/kg of yohimbine was required to antagonize clonidine-induced hyperphagia in monkeys (Schlemmer et al., 1981); and 2.5 mg/kg i.v. of yohimbine was needed to produce an alpha$_2$-NE-mediated change in prolactin plasma levels in stumptail macaques (Gold et al., 1979).

An additional impression of noradrenergic mediation has been gained from experiments using antipsychotic drugs to antagonize amphetamine-induced behavioral changes in the studies reported above. In some monkeys, haloperidol and pimozide restored amphetamine-disrupted social grooming to levels significantly *above* baseline levels, an effect not seen with chlorpromazine (Schlemmer and Davis, 1983). Coincidentally, haloperidol and pimozide are more selective dopamine antagonists and produce less antagonism of noradrenergic receptors, whereas chlorpromazine is a potent antagonist at both catecholamine receptors (Peroutka and Snyder, 1980). Since amphetamine is an indirect agonist of dopamine and norepinephrine systems, it is possible that the increase in social grooming with the amphetamine-haloperidol and amphetamine-pimozide combinations may have resulted from stimulation of the noradrenergic receptors left unprotected by these antipsychotics. But it is important to note that this response only occurred after dopamine blockade, since amphetamine alone consistently reduced social grooming.

In other studies, one dose of the alpha-adrenergic antagonist aceperone (10 mg/kg) and the beta-adrenergic antagonist propranolol (20 mg/kg) failed to induce significant behavioral changes in marmoset groups (Scraggs and Ridley, 1979). Only movement scores were slightly decreased by aceperone. Studies on the effects of other noradrenergic antagonists on primate social behavior have not been reported.

In summary, the role of the noradrenergic system in the mediation of behavior in group-housed monkeys has not been well studied. Those studies available suggest that an agonist effect at alpha_2-NE receptors stimulates food intake in one primate species. In addition, stimulation of noradrenergic

TABLE 12.3b. Effect of preferential agonists and antagonists on target behaviors in stumptail macaques

		Checking	Locomotion	Subm. Gest.
DA				
Apomorphine	Base	24.25 ± 0.42	3.34 ± 0.44	2.74 ± 0.64
	Tx	84.25 ± 8.37**	11.50 ± 5.77*	7.44 ± 1.85**
Haloperidol	Base	47.08 ± 2.86	4.40 ± 0.66	2.72 ± 0.64
	Tx	14.52 ± 2.52	0.60 ± 0.15	0.84 ± 0.32**
α_2-*NE*				
Clonidine	Base	47.94 ± 2.11	7.44 ± 1.99	3.75 ± 0.65
	Tx	33.75 ± 4.91	7.00 ± 1.47	2.00 ± 1.68
Yohimbine	Base	42.73 ± 3.03	5.75 ± 0.74	3.64 ± 1.11
	Tx	34.50 ± 2.25	5.00 ± 1.00	5.00 ± 1.08
5-HT				
Quipazine	Base	38.73 ± 1.58	5.65 ± 0.89	3.2 ± 0.54
	Tx	66.38 ± 4.03**	2.13 ± 2.13*	5.13 ± 2.20
Cinanserin	Base	47.08 ± 2.86	4.40 ± 0.66	2.72 ± 0.64
	Tx	34.56 ± 4.00**	2.68 ± 0.56	1.20 ± 0.23
ACh				
Physostigmine + MS	Base	38.80 ± 1.43	4.04 ± 0.70	1.88 ± 0.54
	Tx	37.20 ± 4.35	1.20 ± 0.64	2.40 ± 1.04
Atropine	Base	38.80 ± 1.43	4.04 ± 0.70	1.88 ± 0.54
	Tx	31.50 ± 3.13	3.20 ± 1.46	1.60 ± 0.53

Each value represents the mean + SEM frequency for each behavior for 4 to 5 monkeys. Doses for each drug are: apomorphine, 0.5 mg/kg; haloperidol, 0.1 mg/kg; clonidine 0.1 mg/kg; yohimbine, 5 mg/kg; quipazine, 1.6 mg/kg; cinanserin, 5 mg/kg; physostigmine, 0.06 mg/kg, + methscopolamine (MS), 0.01 mg/kg; and atropine, 0.2 mg/kg. All drugs were administered intramuscularly.

Statistical significance is shown as:
*$p < 0.05$ when treatment (Tx) is compared with the respective baseline (Base).
**$p < 0.01$ when treatment (Tx) is compared with the respective baseline (Base).

systems may be involved in the mediation of social grooming in primates; but more direct evidence is necessary to confirm this impression.

Serotonin

The role of 5-hydroxytryptamine (5-HT, serotonin) in the mediation of primate social behavior has been studied in several laboratories. The most exhaustive studies have been conducted using vervet monkeys in the laboratory of Raleigh, McGuire, Brammer, and coworkers. This group has compared the effects of three drugs which facilitate 5-HT transmission by different mechanisms: quipazine, a direct 5-HT receptor agonist; tryptophan, a 5-HT precursor; and fluoxetine, a relatively selective 5-HT uptake inhibitor (Raleigh et al., 1985). All three 5-HT agonists elicited similar behavioral changes in 15 vervets. Social grooming, approaching other monkeys, resting, and eating were all increased by each agonist, while locomotion, avoid other monkeys, be vigilant, and be solitary were decreased following administration of the 5-HT agonists. Dominant animals in the groups had a greater response than subordinate monkeys which is in agreement with a previous report by these investigators linking serotonin function to dominance in vervets (Raleigh et al., 1983).

Quipazine has also been tested in a colony of stumptail macaques in our laboratory (Table 12.3). Acute doses of quipazine, 0.1-3.0 mg/kg, were administered i.m., to four monkeys 15 minutes prior to one-hour behavioral observations (described above). Only one monkey per day received drug treatment. In this species, quipazine significantly decreased social grooming and self-grooming, slightly increased distancing (non-significantly), and significantly elevated checking from baseline levels. These findings are dissimilar to the behavioral changes induced by comparable quipazine doses in vervets (Raleigh et al., 1985). Quipazine, however, did significantly decrease locomotion in both species. Quipazine also induced limb jerks and body shakes (behaviors associated with hallucinogens in stumptail macaques), but these emergent behaviors were not reported in vervets. An explanation for these disparate findings is not apparent at this time.

Two types of serotonin antagonists, a receptor antagonist and a synthesis inhibitor, have been studied in primate social groups. Acute or chronic administration of four serotonin receptor antagonists—cinanserin (5 mg/kg), metergoline (0.3 mg/kg), methysergide (1 mg/kg), and cyproheptadine (1 mg/kg)—failed to induce major overt behavioral changes in members of stumptail macaque social colonies; yet, each of the 5-HT blockers antagonized hallucinogen-induced behavior at the same doses (Schlemmer and Davis, 1986). The only exception was a slight, but statistically significant reduction in checking by some, but not all agents (e.g., cinanserin, Table 12.3). Similarly, depletion of brain 5-HT by the serotonin synthesis inhibitor PCPA did not significantly affect initiated or passive social activity in this species (Redmond et al., 1971a); nor did this depletion adversely effect the separation syndrome in rhesus monkeys (Kraemer and McKinney, 1979). Conversely, PCPA did induce significant behavioral changes when administered chronically to both crab-eating macaques (Boelkins, 1979) and vervets (Raleigh et al., 1980). Boelkins found that PCPA (330–735 mg/kg) which is 2 to 5 times the dose used by Redmond et al., (1971b) for 7 to 17 days decreased sexual interaction and increased passive social interaction in immature macaques. In vervets, chronic PCPA (80 mg/kg) increased locomotion, being vigilant, avoiding other monkeys, being solitary, and initiated and received aggression (Raleigh et al., 1980). Initiated social grooming, huddling, approaches, resting, and eating were decreased by this regimen. These data are in good agreement with data reported above for behavioral changes induced by 5-HT agonists in vervets.

In summary, it appears that the role of serotonin systems in the mediation of primate social behavior may vary between species. Serotonin systems play a particularly important role in the mediation of social and individual behavior and hierarchical status in vervets, but may not serve a similar role in larger macaque monkeys such as stumptail macaques.

Acetylcholine

Cholinergic systems (ACh) have not been extensively studied in primate social colonies. In the authors' laboratory, the cholinesterase inhibitor, physostigmine, and the antimuscarinic, atropine, have been used to examine the effect of a centrally active cholinergic agonist and antagonist in members of an adult stumptail macaque social colony. Physostigmine (0.06 mg/kg) was administered intramuscularly 45 minutes prior to behavioral observation (detailed above). The peripheral antimuscarinic, methscopolamine (0.01 mg/kg), was administered intramuscularly 75 minutes prior to observation to minimize peripheral effects of the cholinesterase inhibitor. In all, five monkeys received treatment for three consecutive days. However, only two or three monkeys were treated per day. In general, physostigmine-treated animals were hypoactive compared to baseline activity scores. Both initiated social grooming and self-grooming were significantly decreased while distancing scores were increased (Table 12.3). Locomotion was reduced, but checking remained unchanged. Interestingly, these changes are consistent with target behavioral changes induced by inhibitors of the catecholamine system, reserpine, and AMPT. Postural changes, however, were not noted with physostigmine.

Atropine (0.2 mg/kg) was administered intramuscularly 30 minutes prior to behavioral observation in a manner similar to that described for physostigmine. Atropine did not induce significant behavioral changes at this dose (Table 3), although the same dose was effective in antagonizing the behavioral changes induced by physostigmine.

Opiates

Opiates have also been administered to primates living in social groups to assess the drugs' effects on social and individual behavior. Both opiate agonists and antagonists have been reported to alter social behaviors of association, dominance, submission, and sexuality, as well as individual behaviors.

Crowley et al. (1974) reported that adult male pigtail macaques receiving acute doses of morphine (0.05-0.02 mg/kg) typically exhibited a decrease in sexual behavior, while maintaining baseline scores for both agonistic behavior and locomotor activity. Following the administration of morphine (0.5-2.0 mg/kg) to juvenile squirrel monkeys, distancing scores increased and social contact between treated juveniles and their mothers was reduced (Miczek et al., 1981). Methadone (10-20 mg/kg) decreased several social and individual behaviors, including social grooming, social proximity, locomotion (night only), sleeping, eating, and food foraging in bonnet macaques (Crowley et al., 1975). In the same study, locomotion was increased following the administration of methadone. Agonistic behavior was not affected by methadone.

The opiate antagonists naloxone and naltrexone appear to enhance social behavior in primates. Both naloxone (0.5 mg/kg) and naltrexone (0.25-1.0 mg/kg), increased social grooming in pairs of adult talapoin monkeys, but had no apparent effect on self-grooming (Fabre-Nys et al., 1982). Both antagonists were associated with decreases in locomotion and scratching. Naltrexone (0.5 mg/kg) also increased sexual behaviors such as inspections, mounts, and ejaculations, as well as agonistic behavior in high-ranking male talapoin monkeys (Meller et al., 1980).

In summary, opiate systems appear to play a role in the mediation of social grooming, as well as, motor, sexual, and feeding behavior in primates. Opiates tended to decrease these behaviors with the exception of locomotion, while opiate antagonists increased grooming, sexual behavior, and feeding, but increased locomotion.

Summary

Although the number of drug studies which have sought to determine the mediation of primate social behavior is considerably fewer than those for solitary behavior in lower species (e.g., rats), a preliminary picture is beginning to emerge. The picture is certainly clearer for some behaviors than for others, which may, in part, result from attempts to unravel the mediation of social behavior requiring well-integrated, complex CNS systems that must be intact for appropriate function. Disruption of the function of one of several component systems could interfere with the proper expression of this behavior, potentially suggesting a misleading pattern of neurotransmitter mediation. Moreover, caution must be exercised in interpreting these findings since the addition of a social component increases a number of uncontrolled variables in these experiments and the number of animals tested is relatively small. Therefore, any conclusions should be viewed with appropriate caution.

Of particular interest are the target behaviors for primate models of major mental disorders. A summary of the effect of preferential agonists and antagonists of key neurotransmitter systems on these behaviors across several species is presented in Table 12.4.

Initiated social grooming, which is significantly reduced in models of both disorders, appears to be influenced by each neurotransmitter system studied in monkeys which probably reflects highly integrated mediation of this behavior. Most agents reduced social grooming, but opiate antagonists and serotonin agonists (in one species) facilitated heterogrooming. In addition, indirect evidence from amphetamine studies suggests that an increase in norepinephrine activity may increase social grooming as well. Therefore, catecholamine, serotonin, and opiate systems may be particularly important in mediating social grooming. On the other hand, self-grooming in social groups was only disrupted by some agonists and antagonists. However, self-grooming was significantly elevated by amphetamine, suggesting that simultaneous activation of both norepinephrine and dopamine systems enhance this behavior.

Spatial isolation, or an increase in distancing from other colony members, has been reported in models of depression and psychosis and may be related to autistic behavior present in a number of human psychopathologies. Distancing appears to be influenced by several systems; however, stimulation of dopamine systems by agonists such as apomorphine and amphetamine appears to induce the most pronounced spatial isolation.

TABLE 12.4. Summary of behavioral changes induced in primates by agonists and antagonists of neurotransmitter systems

	Social Groom	Self-Groom	Distance	Checking	Locomotion	Subm. Gest.	Scratch
DA							
Agonists	↓	↓	↑	↑	↑	↑	0-↑
Antagonists	↓	↓	0-↑	↓	↓	0-↓	0
α₂-NE							
Agonists	↓	↓	0	↓	0	0	0
Antagonists	↓	↓	0	↓	0	0	0
5-HT							
Agonists	↓-↑	↓	↓-↑	↓-↑	↓	0	0
Antagonists	0	0	0	↑-↓	0	0	0
ACh							
Agonists	↓	↓	↑	0	↓	0	↓
Antagonists	0	0	0	0	0	0	0
Opiate							
Agonists	↓	N.R.	↑	N.R.	↓-↑	0	↑
Antagonists	↑	0	N.R.	N.R.	↓	N.R.	↓

↑ = increase; ↓ = decrease; 0 = no change; N.R. = Not reported.

Checking, which is a form of visual exploration of the cage environment, is decreased in depression models and increased in models of psychosis. Checking can also be affected by cage activity. For example, checking is incresed by conflicts within the cage and is usually decreased during huddling with other monkeys. Checking appears to be influenced by monoaminergic systems. Dopamine systems appear to be particularly important in the mediation of checking. Serotonin mediation of the checking response may differ between primate species. Checking was increased by disruption of serotonin transmission and decreased by facilitation of serotonin transmission in vervets, but increased by serotonin agonists and decreased by serotonin antagonists in stumptail macaques.

Locomotion (motor exploration of the environment) is decreased in models of depression. Dopamine systems also appear to play a significant role in the mediation of primate locomotion. Locomotion was increased by stimulation of dopamine systems by agonists and decreased by dopamine antagonists. Opiate systems also appear to play a role in the regulation of locomotion since opiate agonists increased and opiate antagonists decreased locomotion. Serotonin and cholinergic agonists reduced locomotion, but antagonists of these systems failed to significantly affect locomotion. Changes in submissive behavior and scratching are important in models of psychosis. Only the dopamine system appeared to profoundly influence submissive behavior. Dopamine agonists induced significant increases in submissiveness. The mediation of aggression is less clear, although serotonin systems appear to be involved in the determination of dominance in male vervets. Scratching may be influenced by dopamine, opiate, and cholinergic systems, although none of the preferential agonists or antagonists induced the intensity of scratching that was seen with amphetamine.

Bizarre behavior is an important component of psychosis and should be one feature which can be replicated in animal models. Grossly abnormal or bizarre behavior may be induced on occasion by a number of pharmacologic agents. However, amphetamine and dopamine agonists appear to most consistently induce bizarre behavior across several species. Although this behavior often takes the form of stereotypy, non-stereotyped aberrant behavior is also seen in stimulant-treated monkeys.

Conclusions

Two major criticisms leveled against animal models of mental disorders are that these models have not been very useful in (1) predicting clinical efficacy of new therapeutic agents or (2) generating new hypotheses concerning the etiology of psychopathologies (Sitaram and Gershon, 1983). Part of the problem may stem from the inability to accurately model a very complex *group* of human behaviors in animals. Until recently, there has been a tendency to attempt to model a particular disorder rather than a particular symptom or symptoms of the disorder. Focusing attention on modeling specific aspects of psychopathology may prove to be a more fruitful and realistic approach to studying these disorders in animals. Non-human primates offer a significant advantage for such an approach because of the wide variety of behaviors included in their behavioral repertoire. Furthermore, social behavior should be studied as well since inappropriate social interaction is an integral part of depressive, manic, and psychotic behavior. Therefore, primate social colonies offer a particularly relevant paradigm for modeling psychopathology where a wider range of potentially relevant behaviors can be evaluated.

Experiments directed at developing models of mental disorders in primate social colonies have produced some behavioral changes which bear greater similarity to the human condition, such as the depressive posture of AMPT-treated monkeys. Primate social colony models have not lead to the formulation of new hypotheses concerning these disorders, and these models have not been frequently used to predict efficacy of new drugs. However, the social colony paradigm has clearly found utility in testing existing hypotheses of affective disorders and schizophrenia (e.g., McKinney et al. 1971; Redmond et al., 1971b; Garver et al., 1975). The primate social colony models recently have been included in the preclinical screen of compounds for antipsychotic properties (Schlemmer and Davis, 1985; New et al., 1988). Understanding the mediation of relevant primate behavioral changes (target behaviors) and ways to prevent or reverse these changes could potentially provide new insights into the etiology of psychiatric illness, which could ultimately lead to the development of new treatment modalities. Moreover, monitoring the reversal of target behaviors only induced in primates (e.g., increased submissive responses) could reveal therapeutic properties of compounds that would otherwise go undetected in models using lower species.

Presently, primate social colony models of depression and psychosis are in general agreement with the current hypotheses for both disorders. Models of depression appear to be linked with disruption of catecholamine transmission. Alternatively, facilitation of catecholamine transmission, particularly dopamine, elicits psychotic-like behavior in monkeys; and dopamine antagonists reverse these behavioral changes. Data from studies of the effect of preferential agonists and antagonists on primate social behavior are suggestive, but not conclusive. However, sufficient information has been generated from this work to offer direction for future studies.

Future Directions

There is a need for the development of new, innovative models of mental disorders to complement the present models. One possibility comes from the work of Kalin (1985; also, this volume) with corticotropin-releasing factor (CRF). Intravenous or intraventricular administration of CRF to rhesus monkeys induced changes in solitary behavior which are consistent with a model of depression. It would be of interest to determine if similar behavioral changes along with detriments in social interaction occurred in the social setting. Another potential model is the withdrawal reaction following discontinuation of stimulant drug administration. Severe depressive episodes have been described upon the monkeys' withdrawal from cocaine and amphetamine. Indeed, one report suggests that the tricyclic antidepressant desipramine was effective in diminishing the severity of cocaine and amphetamine withdrawal in several users (Tennant and Rawson, 1983). It would be of both theoretical and practical significance if this syndrome could be reproduced and tested in monkeys.

More studies are needed to determine the role of neurotransmitter systems in the mediation of primate social behavior. Other neurotransmitters, such as neuropeptides must be included, and previous findings should be extended to look at the role of the individual receptor subtypes in the mediation of primate behavior. In the dopamine system, for example, the behavioral effects of the selective D1 and D2 agonists (SKF38393) and l-quinpirole (LY171555) could be compared. Similarly, the effects of D1 and D2 antagonists on the amphetamine syndrome could be tested. The increased selectivity of these compounds may prove useful in limiting the number of behavioral changes induced by one agent, thereby reducing the number of interfering variables which must be considered.

Acknowledgment

The authors wish to acknowledge the contributions of Jane Ann Retallack, Shawn Montell, Doris McGinness, Dr. Corinne Tyler, and Dr. William J. Heinze to this work.

References

Angrist BM, Sathananian G, Wilk S, Gershon S (1974): Amphetamine psychosis: Behavioral and biochemical aspects. *J Psychiatr Res* 11:13-23

Bellarosa A, Bedford J, Wilson MC (1980): Sociopharmacology of d-Amphetamine in *Macaca arctoides. Pharmacol Biochem Behav* 13:221-228

Boelkins RC (1979): Effects of parachlorophenylalanine on the behavior of monkeys. In: *Serotonin and Behavior,* Barchas J, Usdin E, eds. New York: Academic Press

Burki HR, Eichenberger E, Sayers AC (1975): Clozapine and the dopamine hypothesis of schizophrenia: A critical appraisal. *Pharmakopsychiatr Neuropsychopharmakol* 8:115-121

Connell PH (1958): Amphetamine psychosis. *Maudsley Monographs* No.5. London Oxford University

Crowley TJ, Stynes AJ, Hydinger M, Kaufmann IC (1974): Ethanol, methamphetamine, pentobarbital, morphine, and monkey social behavior. *Arch Gen Psychiatry* 31:829-838

Crowley TJ, Stynes AJ, Hydinger M, Feiger A (1975): Monkey motor stimulation and altered social behavior during chronic methadone administration. *Psychophar-macology* 43:135-144

Dubach MF, Bowden DM (1983): Response to intracerebral dopamine injection as a model of schizophrenic symptomatology. In: *Ethopharmacology: Primate Models of Neuropsychiatric Disorders,* Miczek KA, ed. New York: Alan R. Liss

Fabre-Nys C, Meller RE, Keverne EB (1982): Opiate antagonists stimulate affiliative behaviour in monkeys. *Pharmacol Biochem Behav* 16:653-659

Freis ED (1954): Mental depression in hypertensive patients treated for long periods with large doses of reserpine. *New Engl J Med* 251:1006

Garver DL, Schlemmer RF, Jr, Maas JW, Davis JM (1975): A schizophreniform behavioral psychosis mediated by dopamine. *Am J Psychiatry* 132:33-38

Gold MS, Donabedian RK, Redmond DE Jr (1979): Further evidence for alpha$_2$-adrenergic receptor mediated inhibition of prolactin secretion: The effect of yohim-bine. *Psychoneuroendocrinology* 3:253-260

Haber S, Berger PA, Barchas PR (1979): The effects of amphetamine on agonistic behaviors in nonhuman primates. In: *Catecholamines: Basic and Clinical Frontiers,* Usdin E, Kopin IJ, Barchas J, eds. New York: Pergamon Press

Kalin NH (1985): Behavioral effects of ovine corticotropin-releasing factor admin-istered to rhesus monkeys. *Fed Proc* 44:249-253

Kjellberg B, Randrup A (1973): Disruption of social behavior of vervet monkeys (*Cercopithecus*) by low doses of amphetamines. *Pharmakopsychiatria* 6:287-293

Kornetsky C (1977): Animal models: Promises and problems. In: *Animal Models in Psychiatry and Neurology,* Hanin I, Usdin E, eds. Oxford: Pergamon Press

Kraemer GW, Ebert MH, Lake CR, McKinney WT (1983): Amphetamine challenge: Effects in previously isolated rhesus monkeys and implications for animal mod-els of schizophrenia. In: *Ethopharmacology: Primate Models of Neuropsychiatric Disorders,* Miczek KA, ed. New York: Alan R. Liss

Kraemer GW, McKinney WT (1979): Interactions of pharmacological agents which alter biogenic amine metabolism and depression. *J Affective Disord* 1:33-54

Machiyama Y, Utena H, Kikuchi M (1970): Behavioural disorders in Japanese monkeys produced by the long-term administration of methamphetamine. *Proc Japan Acad* 46:738-743

McKinney WT Jr (1974): Primate social isolation. *Arch Gen Psychiatr* 31:422-426

McKinney WT Jr, Eising RG, Moran EC, Suomi SJ, Harlow JF (1971): Effects of reserpine on the social behavior of rhesus monkeys. *Dis Nerv Syst* 32:735-741

Meller RE, Keverne EB, Herbert J (1980): Behavioural and endocrine effects of naltrexone in male talapoin monkeys. *Pharmacol Biochem Behav* 13:663-672

Miczek KA (1983): *Ethopharmacology: Primate Models of Neuropsychiatric Disorders.* New York: Alan R. Liss

Miczek KA, Woolley J, Schlisserman S, Yoshimura H (1981): Analysis of am-phetamine effects on agonistic and affiliative behavior in squirrel monkeys (*Saimari sciureus*). *Pharmacol Biochem Behav* 14:103-107

Miczek KA, Yoshimura J (1982): Disruption of primate social behavior by *d*-amphetamine and cocaine: Differential antagonism by antipsychotics. *Psy-chopharmacology* 76:163-171

Miller MH, Geiger E (1976): Dose effects of amphetamine on macaque social behavior: Reversal by haloperidol. *Res Commun Psychol Psychiat Behav* 1:125-142

Miller RE, Levine JM, Mirsky IA (1973): Effects of psychoactive drugs on nonverbal communication and group behavior in monkeys. *J Person Soc Psychol* 28:396-405

New JS, Yevich JP, Temple DL, Jr, New KB, Gross SM, Schlemmer RF Jr, Eison MS, Taylor DP, Riblett LA (1988): Atypical antipsychotic agents: Patterns of activity in a series of 3-substituted-2-pyridinyl-1-piperazine derivatives. *J Med Chem* 31:618–624

Nielsen EB, Eison MS, Lyon M, Iversen SD (1983): Hallucinatory behaviors in primates produced by around-the-clock amphetamine treatment for several days via implanted capsules. In: *Ethopharmacology: Primate Models of Neuropsychiatric Disorders*, Miczek KA, ed. New York: Alan R. Liss

Peroutka SJ, Snyder SH (1980): Relationship of neuroleptic drug effects at brain dopamine, serotonin, alpha-adrenergic, and histamine receptors to clinical potency. *Am J Psychiatry* 137:1518–1522

Poignant JC, Avril A (1978): Pharmacological studies on drug acting on the social behavior of the squirrel monkeys. *Arzneimittelforsch* 28:267–271

Raleigh MJ, Brammer GL, McGuire MT (1983): Male dominance, serotonergic systems, and the behavioral and physiological effects of drugs in vervet monkeys (*Cercopithecus aethiops sabaeus*). In: *Ethopharmacology: Primate Models of Neuropsychiatric Disorders*, Miczek KA, ed. New York: Alan R. Liss

Raleigh MJ, Brammer GL, McGuire MT, Yuwiler A (1985): Dominant social status facilitates the behavioral effects of serotonergic agonists. *Brain Res* 348:274–282

Raleigh MJ, Brammer GL, Yuwiler A, Flannery JW, McGuire MT, Geller E (1980): Serotonergic influences on the social behavior of vervet monkeys (*Cercopithecus aethiops sabaeus*). *Exp Neurol* 68:322–334

Raleigh MJ, Buwiler A, Brammer GL, McGuire MT, Geller E, Flannery JW (1981): Peripheral correlates of serotonergically-influenced behaviors in vervet monkeys (*Cercopithecus aethiops sabaeus*). *Psychopharmacology* 72:241–246

Redmond DE Jr, Hinrichs RL, Mass JW, Kling A (1973): Behavior of free-ranging macaques after intraventricular 6-hydroxydopamine. *Science* 181:1256–1258

Redmond DE, Jr, Maas JW, Kling A, Dekirmenjian H (1971a): Changes in primate social behavior after treatment with alpha-methyl-paratyrosine. *Psychosom Med* 33:97–113

Redmond DE, Jr, Maas JW, Kling A, Graham CW, Dekirmenjian H (1971b): Social behavior of monkeys selectively depleted of monoamines. *Science* 174:428–431

Ridley RM, Baker HF (1983): Is there a relationship between social isolation, cognitive inflexibility, and behavioral stereotype? An analysis of the effects of amphetamine in the marmoset. In: *Ethopharmacology: Primate Models of Neuropsychiatric Disorders*, Miczek KA, ed. New York: Alan R. Liss

Schiorring E (1977): Changes in individual and social behavior induced by amphetamine and related compounds in monkeys and man. In: *Cocaine and Other Stimulants*, Ellinwood EH, ed. New York: Plenum Press

Schlemmer RF Jr, Casper RC, Narasimhachari N, Davis JM (1979): Clonidine-induced hyperphagia and weight gain in monkeys. *Psychopharmacology* 61:233–234

Schlemmer RF Jr, Casper RC, Elder JK, Davis JM (1981): Hyperphagia and weight gain in monkeys treated with clonidine. In: *Psychopharmacology of Clonidine*, Lal H, Fielding S, eds. New York: Alan R. Liss

Schlemmer RF, Jr, Davis JM (1981b): Evidence for dopamine mediation of submissive gestures in the stumptail macaque monkey. *Pharmacol Biochem Behav* 14(Suppl. 1):95–102

Schlemmer RF, Jr, Davis JM (1981a): Similarities in the behavioral effects of *d*-

260   R. Francis Schlemmer, Jr., Jennifer E. Young, and John M. Davis

amphetamine and apomorphine in selected members of a primate social colony. In: *Recent Advances in Neuropsychopharmacology, Advances in the Bio-Sciences,* vol. 31, Angrist B, Burrows GD, Lader M, Lingjaedre O, Sedvall G, Wheatley D, eds. Oxford: Pergamon Press

Schlemmer RF, Jr, Davis JM (1983): A comparison of three psychotomimetic-induced models of psychosis in nonhuman primate social colonies. In: *Ethopharmacology: Primate Models of Neuropsychiatric Disorders,* Miczek KA, ed. New York: Alan R. Liss

Schlemmer, RF Jr, Davis JM (1985): Antagonism of amphetamine-induced behavioral changes by BMY13859-1 in a primate model of psychosis. *Neurosci Abst* 11:114

Schlemmer RF Jr, Davis JM (1986): A primate model for the study of hallucinogens. *Pharmacol Biochem Behav* 24:381-392

Schlemmer RF Jr, Jackson JA, Preston KL, Bederka JP Jr., Garver DL, Davis JM (1978): Phencyclidine-induced stereotyped behavior in monkeys. *Eur J Pharmacol* 52:379-384

Schlemmer RF Jr, Narasimahachari N, Davis JM (1980): Dose dependent behavioral changes induced by apomorphine in selected members of a primate social colony. *J Pharm Pharmacol* 32:285-289

Scraggs PR, Baker HF, Ridley RM (1979): Interaction of apomorphine and haloperidol: Effects on locomotion and other behavior in the marmoset. *Psychopharmacology* 66:41-43

Scraggs PR, Ridley RM (1979): The effect of dopamine and noradrenaline blockade on amphetamine-induced behaviour in the marmoset. *Psychopharmacology* 62:41-45

Sitaram N, Gershon S (1983): From animal models to clinical testing-promise and pitfalls. *Prog Neuropsychopharmacol Biol Psychiatry* 7:227-228

Smith EO, Byrd LD (1983): Contrasting the effects of *d*-amphetamine on affiliation and aggression in monkeys. *Pharmacol Biochem Behav* 20:255-260

Snyder SH (1972): Catecholamines in the brain as mediators of amphetamine psychosis. *Arch Gen Psychiatry* 27:169-179

Tennant FS Jr, Rawson RA (1983): Cocaine and amphetamine dependence treated with desipramine. In: *Problems of Drug Dependence, 1982, NIDA Research Monograph 43,* Harris LS, ed. Rockville: PHS

Wilson MC, Bailey L, Bedford JA (1983): Effects of subacute administration of amphetamine on food competition in primates. In: *Ethopharmacology: Primate Models of Neuropsychiatric Disorders,* Miczek KA, ed. New York: Alan R. Liss

Woolverton WL, Goldberg LI, Ginos JZ (1984): Intravenous self-administration of dopamine receptor agonists by rhesus monkeys. *J Pharmacol Exp Ther* 230:678-683

Young JE, Verlangieri AJ, Wilson MC (1986): The effects of clonidine on food consumption and food competition in male stumptail macaques *(Macaca arctoides). Pharmacol Biochem Behav* 24:1567-1572

13

The Neuropharmacology of Serotonin and Sleep: an Evaluation

JOHN D. FERNSTROM AND ROSS H. PASTEL

Introduction

Depression is often associated with disruption of the normal sleep cycle. In particular, though patients frequently show disturbances in sleep continuity and a reduction in slow wave sleep, the most commonly observed effects are a shortened latency to the first episode of rapid eye movement (REM) sleep after sleep onset (i.e., a reduced REM latency), and an increase in REM density in the first REM periods during the night (Kupfer, 1982). The administration of antidepressant drugs typically suppresses most measures of REM sleep (see Kupfer, 1982), and it is an issue of general interest whether there is a connection between the pharmacologies of REM suppression and of antidepressant treatment. Several of the antidepressants in past, current, and experimental use are active at serotonin (5HT) synapses. The antidepressant action of such drugs is commonly thought to be associated with their ability to enhance transmission across 5HT synapses (see Willner, 1985). If this is the case and if there is a connection between a drug's effects on sleep and depression, then these agents should have predictable actions on sleep, as should other drugs that affect 5HT synaptic function. The purpose of this review is to evaluate this latter proposition. To this end, a discussion is presented of the effects that each of several classes of agents that modify 5HT transmission has on sleep in mammals. These are broadly classed as agents that decrease or increase synaptic transmission. In the former category are compounds like para-chlorophenylalanine, which depletes the brain of 5HT and 5HT neurotoxins and antagonists (Figure 13.1). In the latter category are included direct-acting 5HT receptor agonists, releasers, reuptake blockers, and precursors. The hope in such a review would be that at the end a clear, simple view would emerge concerning the role of 5HT neurons in sleep. As will become apparent, such is not the case. Even the most specific 5HT drugs sometimes produce conflicting results; there appear to be species differences in the actions of many 5HT drugs on sleep, making the use of animal models problematic; and it has become apparent that the phenomenon of sleep is perhaps too complex a set of processes to elaborate

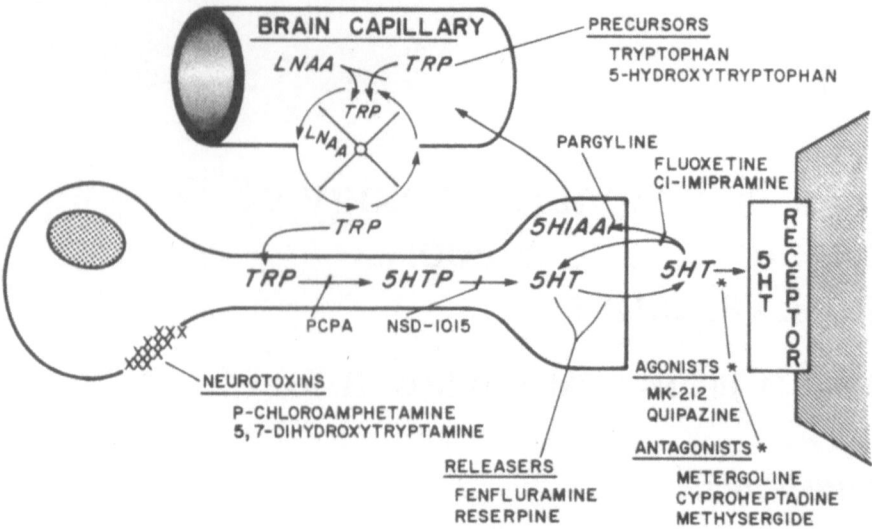

FIGURE 13.1. Neuropharmacologic agents that affect the serotonin neuron in brain. Shown are the general categories of drugs, with examples, that influence the different aspects of the life history of serotonin within and around the neuron. Abbreviations: TRP = tryptophan; LNAA = large, neutral amino acids; 5HTP = 5-hydroxytryptophan; 5HT = 5-hydroxytryptamine (serotonin); 5HIAA = 5-hydroxyindoleacetic acid, the principal metabolite of 5HT.

simple state changes in response to the administration of a drug, however specific.

Drugs that Diminish Serotonin Neurotransmission

Para-chlorophenylalanine (PCPA)

This review begins with a perspective on para-chlorophenylalanine, because it has been the principal agent used to propound a 5HT-sleep connection. Prior to 1966, there appears to have been no pharmacologic agent generally regarded as both potent and specific in modifying transmission across 5HT synapses. This situation changed with the discovery of PCPA. Para-chlorophenylalanine is an inhibitor of tryptophan hydroxylase (Koe and Weissman, 1966), the enzyme catalyzing the rate-limiting step in 5HT synthesis (Lovenberg et al., 1968) (Figure 13.1). Because this enzyme is specific to the 5HT biosynthetic pathway (Lovenberg et al., 1968), the use of PCPA to block tryptophan hydroxylase also guarantees a specific neurochemical action of the drug in brain: viz., to block 5HT synthesis and thus to deplete the brain of this amine. (PCPA also inhibits hepatic phenylalanine hydroxylase [Lipton et al., 1967], but it is unknown if this action is important in eliciting behavioral and physiological alterations typically ascribed to effects on

5HT synthesis.) Para-chlorophenylalanine was therefore embraced by psychopharmacologists as an agent that would quickly allow a determination of how 5HT neurons were involved in the control of particular brain functions. One of these functions was sleep. In fact, sleep pharmacologists appear to have been (and still are) particularly enthused by PCPA: published articles on the drug's effects on sleep appeared almost immediately (Jouvet, 1969) after the main publication of the biochemical actions of PCPA (Koe and Weissman, 1966). A steady stream continues to the present.

What are the actions of PCPA on sleep that make this drug so interesting? The bulk of the work has been in cats and rats. Typically, the drug is administered I.P. in doses ranging between 50 and 800 mg/kg. The effects on sleep do not become apparent for 24 to 36 hr, which is the approximate time required for the drug to induce maximal depletion of brain 5HT content. At this time, non-REM (NREM) sleep and REM sleep decline substantially, reaching a nadir between 24 to 40 hr post-PCPA injection. Thereafter, sleep returns gradually to normal within about 200 hours (Koella et al., 1968; Mouret et al., 1968; Jouvet, 1969). Because the specific biochemical action of the drug in brain (to deplete 5HT) coincided temporally with a clear behavioral effect (the disruption of normal sleep), the conclusion typically drawn was that 5HT neurons must be involved in various aspects of sleep regulation. This view was reinforced by the additional observation in cats experiencing PCPA-induced insomnia that a single injection of 5-hydroxytryptophan (5HTP, 3-5 mg/kg I.P. the immediate precursor of 5HT) almost immediately induced a short period of both normal NREM and REM sleep (Jouvet, 1972). This latter drug should be able to restore sleep, if the depletion of 5HT was important for PCPA's effects, because 5HTP would immediately replenish the brain's store of this amine. (5HTP is the product of tryptophan hydroxylation; its administration should thus lead to its immediate conversion to 5HT by the enzyme catalyzing the second and final step in 5HT synthesis, aromatic L-amino acid decarboxylase [Lovenberg et al., 1962].) Hence, using PCPA alone and in combination with 5HTP, early sleep pharmacologists could reasonably conclude that 5HT neurons must participate in the sleep-waking process.

Over the years, this basic paradigm (PCPA administration suppresses sleep; a single injection of 5HTP acutely restores it) has been studied extensively, in an attempt to gain a more precise insight into sleep regulation. Unfortunately, as more work has appeared, the picture has become less certain. The earliest hypothesis that derived from the data on PCPA and sleep was put forth by Jouvet and his associates, who posited that 5HT neurons are important in the "executive functions" of NREM sleep (Jouvet, 1972). The term "executive" followed from the observation that when 5HT is depleted, NREM sleep is substantially compromised; the functioning of the 5HT neuron was thus viewed as necessary for the occurrence of NREM sleep (it thus served an executive role). While some data support this viewpoint (Mouret et al., 1968; Pujol et al., 1971; Ursin, 1972, 1980; Weitzman et al., 1968), not all do. For example, several investigators have noted that the timecourse of brain 5HT restoration after its depletion by PCPA injection lags well behind the actual return of NREM sleep (Dement et al., 1972; Mouret et al., 1968; Rechtschaffen et al., 1973). If 5HT is an important factor for inducing NREM sleep, its reappearance should precede, not lag behind, the recurrence of NREM sleep. In addition, it is possible to administer PCPA in such a manner (several small, daily doses spread over a several-day period) as to cause almost total depletion of brain 5HT (90+%) without producing any changes in NREM sleep (Dement et al., 1972; Rechtschaffen et al., 1973). This result seems to be substantially at odds with the prime importance of the 5HT neuron in executing NREM sleep. And finally, NREM sleep can be induced in PCPA-treated rats by sleep deprivation (Tobler and Borbely, 1982). This result suggests that the induction of NREM sleep may be (or can be) independent of the presence and activity of 5HT neurons. (However, it should be noted that the opposite result has also been obtained [Sallanon et al., 1983]).

Second, the PCPA data have often been cited by Jouvet as supporting an important role for the 5HT neuron in the "priming functions" of REM sleep (Jouvet, 1969, 1972). This priming role seems to mean that the 5HT neuron modulates the frequency of REM sleep occurrence but does not create REM sleep itself. For example, PCPA depletes brain 5HT and greatly diminishes the rate of REM sleep episode occurrence (Mouret et al., 1968; Kaufman, 1983). One problem with this priming function of 5HT neurons in REM sleep, in the context of PCPA studies, is that it has only rarely been cleanly separated from the NREM sleep effects of 5HT. That is, it is widely held that there is normally a progression from waking into NREM sleep and only thereafter into REM sleep, that REM sleep is almost always preceded by NREM sleep, and that very rarely does a REM sleep episode follow waking directly (Jouvet, 1969; Ursin, 1972). This temporal relationship is relevant to the PCPA studies, since it appears that the degree of REM sleep suppression by the drug correlates well with the degree of NREM sleep suppression (Koella et al., 1968; Ursin, 1980). If so, then the occurrence of REM sleep can simply be tied to 5HT depletion through the depression in NREM sleep; no separate priming mechanism need be involved. In another study (Sallanon et al., 1983), in which cats received PCPA during a period of sleep deprivation, a post-sleep deprivation rebound was observed in REM sleep in the absence of almost any NREM sleep at all. While these data provide an example showing that REM sleep *can* occur without NREM sleep in a PCPA paradigm (i.e., REM sleep is not depressed secondary to a reduction in NREM sleep), they present the interesting result that REM sleep is occurring at a time when 5HT is almost fully depleted. This finding suggests that REM sleep can occur quite independently of the normal functioning of 5HT neurons (an implication recognized by the investigators).

Third, the argument has also been made that PCPA-induced insomnia is not the consequence of a direct, 5HT-related effect on the neural substrates of sleep. Instead, the reduction in serotonin neuronal function leads to a state of hyperresponsivity, thereby preventing the animal from damping the input of external sensory information, which is viewed as a necessary prelude to sleep (Mouret et al., 1968; Tobler and Borbely, 1982). This notion seems more relevant to studies in rats than in cats, for rats show clear hyperreactivity and agitation after PCPA administration (Mouret et al., 1968; Tobler and Borbely, 1982), while cats apparently do not (Ursin, 1980). This possibility has been countered in part by Laguzzi (1982), who reported that when rats were made hyperactive by hippocampal lesions, sleep was unaffected, and PCPA still induced insomnia. According to Laguzzi, if hyperactivity *per se* were important, the lesion should have disrupted sleep, and PCPA should not have produced sleep disruption in these already hyperactive animals. Though this observation is interesting, it does not exclude the possibility that PCPA acts indirectly on sleep via a hyperreactivity mechanism. It only shows that hyperactivity induced by one particular method (hippocampal lesioning) does not disrupt sleep or PCPA's effects on sleep. The concern about indirect hyperreactivity effects of PCPA on sleep is therefore still valid.

The above analysis demostrates that the putatively simple, clear effects of PCPA on sleep are neither simple nor clear. Perhaps the initial depletion of 5HT by PCPA does disrupt sleep. If so, the question is whether the effect follows from the elimination of (a) a set of neurons that dampens the flow of sensory stimuli to the forebrain or moderates motor responses to sensory inputs or (b) a set of neurons that directly controls sleep (or both). Apart from this issue, it is also apparent that the recurrence of sleep after several days, despite continuing low levels of 5HT, presents a problem for the serotonin-sleep hypothesis. In this regard, there are now so many data demonstrating this phenomenon that the question should no longer be one of conforming the results to fit the 5HT hypothesis, but rather the development of new hypotheses for explaining the phenomenon (i.e., what happens neurochemically or anatomically in brain either to diminish the rat's reactivity to stimuli or to restore directly his ability to sleep? What is the adaptive response?). If an answer could be found regarding the re-emergence of sleep despite the continued depletion of 5HT, perhaps it would help illuminate the mechanism by which depressed patients show a restoration of REM sleep several weeks after initiation of antidepressant therapy and the concomitant suppression of REM sleep (Kupfer, 1982).

Serotonin Neurotoxins

Two serotonin analogs, 5,6-dihydroxytryptamine (5,6-DHT) and 5,7-dihydroxytryptamine (5,7-DHT) have been found to be potent and specific toxins for 5HT neurons. When administered into the brain, they cause a rapid and permanent depletion of 5HT, and of 5HT neurons and nerve terminals (Bjorklund et al., 1974). These compounds have been employed as a further model to study the effects of 5HT depletion on sleep.

Pujol et al. (1978) injected 1 mg of 5,6-DHT into the lateral ventricles of cats, and observed a specific depletion of 5HT in brain (norepinephrine [NE] and dopamine [DA] levels were unaffected). In these animals, a significant increase in waking and a significant reduction in NREM sleep occurred. A reduction in REM sleep was also noted, which was variable for a few days after 5,6-DHT injection, but then stabilized at about 50% normal values. The conclusion drawn from the data was that the results paralleled those with PCPA; therefore, 5HT neurons serve an executive function in the generation of NREM sleep.

Kiianmaa and Fuxe (1977) gave 8 micrograms of 5,7-DHT directly into the mesencephalic tegmentum of rats. This treatment caused a significant reduction in brain 5HT levels; it also increased total waking time, and reduced NREM sleep. In this study, chlorimipramine was coadministered with 5,7-DHT to some rats as a control. The expectation was that 5,7-DHT would not reduce 5HT in animals pretreated with this drug (and thus presumably also not affect sleep). However, in this study chlorimipramine protected NE neurons from transmitter depletion, but not the 5HT neurons. (5,6- and 5,7-DHT exert their neurotoxic effects after they are transported into the 5HT neuron by the 5HT reuptake carrier [Bjorklund et al., 1974]. Hence, their destructive effects can be blocked by pretreatment with a 5HT reuptake blocker. Chlorimipramine is active as a 5HT reuptake blocker, and thus the neurochemical results of this study are surprising.) The authors concluded from their study that, despite the unusual neurochemical results their data supported the role of 5HT neurons in the control of NREM sleep and the Jouvet hypothesis that functioning 5HT neurons are required for normal NREM sleep. It is interesting to note that Kiianmaa and Fuxe (1977) also observed that animals treated with 5,7-DHT, like those treated with PCPA, showed a recovery of normal sleep 14 to 21 days after neurotoxin injection, despite permanently low 5HT levels.

Ross et al. (1976) also performed sleep studies in rats using 5,7-DHT. They administered the drug (0.2 mg) into the lateral ventricles. All animals were pretreated with desmethylimipramine, to protect NE neurons. Substantial and selective reductions in forebrain 5HT resulted from this treatment, yielding levels that were 67% to 95% below normal. In these animals, *no* change in NREM sleep was noted over several days, though a small rise in REM sleep occurred. Hence, just as with the PCPA literature (Dement et al., 1972), data exist for 5HT neurotoxins that dissociate brain 5HT from the control of sleep states (Ross et al., 1976).

Serotonin Receptor Antagonists

A number of compounds show considerable specificity in blocking 5HT receptors (e.g., methysergide, cyproheptadine, metergoline, methiothepin). Several have been tested for their effects on sleep. Methysergide has been tested in rabbits, rats, and humans. In the rabbit, the drug (1 or 4 mg/kg I.V.) reduced both total REM and NREM sleep times, with effects on REM sleep being the largest, and persisting for as long as 24 hr postinjection (Tabushi and Himwich, 1971). In the rat, methysergide (1 or 5 mg/kg I.P.) reportedly had no effect on sleep, other than to increase the REM sleep latency and to produce an insignificant reduction in REM sleep (Fornal and Radulovacki, 1982). In humans, treatment with methysergide (8 mg/day × 2 days) did not modify total sleep time (TST), but did reduce REM sleep time. A small rise in NREM sleep was also noted, with a shifting from stage 4 to stages 2-3 (Mendelson et al., 1975). (The stages of NREM sleep in humans include stages 1 and 2, containing comparatively few cortical slow waves, and stages 3 and 4, containing considerable slow wave activity; stages 3 and 4 together are often referred to as slow wave sleep [SWS] in humans.) There thus appear to be species differences in the response to this drug.

A limited amount of data has also been collected on other 5HT antagonists. For example, metergoline has been administered to rats; a dose of 1 mg/kg I.P. was reported to reduce total REM sleep time and the number of REM sleep episodes. Total waking and NREM sleep time were unaffected (Fuxe and Kiianmaa, 1978). In cats, the injection of metergoline (0.5-8 mg/kg) induced an acute, total insomnia at all doses tested (Sallanon et al., 1982). Methiothepin, another 5HT antagonist has also been tested (Sallanon et al., 1982). It reduced REM and NREM sleep times at a low dose (1 mg/kg I.P.), and caused total insomnia at a higher dose (5 mg/kg I.P.). Given intraventricularly, methiothepin (20 micrograms) also suppressed REM sleep, but not NREM sleep (Sallanon et al., 1982). Finally, in cats made insomniac by injection of PCPA, 5HTP was observed to restore acutely both REM and NREM sleep. Pretreatment with either metergoline or methiothepin suppressed the restoration of NREM sleep by 5HTP, and delayed the recurrence of REM sleep by several hours (Sallanon et al., 1982).

Ideally, for the pharmacology to be consistent among different drug classes, the 5HT antagonists should produce effects on sleep that are similar to those produced by chronic 5HT depleters (like PCPA, though with a shorter duration of action). In cats, metergoline and methiothepin produced effects similar to those seen with PCPA (Sallanon et al., 1982). The findings with methysergide in humans and rabbits are also similar to those with PCPA in the same species (Tabushi and Himwich, 1971; Wyatt et al., 1969). In rats, however, methysergide (Fornal and Radulovacki, 1982) and metergoline (Fuxe and Kiianmaa, 1978) did not produce PCPA-like effects. Dose might explain the absence of effects for metergoline (perhaps it was too small), but probably not in the case of methysergide. More generally, there have been too few studies of 5HT antagonists in rats to allow a clear conclusion. Hopefully, more will be forthcoming.

Drugs That Enhance Serotonin Neurotransmission

Serotonin Agonists and Reuptake Blockers

Among the drugs that can stimulate post-synaptic 5HT receptors are direct-acting receptor agonists and indirect-acting presynaptic reuptake blockers. The agonist studied in the greatest detail for its effects on sleep is quipazine. In one early study, when administered to cats intravenously (0.3-3 mg/kg), it induced cortical slow waves, but also a constellation of abnormal behaviors (e.g., fixed gaze, catatonia, protrusion of the claws, arching of the back) (Rodriguez et al., 1973). Given directly into the pontine raphe in small doses, however, it induced apparently normal sleep (Rodriguez et al., 1973). In rats, quipazine (5,10 mg/kg I.P.) was found to suppress both NREM and REM sleep, and also to induce headshaking behavior for a period of 4 to 6 hr (Fornal and Radulovacki, 1981; Mogilnicka, 1981). The effects of quipazine on NREM sleep and head-shaking, but not REM sleep, were blocked by pretreating the animals with metergoline (2.5 mg/kg I.P.) (Fornal and Radulovacki, 1981) or methysergide (5 mg/kg I.P.) (Fornal and Radulovacki, 1982), both 5HT receptor antagonists.

The work in cats is difficult to evaluate; there appears to be but a single published report, and in it quipazine was reputed to induce abnormal behaviors (Rodriguez et al., 1973). The effects of centrally administered quipazine, though more normal, are nonetheless also difficult to interpret since the compound was applied to the raphe, a site of serotonin cell bodies. The drug may therefore have inhibited 5HT neuronal firing, which would inhibit endogenous 5HT release (the firing of 5HT neurons is inhibited by stimulation of autoreceptors: see Aghajanian and Wang [1978]). Hence, in this paradigm, it is impossible to conclude if quipazine affected sleep by enhancing 5HT transmission (as a direct-acting agonist on some postsynaptic site), or by inhibiting it (via stimulation of autoreceptors on 5HT cell bodies, thereby shutting off the neurons and reducing transmission). The studies in rats are no less difficult to interpret. The investigators noted excellent correlations between the induction of headshaking behavior and the suppression of NREM and REM sleep by quipazine. They interpret the head-shaking behavior as indicative of 5HT receptor stimulation (Matthews and Smith, 1980). Since the 5HT antagonists prevented quipazine-induced head-shaking and NREM sleep suppression, but not REM sleep suppression, the data were said to support a 5HT-mediated effect on NREM sleep, but not REM sleep. But it is also possible that quipazine disrupted both NREM and REM sleep because it induced head-shaking. In such studies, the reported head-shaking frequency is upwards of 3/minute (Fornal and Radulovacki, 1982), or once every 20 seconds, if evenly distributed. It is difficult to imagine that a rat could sleep under such conditions. Moreover, the absence of a blockade of the REM sleep-suppressing action of quipazine by 5HT antagonists cannot be used as a bias for excluding a 5HT effect on REM sleep, since the antagonists by themselves appear to suppress REM sleep (Fuxe and Kiianmaa, 1978). One can only hope that new studies of 5HT agonists will appear. It seems reasonable to expect that a dose range can be identified for some of the many agonists available that will produce EEG sleep effects in the absence of gross motor changes.

Several studies have also been carried out with fenfluramine, an analog of amphetamine that has specific actions at serotonin synapses. This drug releases serotonin from the nerve terminal and also blocks its reuptake and storage (Reuter, 1975). Acutely, therefore, it should greatly enhance 5HT receptor stimulation, much as a direct-acting agonist would. In humans, fenfluramine administration (80, 120 mg/day) caused a rapid suppression of REM sleep; with continued treatment, tolerance to this effect developed (Lewis et al., 1971). An enhancement of slow wave sleep (i.e., NREM sleep stages 3 and 4) was also noted at doses of 40 and 80 mg/day. In cats, single doses of fenfluramine (0.5-7.5 mg/kg I.P.) reduced REM sleep and increased REM latency, increased NREM sleep time and total sleep time, but did not alter sleep latency. With repeated administration, tolerance was found for the REM sleep effects (Johnson et al., 1971, 1972; Zolovick et al., 1973). This drug is also reputed to elevate NREM sleep in cats previously treated with PCPA, and this effect can apparently be blocked by a 5HT receptor antagonist (Johnson et al., 1972). These latter findings on NREM sleep strongly link the effect of fenfluramine to a 5HT synapse and suggest that the drug either releases a biologically potent residuum of 5HT in the PCPA-treated cat or perhaps acts as a direct agonist (a possibility entertained by Fornal and Radulovacki [1983b] in their studies in rats: see below).

Fornal and Radulovacki (1983a,b) have studied fenfluramine in the rat. They reported a dose-dependent decline in REM and NREM sleep after acute administration of the drug (1-10 mg/kg I.P.). As with the quipazine studies, the reduction in sleep was accompanied by head-shaking behavior. The effects of fenfluramine on NREM sleep and head-shaking were blocked by metergoline pretreatment, while those on REM sleep were not. In contrast, pretreatment with fluoxetine (to block fenfluramine entry into 5HT neurons) or PCPA (to deplete the 5HT neuron of its amine) did not block the actions of fenfluramine on head-shaking and sleep. These results were interpreted as indicating that fenfluramine may directly stimulate 5HT receptors.

But the results with fenfluramine in rats are difficult to clothe with a simple interpretation regarding 5HT and sleep. As with quipazine, it cannot be determined if the disruption of sleep was mediated indirectly by the head-shaking or directly by an action on sleep processes. Even if the effects were specifically tied to a 5HT synapse, a tenable conclusion is that the drug suppressed a specific, 5HT-mediated effect on head-shaking, not on sleep. In the metergoline study, in which the REM sleep effects of fenfluramine were not blocked by the antagonist, this result probably reflected the fact that metergoline itself diminished REM sleep (Fuxe and Kiianmaa, 1978). It is thus incorrect to conclude that since metergoline did not block the suppressive effect of fenfluramine on REM sleep, serotonin neurons are not involved in REM sleep mechanisms. Additionally, if the head-shaking is the source of the NREM sleep-suppressive effects of fenfluramine in rats, this phenomenon could account for the difference between the effects of the drug in cats (to raise NREM sleep) and rats (to reduce NREM sleep). The cats apparently did not develop head-shaking. The investigators discount the importance of head-shaking in discussing their effects on sleep by indicating that the animals did not show any entry into NREM sleep in the intervals between head-shakes. This statement, however, is not convincing.

Whereas fenfluramine has a multiplicity of effects on the 5HT synapse, there are other drugs more focused in their biochemical mode of action. Some have been tested for their effects on sleep, and the specific 5HT reuptake blockers fluoxetine and zimelidine provide interesting examples. Zimelidine is a highly specific blocker of 5HT reuptake, showing little effect on catecholamine or cholinergic neurons (Heel et al., 1982). It has been used as an antidepressant, and has clear effects on sleep in both humans and animals. In humans, the drug's main effect seems to be on REM sleep. It increases REM sleep latency, reduces total REM sleep time, and reduces the number of REM sleep periods. It also causes smaller effects on other sleep parameters: Sleep latency showed some increase; the number of awakenings rose somewhat; and the subjects exhibited lighter NREM sleep (Shipley et al., 1984). Similar effects have been noted in rats (Fuxe and Kiianmaa, 1978; Reyes et al., 1983), particularly that the drug significantly reduces total REM sleep time and the number of REM sleep episodes. It also reputedly enhances the total time spent in SWS1 (a sub-stage of NREM sleep in rats containing relatively few cortical slow waves) and the number of SWS1 episodes (Fuxe and Kiianmaa, 1978), but this latter effect is not obtained by all investigators (Reyes et al., 1983, 1986).

Another specific blocker of 5HT reuptake, fluoxetine (0.5, 1.25, 2.5, 5 mg/kg), also decreased REM sleep in cats (Slater et al., 1978). The authors have recently looked at the effect of fluoxetine on sleep in rats. In their studies (Pastel and Fernstrom, 1987), the animals received one of several doses of the drug (0.625, 1.25, 2.5, or 5 mg/kg I.P.) at light onset, and the EEG was monitored for several hours thereafter. They noted that a clear suppression of REM sleep time in the rat could be obtained for 2 to 4 hours with a dose as low as 1.25 mg/kg. No effects were observed on NREM sleep or waking at the doses tested below 5 mg/kg. The 5 mg/kg dose produced a transient suppression of NREM sleep (first 2 hr) as well as the expected effect on REM sleep. No behavioral effects of fluoxetine were noted that might influence sleep (e.g., head-shaking). Hence, the results in the rat are compatible with those in the cat, and suggest that at the lower doses fluoxetine's effects are specific to REM sleep. Similar results have also been obtained for two other 5HT reuptake blockers, indalpine (Kafi-de St Hilaire et al., 1984) and citalopram (Hyttel, 1982).

Taken together, many of the studies reviewed in this section are sufficiently flawed to give one concern about embracing their conclusions. However, there are several studies that seem relatively unambiguous: (a) those in humans and cats using fenfluramine (Johnson et al., 1971, 1972; Zolovick et al., 1973), in which the drug reduced REM sleep time and latency and increased NREM sleep; and (b) those in humans, cats and rats using zimelidine, fluoxetine (Fuxe and Kiianmaa, 1978; Pastel and Fernstrom, 1987; Reyes et al., 1983, 1986; Shipley et al., 1984; Slater et al., 1978), or other 5HT reuptake blockers in which this type of drug consistently reduced REM sleep time and the number of REM sleep episodes. These results suggest at least that activation of 5HT synapses consistently suppresses REM sleep. The mechanism by which fenfluramine enhances NREM sleep is not presently known.

Serotonin Precursors (5-hydroxytryptophan and tryptophan)

5-Hydroxytryptophan and tryptophan are the precursors of 5HT. Tryptophan is the physiologic precursor of 5HT and is converted to 5HTP in a reaction catalyzed by tryptophan hydroxylase (Figure 13.1). 5-Hydroxytryptophan is then decarboxylated to 5HT by the enzyme aromatic-L-amino acid decarboxylase (AAAD). Tryptophan hydroxylase is highly localized to 5HT-producing neurons (and other 5HT-producing cells) (Aghajanian and Asher, 1971). It is normally unsaturated with substrate, and thus, when brain tryptophan levels are increased, such as following an injection of L-tryptophan, 5HT synthesis and levels increase rapidly and specifically in those neurons that normally make it (Fernstrom, 1983). As a pharmacologic agent, therefore, tryptophan provides a unique level of specificity in affecting the 5HT neuron. In contrast, AAAD is not localized to the 5HT neuron; in fact, the enzyme is ubiquitous (Lovenberg et al., 1962). The K_m of AAAD for 5HTP is so large that the enzyme is probably never saturated with this substrate (Lovenberg et al., 1962). Consequently, an injection of 5HTP will stimulate 5HT synthesis, but in virtually all parts of the body, including neurons that normally do not make 5HT (Moir and Eccleston, 1968). In contrast to tryptophan, therefore, 5HTP is a pharmacologic agent that does *not* produce highly specific effects on the 5HT neuron, nor at the 5HT synapse. [For example, 5HTP can be taken up by DA neurons, converted to 5HT and lead to the release of DA [Ng et al., 1972]. It can probably also be converted to 5HT and then released by non-5HT neurons [or even glia], and presumably act at the myriad of 5HT "receptors" found throughout the brain at sites that do not contain 5HT terminals [Aghajanian and Wang, 1978].)

Because these two 5HT precursors can "drive" 5HT synthesis, both have been employed to study the effects of 5HT neurons on sleep (despite the fact that 5HTP is non-specific). 5-hydroxytryptophan was used in the early PCPA studies and was reported in low doses (2-5 mg/kg) to reverse acutely the NREM and REM sleep suppression produced by PCPA in cats (Jouvet, 1972). At very high doses (30-50 mg/kg), 5HTP restored NREM sleep but the return of REM sleep was delayed (Koella et al., 1968; Jouvet, 1972). The amino acid also was administered in low doses (5 mg/kg) to cats made insomniac by raphe lesioning and was observed to have no effects on sleep (Pujol et al., 1971). At a high dose, 5HTP (40 mg/kg) induced cortical slow waves in raphe-lesioned cats, but the animals were not asleep; no return of REM sleep was noted.

5-Hydroxytryptophan has also been administered to normal animals and humans. In cats, Pujol et al. (1971) observed that a small dose of 5HTP (5 mg/kg) was without effect on sleep. Injection of a high dose (40-50 mg/kg) produced a slow-wave-like sleep that was considered abnormal by the investigators. REM sleep was also suppressed for 6 hr by this dose of 5HTP and was followed thereafter by a rebound (Jouvet, 1969; Pujol et al., 1971). These findings are almost identical to those reported by Polc et al. (1979), except that in cats receiving 50 mg/kg 5HTP the induced NREM sleep was *not* noted to be unusual. However, Ursin (1976) reported finding no effects of 40 mg/kg 5HTP during the 24-hour period following injection on total sleep time, total awake time, or total NREM sleep and only an insignificant increment in REM sleep. Nevertheless, the animals seemed to spend a considerable amount of waking time in a state of "drowsiness," and no REM sleep occurred during the first 6 hr post-injection. Finally, REM sleep suppression can also be obtained in cats with a very small, ineffective dose of 5HTP (0.5 mg/kg) if it is combined with an ineffective dose of a 5HT reuptake blocker (fluoxetine, 0.5 mg/kg) (Slater et al., 1978). Together, these results suggest that the most consistent effect obtained in these studies is REM sleep suppression. It should be noted, however, that the administration of 5HTP does produce vomiting and diarrhea in cats (Pujol et al., 1971; Ursin, 1976), which could be a complicating factor. Only in the study of Polc et al. (1979) were such effects absent, probably because their animals were pretreated with a peripheral inhibitor of AAAD.

In normal humans, 5HTP ingestion (200, 600 mg) increased REM sleep in one study, while total sleep time did not change, and a small reduction in NREM sleep was noted (Wyatt et al., 1971). In another study (Autret et al., 1976), subjects ingested 900 mg 5HTP/day chronically, either alone or with RO4-4602 (an antagonist of aromatic L-amino acid decarboxylase which at the dose employed, works selectively outside of the brain); and were studied for several weeks. While several "trends" in sleep were evident in this study, it is noteworthy that most of the changes reported to follow the administration of 5HTP alone (i.e., without RO4-4602) did not achieve statistical significance. When 5HTP was coadministered with RO4-4602, the latency to REM sleep was increased, and NREM sleep was reduced. In a third study, no effects of 5HTP (200 mg) were noted on any sleep parameters, except when subjects were first deprived of sleep. When 5HTP was administered after sleep deprivation, the rebound in NREM sleep was blocked and did not occur on succeeding, drug-free nights (Nakazawa et al., 1980). Probably the clearest conclusion that can be drawn from these results is either that there are insufficient data in the literature with which to evaluate the true effect of 5HTP on sleep, or that there is no robust effect of 5HTP on sleep in humans. No one group confirmed another's findings with 5HTP. In contrast, the data in animals suggest a REM-suppressant effect of this amino acid. The explanation for this species difference is presently unknown.

The results of studies with tryptophan are no more satisfying, in humans or animals. In rats, tryptophan administration appears to have no effects on sleep other than at several doses (30, 450, 600 mg/kg) to reduce sleep latency (Hartmann and Chung, 1972), REM sleep latency (Hill and Reyes, 1978), or both (Wojcik et al., 1980). In this latter study, however, a dose of 120 mg/kg did not alter sleep or REM sleep latencies but did produce a small reduction in REM sleep time during the second hour after trypto- phan injection and a small increase in waking time. In cats, Radulovacki (1982) noticed no effect of tryptophan (30 mg/kg) on any sleep param- eter. However, if the cats also received the 5HT antagonist methysergide (0.5 mg/kg) which itself greatly increased sleep and REM sleep latencies and reduced REM and NREM sleep times, tryptophan administration then produced small reductions in sleep latency and waking time and modest in- crements in TST and REM sleep time. These results certainly do not confirm the results of 5HTP studies (to suppress REM sleep) and at best suggest the possibility that tryptophan reduces sleep latency. Before a firm conclusion can be drawn, however, additional data on tryptophan's effects should be collected, both alone and in combination with other 5HT drugs.

In humans, the results are also inconsistent. In an early, comprehensive study in humans, oral tryptophan administration (7.5 gm) to normal subjects produced an increase in TST and NREM sleep time and reduced REM sleep time (Wyatt et al., 1970). There was no change in sleep latency, but REM sleep latency increased. In insomniac patients, this same treatment also increased TST (an effect also observed by Hartmann et al. [1971]) and NREM sleep time and reduced awakenings and early morning wakefulness. Slow wave sleep (i.e., NREM stages 3 and 4) time and REM sleep time were not significantly affected. And, in a group of patients receiving PCPA as a treatment for carcinoid tumor, in which REM sleep (but not NREM sleep) was depressed, tryptophan administration (5-14 gm) increased NREM sleep and decreased REM sleep. The investigators concluded that tryptophan could not be causing its effects via conversion to 5HT, since (a) the results are opposite to those they had obtained with 5HTP (Wyatt et al., 1971) and similar to those obtained with PCPA and (b) PCPA did not compromise the effect of tryptophan on sleep (Wyatt et al., 1970). In another, similar study (Griffiths et al., 1972), normal subjects receiving the same dose of tryptophan as in the above study (7.5 gm) showed an increase in NREM sleep time but no change in REM sleep. The tryptophan load also made the subjects drowsy, but sleep latency was not significantly affected. At a higher dose (12 gm), opposite effects were obtained: REM sleep time was dramatically increased, while NREM sleep time was unaffected. Both sleep and REM sleep latencies were also reduced. And, in several other studies, the administration of tryptophan (1-15 gm) appeared only to reduce sleep latency (particularly in individuals with long sleep latencies) (Brown et al., 1979; Hartmann, 1982/83; Hartmann and Spinweber, 1979; Nicholson and Stone, 1979) or to have no effect at all (Nicholson and Stone, 1979).

Overall, the results of sleep studies to date involving tryptophan in both humans and animals have been thoroughly inconsistent. This may mean that tryptophan is without significant effects on mammalian sleep processes, perhaps because brain 5HT neurons adjust 5HT release downward to compensate for increased 5HT synthesis and release following precursor loading (Gallager and Aghajanian, 1976; Trulson and Jacobs, 1976). (This possibility seems relatively unlikely, however, since tryptophan administration does readily influence other brain functions thought to be tied to 5HT neurons [e.g., growth hormone secretion (Arnold and Fernstrom, 1981) and blood pressure regulation (Sved et al., 1982)].) Or more likely, the effects of tryptophan on sleep are relatively modest and therefore easily missed. It must be recalled that while tryptophan is often used like a drug, it is in reality an essential dietary nutrient whose supply to the body and brain varies physiologically (Fernstrom, 1983). It therefore seems unlikely that the body would be designed to be exposed to vagaries in tryptophan supply if this amino acid were a highly potent pharmacologic agent. It might, however, be subject to precursor-induced changes in serotonin synthesis and release if such changes were relatively modest and useful. Hence, investigators exploring for effects of tryptophan on sleep (or other brain-mediated processes) should be prepared to search for such small effects, not large pharmacologic ones.

General Discussion and Conclusions

There are a number of unresolved issues that emerge from the present review. First and foremost is that available data, assembled together, do not present a clear, simple relationship between 5HT neuronal function and sleep. If we begin dissecting subsets of the data, however, some interesting issues emerge.

First, in studies involving *chronic* impairment (PCPA) or permanent destruction (neurotoxins) of 5HT neurons, it may be useful to consider the evolution of the sleep disruption as two stages. The first is the initial depletion state, in which the drop-off in 5HT is accompanied by a reduction in NREM and REM sleep. In this sense, the shutting-down of 5HT neurons may clearly precipitate sleep disruption. But an important, unresolved issue is whether this disruption represents a turning off of (a) a set of neurons directly involved in the control of sleep circuits or (b) a set of neurons that restricts the flow of sensory information to higher brain centers and/or the motor responsivity to sensory input. The second stage is the recovery phase, in which sleep returns to normal in the presence of either a permanent 5HT deficit or a very slow restoration of normal 5HT levels. It would seem that there is no easy way to conform this set of changes to fit a direct 5HT-sleep hypothesis. If this view is correct, then it is time to begin searching for the mechanism(s) of this adaptive response. That is, in the absence of the 5HT neuron and its normal function in sleep processes, how does the brain change to allow normal sleep to return?

As a second issue, focusing on the *acute* affects of 5HT drugs on sleep, it does appear that a common feature of many of the drugs that enhance 5HT neurotransmission (e.g., fenfluramine, zimelidine, fluoxetine, 5HTP) is to suppress REM sleep. However, the results of studies with 5HT antagonists suggest that reducing 5HT synaptic transmission *also* suppresses REM sleep. But it is premature to become perplexed by this putative effect of 5HT antagonists: there are too few data in the literature describing it, and these are relatively incomplete. Increasingly thorough studies of the effects of 5HT antagonists are therefore needed. If these show that the antagonists promote REM sleep, then the straightforward conclusion would be that the short-term effect of 5HT neuronal firing is to inhibit REM sleep. If the result is to establish that 5HT antagonists, like drugs that stimulate 5HT neurotransmission, inhibit REM sleep, then future studies will have to begin to grapple with the possibility of a control system in which the functioning of a single set of neurons can be the production of the same effect at both very high and very low levels of activity (or one in which the 5HT neuron is not involved).

As a third issue, it is interesting to note that not only do 5HT drugs influence sleep but the activity of 5HT neurons changes with the state of vigilance. For example, McGinty and Harper (1976) and Trulson and Jacobs (1976) noted that 5HT neurons (i.e., raphe units) are most active during waking, slow progressively as the animal proceeds into deeper stages of NREM sleep, and are slowest during REM sleep (firing actually ceases during PGO activity, a REM sleep phenomenon). Low 5HT neuronal activity during REM sleep would fit with the simple control model for REM sleep developed by McCarley (1980), in which the turning off of 5HT neurons is a precondition for REM sleep. Many of the results with such drugs as fenfluramine, zimelidine, fluoxetine, and 5HTP discussed in the present review could readily be fit into this context, for these agents stimulate 5HT transmission (at least acutely) and suppress REM sleep. (Of course, this analysis runs counter to the REM "priming" function postulated for 5HT neurons by Jouvet [1969, 1972].)

There is also evidence that during REM sleep deprivation (Dement et al., 1972; Hery et al., 1970) and total sleep deprivation (Toru et al., 1984), 5HT turnover (and presumably release) is increased (though one report disagrees [Borbely et al., 1980].) This raises an important question for future investigation. Do these results simply sustain the view that 5HT neurons are normally active during periods of waking, as the neurophysiological results show (McGinty and Harper, 1976; Trulson and Jacobs, 1979)? Or might they imply that the 5HT neurons are firing at supranormal rates in their capacity as dampeners of sensory input (Ross et al., 1976) as part of the developing "pressure" for sleep? There appear in general to be many such scattered threads and issues in the serotonin literature that if brought together and developed, might lead to new insights into the functioning of 5HT neurons in sleep processes.

Finally, in the context of the affective disorders, the authors have attempted to provide some illumination of past data commonly used to support a serotonin-sleep connection. It should be apparent from the analysis that, although existing data provide interesting leads for future investigation into the role of 5HT neurons in sleep processes, the relationship between 5HT and sleep is by no means simple or clear at present. Hence, when antidepressant drugs like zimelidine and chlorimipramine, which enhance transmission across 5HT synapses, are seen to suppress REM sleep, it is premature to attribute the effect to a simple biochemical mechanism involving the 5HT neuron. Moreover, though depression-associated sleep changes and their dissipation with successful antidepressant treatment may indicate a causal connection between sleep and depression, care should be taken in linking the effects to 5HT neurotransmission when 5HT reuptake blockers are used as the antidepressant therapy.

Acknowledgments. Some of the studies described in this article were supported by a grant from the National Institutes of Health (MH38178; HD24730). J.D.F. is the recipient of an NIMH Research Scientist Development Award, Level II (MH00254).

Further Reading

Aghajanian GK, Asher IH (1971): Histochemical fluorescence of raphe neurons: selective enhancement by tryptophan. *Science* 172:1159-1161

Aghajanian GK, Wang RY (1978): Physiology and pharmacology of central serotonergic neurons. In: *Psychopharmacology: A Generation of Progress,* Lipton MA, DiMascio A, Killam KF, eds. New York: Raven Press, pp. 171–183

Arnold MA, Fernstrom JD (1981): L-tryptophan injection enhances pulsatile growth hormone secretion in the rat. *Endocrinology* 108:331-335

Autret A, Minz M, Bussel B, Cathala HP, Castaigne P (1976): Human sleep and 5-HTP. Effects of repeated high doses and of association with benserazide (RO.04.4602). *Electroencephalogr Clin Neurophysiol* 41:408-413

Bjorklund A, Baumgarten HG, Nobin A (1974): Chemical lesioning of central monoamine axons by means of 5,6-dihydroxytryptamine and 5,7-dihydroxytryptamine. In: *Advances in Biochemical Psychopharmacology,* Volume 10, Costa E, Gessa GL, Sandler M, eds. New York: Raven Press, pp. 13–33

Borbely AA, Steigrad P, Tobler I (1980): Effect of sleep deprivation on brain serotonin in the rat. *Behav Brain Res* 1:205-210

Brown CC, Horrom NJ, Wagman AM (1979): Effects of L-tryptophan on sleep onset insomniacs. *Waking Sleeping* 3:101-108

Dement WC, Mitler MM, Henriksen SJ (1972): Sleep changes during chronic administration of parachlorophenylalanine. *Rev Can Biol* 31:239-246

Fernstrom JD (1983): The role of precursor availability in the control of monoamine biosynthesis in the brain. *Physiol Rev* 63:484-546

Fornal C, Radulovacki M (1981): Sleep suppressant action of quipazine: relation to central serotonergic stimulation. *Pharmacol Biochem Behav* 15:937-944

Fornal C, Radulovacki M (1982): Methysergide blocks the sleep suppressant action of quipazine in rats. *Psychopharmacology* 76:255-259

Fornal C, Radulovacki M (1983a): Sleep suppressant action of fenfluramine in rats. I. Relation to postsynaptic serotonergic stimulation. *J Pharmacol Exp Ther* 225:667-674

Fornal C, Radulovacki M (1983b): Sleep suppressant action of fenfluramine in rats. II. Evidence against the involvement of presynaptic serotonergic mechanism. *J Pharmacol Exp Ther* 225:675-681

Fuxe K, Kiianmaa K (1978): 5-hydroxytryptamine neurons and the sleep-wakefulness cycle. Effects of metergoline and zimelidine. *Neurosci Lett* 8:55-58

Gallager DW, Aghajanian GK (1976): Inhibition of firing of raphe neurones by tryptophan and 5-hydroxytryptophan: blockade by inhibiting serotonin synthesis with RO4-4602. *Neuropharmacology* 15:149-156

Griffiths WJ, Lester BK, Coulter JD, Williams HL (1972): Tryptophan and sleep in young adults. *Psychophysiology* 9:345-356

Hartmann E (1982/83): Effects of L-tryptophan on sleepiness and on sleep. *J Psychiatr Res* 17:107-113

Hartmann E, Chung R (1972): Sleep-inducing effects of L-tryptophan. *J Pharm Pharmacol* 24:252-253

Hartmann E, Chung R, Chien CP (1971): L-Tryptophane and sleep. *Psychopharmacologia* 19:114-127

Hartmann E, Spinweber CL (1979): Sleep induced by L-tryptophan. Effect of dosages within the normal dietary intake. *J Nerve Ment Dis* 167:497-499

Heel RC, Morley PA, Brogden RN, Carmine AA, Speight TM, Avery GS (1982): Zimelidine: a review of its pharmacological properties and therapeutic efficacy in depressive illness. *Drugs* 24:169-206

Hery F, Pujol JF, Lopez M, Macon J, Glowinski J (1970): Increased synthesis and utilization of serotonin in the central nervous system of the rat during paradoxical sleep deprivation. *Brain Res* 21:391-403

Hill SY, Reyes RB (1978): Effects of L-tryptophan and ethanol on sleep parameters in the rat. *Psychopharmacology* 58:229-233

Hyttel J (1982): Citalopram-pharmacological profile of a specific serotonin uptake inhibitor with antidepressant activity. *Prog Neuropsychopharmacol Biol Psychiatry* 6:277-295

Johnson DN, Funderburk WH, Ward JW (1971): Effects of fenfluramine on sleep-wakefulness in cats. *Psychopharmacologia* 20:1-9

Johnson DN, Funderburk WH, Ruckart RT, Ward JW (1972): Contrasting effects of two 5-hydroxytryptamine-depleting drugs on sleep patterns in cats. *Eur J Pharmacol* 20:80-84

Jouvet M (1969): Biogenic amines and the states of sleep. *Science* 163:32-41

Jouvet M (1972): The role of monoamines and acetylcholine-containing neurons in the regulation of the sleep-waking cycle. *Ergebn Physiol* 64:166-307

Kafi-de St Hilaire S, Merica H, Gaillard JM (1984): The effects of indalpine - a selective inhibitor of 5-HT uptake - on rat paradoxical sleep. *Eur J Pharmacol* 98:413-418

Kaufman LS (1983): Parachlorophenylalanine does not affect pontinegeniculate-occipital waves in rats despite significant effects on other sleep-waking parameters. *Exp Neurol* 80:410-417

Kiianmaa K, Fuxe K (1977): The effects of 5,7-dihydroxytryptamine-induced lesions of the ascending 5-hydroxytryptamine pathways on the sleep-wakefulness cycle. *Brain Res* 131:287-301

Koe BK, Weissman A (1966): p-Chlorophenylalanine: a specific depletor of brain serotonin. *J Pharmacol Exp Ther* 154:499-516

Koella WP, Feldstein A, Czicman JS (1968): The effect of para-chlorophenylalanine on the sleep of cats. *Electroencephalogr Clin Neurophysiol* 25:481-490

Kupfer D (1982): Interaction of EEG sleep, antidepressants, and affective disease. *J Clin Psychiatry* 43:30-35

Laguzzi RF (1982): Effects of serotonin synthesis inhibition on sleep in hippocampectomized rats. *Brain Res* 240:175-177

Lewis SA, Oswald I, Dunleavy DLF (1971): Chronic fenfluramine administration: some cerbral effects. *Br Med J* 3:67-70

Lipton MA, Gordon R, Guroff G, Udenfriend S (1967): p-Chlorophenylalanine-induced chemical manifestations of phenylketonuria in rats. *Science* 156:248-250

Lovenberg W, Jequier E, Sjoerdsma A (1968): Tryptophan hydroxylation in mammalian systems. In: *Advances in Pharmacology*, volume 6A, Garattini S, Shore PA, eds. New York: Academic Press, pp. 21-36

Lovenberg W, Weissbach H, Udenfriend S (1962): Aromatic L-amino acid decarboxylase. *J Biol Chem* 237:89-92

Matthews WD, Smith CD (1980): Pharmacological profile of a model for central serotonin receptor activation. *Life Sci* 26:1397-1403

McCarley RW (1980): Mechanisms and models of behavioral state control. In: *The Reticular Formation Revisited*, Hobson JA, Brazier MAB, eds. New York: Raven Press, pp. 375-403

McGinty DJ, Harper RM (1976): Dorsal raphe neurons: depression of firing during sleep in cats. *Brain Res* 101:569-575

Mendelson WB, Reichman J, Othmer E (1975): Serotonin inhibition and sleep. *Biol Psychiatry* 10:459-464

Mogilnicka E (1981): REM sleep deprivation changes behavioral response to catecholaminergic and serotonergic receptor activation in rats. *Pharmacol Biochem Behav* 15:149-151

Moir ATB, Eccleston D (1968): The effects of precursor loading in the cerebral metabolism of 5-hydroxyindoles. *J Neurochem* 15:1093-1108

Mouret J, Bobillier P, Jouvet M (1968): Insomnia following parachlorophenylalanine in the rat. *Eur J Pharmacol* 5:17-22

Nakazawa Y, Hasuzawa H, Kotorii T, Ohkawa T, Sakurada H, Nonaka K, Dainoson K (1980): Study on the effects of L-5HTP on the stages of sleep in man as evaluated by using sleep deprivation. *Folia Psychiatr Neurol Jpn* 34:83-87

Ng LKY, Chase TN, Colburn RW, Kopin IJ (1972): Release of [3H]Dopamine by L-5-hydroxytryptophan. *Brain Res* 45:499-505

Nicholson AN, Stone BM (1979): L-tryptophan and sleep in healthy man. *Electroencephalogr Clin Neurophysiol* 47:539-545

Pastel RH, Fernstrom JD (1987): Short-term effects of fluoxetine and trifluoromethylphenylpiperazine on electroencephalographic sleep in the rat. *Brain Res* 436:92-102

Polc P, Schneeberger J, Haefely W (1979): Effects of several centrally active drugs on the sleep-wakefulness cycle of cats. *Neuropharmacology* 18:259-267

Pujol JF, Buguet A, Froment JL, Jones B, Jouvet M (1971): The central metabolism of serotonin in the cat during insomnia. A neurophysiological and biochemical study after administration of p-chlorophenylalanine or destruction of the raphe system. *Brain Res* 29:195-212

Pujol JF, Keane P, Bobillier P, Renaud B, Jouvet M (1978): 5,6-dihydroxytryptamine as a tool for studying sleep mechanisms and interactions between monoaminergic systems. *Ann NY Acad Sci* 305:576-589

Radulovacki M (1982): L-Tryptophan's effects on brain chemistry and sleep in cats and rats: a review. *Neurosci Biobehav Rev* 6:421-427

Rechtschaffen A, Lovell RA, Freedman DX, Whitehead WE, Aldrich M (1973): The effect of parachlorophenylalanine on sleep in the rat: some implications for the serotonin-sleep hypothesis. In: *Serotonin and Behavior,* Barchas J, Usdin E, eds. New York, Academic Press

Reuter CJ (1975): A review of the CNS effects of fenfluramine, 780SE and norfen-fluramine on animals and man. *Postgrad Med J* 51(suppl 1):18-27

Reyes RB, Hill SY, Kupfer DJ (1983): Effects of acute doses of zimelidine on REM sleep in rats. *Psychopharmacology* 80:214-216

Reyes RB, Hill SY, Kupfer DJ (1986): Effects of repeated zimelidine administration on sleep parameters in the rat. *Psychopharmacology* 88:54-57.

Rodriguez R, Rojas-Ramirez JA, Drucker-Colin RR (1973): Serotonin-like actions of quipazine on the central nervous system. *Eur J Pharmacol* 24:164-171

Ross CA, Trulson ME, Jacobs BL (1976): Depletion of brain serotonin following intraventricular 5,7-dihydroxytryptamine fails to disrupt sleep in the rat. *Brain Res* 114:517-523

Sallanon M, Buda C, Janin M, Jouvet M (1982): 5-HT antagonists suppress sleep and delay its restoration after 5-HTP in p-chlorophenylalanine-pretreated cats. *Eur J Pharmacol* 82:29-35

Sallanon M, Janin M, Buda C, Jouvet M (1983): Serotonergic mechanisms and sleep rebound. *Brain Res* 268:95-104

Shipley JE, Kupfer DJ, Dealy RS, Griffin SJ, Coble PA, McEachran AB, Grochocinski VJ (1984): Differential effects of amitriptyline and of zimelidine on the sleep electroencephalogram of depressed patients. *Clin Pharmacol Ther* 36:251-259

Slater IH, Jones GT, Moore RA (1978): Inhibition of REM sleep by fluoxetine, a specific inhibitor of serotonin uptake. *Neuropharmacology* 17:383-389

Sved AF, Van Itallie CM, Fernstrom JD (1982): Studies of the antihypertensive action of L-tryptophan. *J Pharmacol Exp Ther* 221:329-333

Tabushi K, Himwich HE (1971). Electroencephalographic study of the effects of methysergide on sleep in the rabbit. *Electroencephalogr Clin Neurophysiol* 31:491-497

Tobler I, Borbely AA (1982): Sleep regulation after reduction of brain serotonin: effect of p-chlorophenylalanine combined with sleep deprivation in the rat. *Sleep* 5:145-153

Toru M, Mitsushio H, Mataga N, Takashima M, Arito H (1984): Increased brain serotonin metabolism during rebound sleep in sleep-deprived rats. *Pharmacol Biochem Behav* 20:757-761

Trulson ME, Jacobs BL (1976): Dose-response relationships between systemically administered L-tryptophan or 5-hydroxytryptophan and raphe unit firing activity in the rat. *Neuropharmacology* 15:339-344

Trulson ME, Jacobs BL (1979): Raphe unit activity in freely moving cats: correlation with level of behavioral arousal. *Brain Res* 163:135-150

Ursin R (1972): Differential effect of para-chlorophenylalanine on the two slow wave sleep stages in the cat. *Acta Physiol Scand* 86:278-285

Ursin R (1976): The effects of 5-hydroxytryptophan and L-tryptophan on wakefulness and sleep patterns in the cat. *Brain Res* 106:105-115

Ursin R (1980): Does para-chlorophenylalanine produce disturbed waking, disturbed sleep or activation by ponto-geniculo-occipital waves in cats? *Waking Sleeping* 4:211-221

Weitzman ED, Rapport MM, McGregor P, Jacoby J (1968): Sleep patterns of the monkey and brain serotonin concentration: effect of p-chlorophenylalanine. *Science* 160:1361-1363

Willner P (1985): Antidepressants and serotonergic neurotransmission: an integrative review. *Psychopharmacology* 85:387-404

Wojcik WJ, Fornal C, Radulovacki M (1980): Effect of tryptophan on sleep in the rat. *Neuropharmacology* 19:163-167

Wyatt RJ, Engelman K, Kupfer DJ, Scott J, Sjoerdsma A, Snyder F (1969): Effects of para-chlorophenylalanine on sleep in man. *Electroencephalogr Clin Neurophysiol* 27:529-532

Wyatt RJ, Engelman K, Kupfer DJ, Fram DH, Sjoerdsma A, Snyder F (1970): Effects of L-tryptophan (a natural sedative) on human sleep. *Lancet* ii:842-846

Wyatt RJ, Zarcone V, Engelman K, Dement WC, Snyder F, Sjoerdsma A (1971): Effects of 5-hydroxytryptophan on the sleep of normal human subjects. *Electroencephalogr Clin Neurophysiol* 30:505-509

Zolovick AJ, Stern WC, Panskepp J, Jalowiec JE, Morgane PJ (1973): Sleep-waking patterns in cats after administration of fenfluramine and other monoaminergic modulating drugs. *Pharmacol Biochem Behav* 1:41-46

Epilogue

Floyd E. Bloom

Attending this meeting as an interested observer and a non-participant in the animal models field, several issues strike me as pertinent to any attempt to summarize these proceedings. These issues transcend any specific presentation, and I will cite specific speakers only to illustrate the points on which I wish to comment briefly.

The Cycle of Advances in Experimental Medicine

Physicians diagnose diseases based upon an evaluation of the symptoms or complaints described by the patient or their family of a change in emotion or overt behavior. When possible signs are sought as more objective indicators of the disorder—ranging from weight loss, skin tone, or sleep patterns to still more objective tests on blood or spinal fluid chemistry to brain electrophysiological or endocrine activity, or the noninvasive methods for assessing brain structure together with metabolic activity and blood flow, and when the symptoms and signs from a consistent reproducible ensemble of findings, a diagnostic entity can be described. Patients with some, most, or all the subjective and objective findings can be ascribed to have the disorder.

At this point, many insights become possible: Patients in the diagnostic category can be followed for the natural history of the disorder to assign specific patterns of geographic, age, sex, or occupational factors that sensitize or protect against the disease and which describe its course. Knowing the frequency of natural remissions and recurrences is obviously critical to the ability to determine whether a specific treatment has made a significant difference. The beautiful long-term studies of affective disorder described by Post are therefore absolutely critical to our understanding of a range of developmental, genetic, and environmental factors that may regulate the appearance of affective psychosis.

A disease that waxes and wanes spontaneously, in which episodic recurrences over many years are far more frequent than a course with a single episode, and in which the interval between episodes decreases as the patient's course continues, obviously represent features that are difficult to establish in animal models without better understanding of the underlying disease process and the means by which it is expressed. The primate studies described by McKinney and Schlemmer tantalize us with the degree to which early childhood experiences or adult drug exposures can lead to predictable behavioral and social signs of depressive symptoms. The seeming similarity between the process of electrophysiological sensitization, or kindling, described by Ehlers, and in the behavioral responses of rodents to cocaine, described by Post, may provide insights into the longer term changes that occur after many years of expressing affective psychosis.

Pathology Specifies Molecular and Cellular Etiologic Factors

An important part of the traditional experimental medicine approach to disease description and analysis is the discovery of the underlying pathology, the area in which psychiatric disorders, including depression, have been traditionally weak. The high excitement and long-term interest in the monoaminergic hypotheses of depression and mania based upon drug effects and indirect assessments of brain aminergic metabolism and responsiveness reflect this paucity of substantive pathology. In many ways our desire to model the depressive signs and symptoms in animals is the direct result of the failure to discover a consistent pathological body of factors that can be construed as etiologic elements in the pathogenesis rather than as sequelae of the expression of disease. The behavioral models critically assessed by Weiss help one to recognize the important neurochemical and behavioral differences, both qualitative and quantitative, that small details of experimental procedure can make. The convergence of at least Weiss' own paradigm on the central noradrenergic function of the nucleus locus coeruleus, a structure of more than passing interest in my own research, makes me feel that better understanding of this and other similar circuit structures may ultimately be useful in solving the possible pathologic processes in depression and other psychoses.

The obvious value of discovering some distinctive and consistent pathology in the patient, whether detectable premortem, by noninvasive studies, or documentable post-mortem, is that it leads one to focus on specific cell structures, thereby providing clues for the earlier diagnosis, and possible causes of sensitization and suggests modes of pathogenesis and outcome that will have implications both for treatment and for prevention. In aligning signs and symptoms in the patient to specific structures with underlying functions in the normal subject, we gain better appreciation of subtle normal functions that may be unperceived in the absence of the disease. Moreover, when similar macroscopic brain components have analogies in animal brains, it seems reasonable at first glance to assume that the same regions will function with the same chemical components in the brains of animals and that there-fore, for example, a hippocampus is a hippocampus is a hippocampus, and a monoamine is a monoamine is a monoamine. Thanks to the recent ex-plosion in methods to evaluate the details of brain circuitry structure and histochemical indices of transmitters in primate brains, I believe this sort of assumption is now much less tenable than one might have expected, espe-cially when one starts with a desire to understand the human structural basis for behavior and extrapolates that function-structure complex exclusively to rodent brains. It is quite clear that when a broad biologic perspective is taken, many specific systems have changed, both in importance and in the details of their operation across the species. These changes range from the transmitter used to regulate neuromuscular or sympathetic autonomic exci-tation, to the emergence of major brain regions in humans that are lacking even in lower mammalian forms.

The human stroke model described by Robinson, shows striking human emotional sequelae and rat locomotor sequelae at least over the short term. In that system, specific depressions or inappropriate affect follow left- or right-sided frontal lobe damage, that can, interestingly, build upon longer term pre-existing frontal lobe injury. But is a right frontal cerebral infarct in a rat even remotely comparable to one in human? If so, or if not, the question needs answers in details that are heuristic.

These human case studies and near-human psychopharmacologic experi-ments impress me in the context of this meeting because they reveal alter-native means to express the sorts of signs and nonverbal symptoms that are expressed by the humans with diagnosed affective disorder. I would view the endocrinopathies described by Kalin as still another alternative means to reveal the same or a highly similar, but specifically attributable, behavioral symptom complex. The differences between the stroke-induced depressions and the endocrinopathic depressions and the true affective psychosis are at least twofold: (1) the affective disorder has no known structural or endocrine cause; (2) the endocrine-related changes that occur are probably of central origin and reflect the disease rather than cause it. These intriguing differ-ences and reflections heighten our attraction to understand the degree to which the brain and behavior can be traced to specific anatomic loci and to understand the degree to which the endocrine organs of the periphery can regulate the activity of the CNS regions that in turn regulate or express affect.

The Inverted U-Shaped Dose-Response Curve

The general explanation that has been used to restore logic here is that the treatment acts upon many components of the responding system and that the effects of higher doses induces complex effects that may neutralize or overcome the lowest active doses. This strikes me as pertinent to the concept of an animal model of a human psychiatric disease because we are modeling human signs and symptoms onto animal signs of behavior, and when there is no behavioral results, the "inverted U-shaped dose-response curve logic" forces us to ask ourselves whether what we have done is too little to produce an effect or too much, in a complex system, to reveal an effect.

This relationship between the dose of some treatment and the resulting response of the animal is an intriguing one that seems at first glance to be counter intuitive. Normally, we might hold that there should be a simple near-linear or natural logarithmic relationship between a treatment and the result. For example, in general the rates at which animals respond to intracranial self-stimulation in the experiments described by Kornetsky or Koob show a consistent threshold, a midrange at which rate is proportional to stimulation current, and then a ceiling-point of no further gain in the maximal response rate. Amphetamine administered to animals in this paradigm yields enhanced sensitivity to steps in current (a leftward dose-response shift). However, when low doses do nothing, and higher doses do something, and still higher doses appear to do less than the intermediate values, we become skeptical as to whether the higher dose treatment is in any way the direct cause of the result that we observed in the middose range.

One final issue that deserves explicit mention is that even if the causative agent and principal locus of the etiology and pathogenesis of affective psychosis were known tomorrow, one would still require animal models of definable elements of the symptoms or objective signs, as predictors of drugs or other treatments. For example, in a robust medical disorder, diabetes mellitus, whose chemical basis and structural etiology are known and well documented, a significant body of animal model research persists to devise experimentally controllable means to assess a variety of counter-complication treatments. The multiple predictive indices of rodent responses to drugs that are useful in man, bypassing almost completely the effects of these drugs in primates, as described by Howard, document this point. Nevertheless, were I forced to predict the sort of animal model that will have the most relevance to depression, my decision now would be drawn to the need to identify those regions of the nonhuman primate brain that are regulating affect, to determine the degree to which the early childhood or adult pharmacologic manipulations can reveal the outward signs and symptoms of depression, and to determine the chemical and functional organization of these regions regulating affect.

Index

Prospective studies, 9
Psychosis models, 241–244

Quipazine
observational models and, 196
sleep effects, 268, 269
social behavior effects, 249, 250, 251

Raphe nucleus, serotonin content, 197–198
Rat
as anhedonia model, 165–177, 178
as behavioral sensitization model, 32–38
depression induction in, 83
as electrophysiological kindling model,
30–32, 38–47
as infant separation model, 99, 100–108
intracranial self-stimulation, 165–177, 178
as locomotor activity model, 141, 142–
146, 147–148
maternal separation and, 147–148
as post-stroke depression model, 83–91
behavior, 84–86
hyperactivity, 84–91, 94
middle cerebral artery ligation, 83,
84–86, 89
as schizophrenia model, 20–21, 22
as sleep disturbance model, 263, 264, 265,
266–267, 270, 272–273
as uncontrollable shock model, 111–134
associative disturbances, 221–223
escape performance, 214–219
reinforcement, 224–225
Receptor
antidepressant drugs and
direct effects, 190–192
sensitivity changes, 192–193
in stress-related depression, 115–118,
126–127, 128
Reinforcement. See also Intracranial self-
stimulation (ICSS)
alternatives, 164
uncontrollable shock and, 223–227
Reinforcer, definition, 163
REM sleep, in depression, 106. See also
Sleep disturbance model
antidepressant drugs and, 261, 276
5-hydroxytryptophan and, 263
methysergide and, 266–267
para-chlorophenylalanine and, 263, 271,
273
serotonin agonists and, 267–268
serotonin neurotoxins and, 265–266
serotonin precursors and, 271, 272–273
serotonin reuptake blockers and, 267–268

Reserpine, as depression cause, 189, 193
in non-human primates, 239–240
Reserpine-antagonism model, 194
Response-outcome association deficits,
221–223
Retrospective studies, 8, 13
Reward. See also Intracranial self-stimulation
(ICSS)
definition, 163
measures of, 163–164
uncontrollable shock and, 223–227

Salbutamol, 192, 193, 195
Schizophrenia models, 20–21, 22
amphetamine-related, 241, 242–243
Scopolamine, 173, 196, 218
Seizure
corticotropin-releasing factor-related, 49
opiate peptide-related, 49–50
Seizure model, 30–32, 38–47
affective illness parallels, 40–41
anticonvulsants and, 41–47
pharmacology of, 43–47
characteristics, 38, 39
as manic-depressive illness model, 31,
38–42, 45–47
spontaneity phase, 38, 39–40, 43, 45
Self-grooming, 245, 246, 247, 251, 253
Sensitization, behavioral, 30–31, 32–38, 49
affective illness parallels, 40–41
conditioning in, 34
diazepam and, 34
haloperidol and, 34, 36
locomotor hyperactivity, 32–35, 37
as manic-depressive illness model, 31, 34,
36–38
sex factors, 33, 34
stereotypy, 34
stress-sensitization and, 37–38
vasopressin and, 34
Separation models, 6–7, 23, 24
amphetamine and, 245
for antidepressant drug development, 196
catecholamines and, 244–245
evaluation, 13
hypothalamic-pituitary-adrenal system
and, 67–68
maternal-infant, 24, 99–100, 106
activity rhythm disturbances, 146–148
plasma cortisol and, 147
peer, 99–110
ACTH and, 66
corticotropin-releasing factor response,
104, 106